Tandem Mass Spectrometry of Lipids

Molecular Analysis of Complex Lipids

New Developments in Mass Spectrometry

Editor-in-Chief:
Professor Simon J Gaskell, *Queen Mary University of London, UK*

Series Editors:
Professor Ron M A Heeren, *FOM Institute AMOLF, The Netherlands*
Professor Robert C Murphy, *University of Colorado Denver, USA*
Professor Mitsutoshi Setou, *Hamamatsu University School of Medicine, Japan*

Titles in the Series:
1: Quantitative Proteomics
2: Ambient Ionization Mass Spectrometry
3: Sector Field Mass Spectrometry for Elemental and Isotopic Analysis
4: Tandem Mass Spectrometry of Lipids: Molecular Analysis of Complex
   Lipids

How to obtain future titles on publication:
A standing order plan is available for this series. A standing order will bring
delivery of each new volume immediately on publication.

For further information please contact:
Book Sales Department, Royal Society of Chemistry, Thomas Graham House,
Science Park, Milton Road, Cambridge, CB4 0WF, UK
Telephone: +44 (0)1223 420066, Fax: +44 (0)1223 420247
Email: booksales@rsc.org
Visit our website at www.rsc.org/books

Tandem Mass Spectrometry of Lipids:

Molecular Analysis of Complex Lipids

Robert C Murphy
University of Colorado Denver, Aurora, CO, USA
Email: robert.murphy@ucdenver.edu

THE QUEEN'S AWARDS
FOR ENTERPRISE:
INTERNATIONAL TRADE
2013

New Developments in Mass Spectrometry No. 4

Print ISBN: 978-1-84973-827-9
PDF eISBN: 978-1-78262-635-0
ISSN: 2044-253X

A catalogue record for this book is available from the British Library

Published by The Royal Society of Chemistry,
Thomas Graham House, Science Park, Milton Road,
Cambridge CB4 0WF, UK

Registered Charity Number 207890

For further information see our web site at www.rsc.org

Printed and bound by CPI Group (UK) Ltd, Croydon, CR0 4YY

Preface

The analysis of lipids by mass spectrometry has been carried out for more than six decades. Fatty acids and simple derivatives such as methyl esters were some of the first lipids investigated in biochemical studies by electron ionization (EI) mass spectrometry. The use of electron ionization and the ancillary technique of gas chromatography for structural studies of simple and complex fatty acids have been extensively covered in a previous mono-graph (*R. C. Murphy, Mass spectrometry of lipids, in Handbook of Lipid Research, ed. F. Snyder, Plenum Press, New York, 1993, vol. 7*). The major limitations of EI and gas chromatography-mass spectrometry (GC-MS) have always been the requirement of volatility of the lipids in order to become ionized within the high vacuum of the ion source. This requirement is not an issue with electrospray ionization (or MALDI) since ionization occurs outside the high vacuum of the mass spectrometer and mass separation of ions is carried out by devices such as quadrupolal electrical fields, magnetic sectors, or time-of-flight sectors under high vacuum conditions. This feature of atmospheric ionization has an enormous advantage to the investigator, as made obvious by the explosive growth of electrospray ionization and the associated technique of liquid chromatography-mass spectrometry (LC-MS) that is now applied to a large number of very diverse lipid structures in biochemical studies. The combined liquid chromatography and mass spectrometer instrument is one established as being able to analyze virtually all types of lipid substances. Unlike the very energetic ionization techniques of electron ionization, electrospray ionization is quite gentle, leading to, in most cases, a single molecular ions species such as $[M - H]^-$, $[M + H]^+$ or $[M + cation]^+$. The mechanism of electrospray ionization has been extensively studied and various reviews should be consulted to obtain a better under-standing of this ionization process, since such details will not be covered in this monograph. MALDI ionization has also been applied to the analysis of

New Developments in Mass Spectrometry No. 4
Tandem Mass Spectrometry of Lipids: Molecular Analysis of Complex Lipids
By Robert C Murphy
© Robert C Murphy 2015
Published by the Royal Society of Chemistry, www.rsc.org

virtually all lipids and the mechanism of ion formation extensively reviewed. Many of the general features of the behavior of electrospray generated molecular ion species in terms of mechanism of carbon–carbon bond cleavages directly apply to MALDI generated ions.

Both techniques generate fairly low energy ions that carry information of the intact molecular weight of the lipid. When a high resolution mass spectrometer, such as ion cyclotron resonance cell, orbitrap, the modern time-of-flight instrument are employed to measure the mass-to-charge ratio of the ion corresponding to the lipid molecular weight, one is able to calculate with excellent accuracy the elemental composition of the observed molecular ion species. However, structural details (unlike that often obtained by electron ionization) require subsequent excitation of the ion to a suffi-ciently high energy level to break covalent bonds. This is typically accom-plished in a tandem mass spectrometer, where collision between a selected ion isolated by a mass separation device takes place with neutral gas mole-cules. Often several collisions are required for a covalent to bond break and resultant product ions are then analyzed by mass. It is the observed mass-to-charge ratios (m/z) and intensity of these product ions that carry attendant structural clues of the precursor ion.

The use of tandem mass spectrometry for the analysis of lipids requires experimental determination of the best instrumental conditions to employ to generate the ion(s) of interest. These conditions include voltages applied to ions for the optimal electrospray ion yield, collision gas and pressure in collision cell, collision energy (laboratory frame of reference), resolution of precursor ion section, and product ion selection to name a few conditions. All of these parameters need to be optimized and instrument companies provide essential guidelines. These topics will not be addressed in this book since they are typically empirically found. That is not to say they are not important, but this book deals with the chemistry of product ion formation. The product ions discussed here are ones that should be observed with any instrument. However, the exact ratio of ions will likely differ between instrument and analysis conditions. The most variable ion abundance is that of the precursor mass or mass selected for collisional activation. One can collisionally activate all precursor ions so that no ion at this mass-to-charge ratio appears in the product ion spectrum. I have chosen to retain a fair amount of precursor ions in the figures generated for this book. This required in some cases to reduce the collision energy for certain lipids. References are provided for all lipids discussed in this monograph and the primary literature should be consulted to gain insight into the optimal mass spectrometer parameters to employ.

In the early 1990s, I completed the monograph *Mass Spectrometry of Lipids*, which covered, in detail, various mass spectrometric approaches to analyze complex lipids from fatty acids to phospholipids. For the vast majority of examples of lipid analysis presented, electron ionization and the closely related ionization technique chemical ionization were presented. Complex lipids that could not pass underivatized through a gas chromatograph were analyzed at that time by fast atom bombardment ionization. Electrospray ionization and MALDI were just in the early phases of development, but the promise was clearly there.

In the intervening two decades we have seen a revolution in biochemistry driven largely by electrospray ionization of proteins and peptides (proteomics) using LC-MS/MS instrumentation. The impact on lipid biochemistry has been equally revolutionary. This monograph will cover this ionization technique applied to the large number of classes of lipids that exist in nature due to enzymatic catalyzed processes. At the heart of understanding electrospray ionization and application to problems in biochemistry, I feel one must come to grips with events taking place with ions while in the mass spectrometer sectors. In other words, to begin to understand the gas phase ion chemistry taking place, not only following collisional activation (collision induced dissociation, CID), but also during the ionization process itself. Therefore, with each lipid class example, I will try to suggest mechanisms by which the observed product ions are formed if mechanisms have not been examined in detail in previous publications. In most cases, these events follow fairly simple chemical rules and it is hoped that by working with specific examples, the reader can expand their facility of understanding ions derived from complex lipids and then apply these principles to more complex lipids and to solve complex biochemical problems. In some cases, these suggestions as to mechanism will be in error but it is hoped that my suggestions will encourage investigators to examine, in detail, the protons being rearranged and bonds broken and formed, then to devise experiments using stable isotope labeling, theoretical calculations, and chemical synthesis of intermediates, so that the correct mechanism can be revealed.

Robert C. Murphy,
University of Colorado,
Aurora, CO, USA

Contents

New Developments in Mass Spectrometry No. 4
Tandem Mass Spectrometry of Lipids: Molecular Analysis of Complex Lipids
By Robert C Murphy
© Robert C Murphy 2015
Published by the Royal Society of Chemistry, www.rsc.org

Acknowledgements

I would like to acknowledge Christopher Johnson, Thomas Leiker, and Charis Uhlson for the many original tandem mass spectra of lipids in this monograph. Their skill and dedication is gratefully appreciated. The tandem quadrupole mass spectrometers employed to generate these tandem mass spectra were AB Sciex API 3200, 4000, and 5500 instruments. The high resolution tandem mass spectra were recorded on a quadrupole time-of-flight instrument, Waters G2S. Many lipid samples were kindly provided by Walter Shaw at Avanti Polar Lipids and eicosanoids by Kirk Maxey at Cayman Chemical Company. The critical review of various chapters by Robert Barkley, Joseph Hankin, Miguel Gijón, Karin Zemski Berry, and Simona Zarini were exceedingly helpful and their professional insight into biochemical mass spectrometry is reflected in this book. Deborah Beckworth, who has been my assistant for over 25 years, endured my dictations and produced the first written draft of the entire book. Her considerable skills make this monograph possible and I thank her for her loyalty and hard work.

Over the course of forty years as a faculty member of the University of Colorado, my research into lipid biochemistry has been supported by the National Institutes of Health, which made possible the acquisition of mass spectrometers and research into the ion chemistry of various lipids. These grants included R01-HL25785-35, P01-HL34303–30, R01-ES022172, and most importantly U54-GM069339-10. I would like to thank Dr. Jean Chin, Program Director at NIH General Medical Science Division for her support, guidance and friendship as the LipidMaps grant administrator.

New Developments in Mass Spectrometry No. 4
Tandem Mass Spectrometry of Lipids: Molecular Analysis of Complex Lipids
By Robert C Murphy
© Robert C Murphy 2015
Published by the Royal Society of Chemistry, www.rsc.org

CHAPTER 1

Fatty Acids

Fatty acids are the simplest and most fundamental lipids found in biology, and they are readily analyzed by mass spectrometry. A rich literature exists concerning the analysis of these structures, primarily using electron ionization (EI); from saturated, straight chain species to more complex unsaturated and alkyl-substituted examples. EI is quite an energetic mode of ionization that leads to substantial ion decomposition from which, when followed by subsequent mass spectrometric analysis, a wealth of structural information can be deduced. The ion chemistry of these events was covered extensively in the first monograph on this topic by this author.[1]

Electrospray ionization (ESI) of free fatty acids (RCOOH) effectively yields molecular ion species at higher efficiency than EI; both positive $[M + H]^+$ and $[M + cation]^+$ as well as negative ions $[M - H]^-$, but in a manner that leads to little, if any, subsequent fragmentation.[2] Therefore, to glean any structural details concerning fatty acids it is necessary to impart internal energy to these fatty acid molecular ion species, sufficient to break covalent bonds. In almost all cases this is accomplished by collisional activation (collision induced dissociation, CID) with a neutral gas molecule in a collision cell, such as Rf-only quadrupole sector, or by increasing the angular momentum of the ion in a Paul-type ion trap or ion cyclotron resonance instrument where it can collide with neutral gas molecules in the cell.

Fatty acids exist in biological matrices as the deprotonated, ionized carboxylic acid (carboxylate anion) or esterified to glycerolipids, glycerophospholipids, wax esters, various saccharolipids, as well as cholesterol esters (steroid alcohol such as ergosterol in fungi) along with other less abundant esters of complex lipids. Many sphingolipids have long chain fatty acyl substituents derived from fatty acids, but exist in this class of

New Developments in Mass Spectrometry No. 4
Tandem Mass Spectrometry of Lipids: Molecular Analysis of Complex Lipids
By Robert C Murphy
© Robert C Murphy 2015
Published by the Royal Society of Chemistry, www.rsc.org

lipids as N-acyl amides that are very stable to base hydrolysis. Analysis of fatty acids can involve electrospray ionization of the free fatty acids or the fatty acids liberated from the various esterified fatty acids after saponification.

The isolation of free fatty acids is very facile and can be carried out by solvent extraction techniques. A typical experiment might involve isolation of all simple and complex fatty acids using techniques such as the Folch or Bligh/Dyer extraction followed by saponification using 0.5 M NaOH at 37 °C for 1 hour.[3,4] This solution is then neutralized and extracted by hexane or ethyl acetate to isolate the very nonpolar fatty acids in the organic layer in high yield. Subsequent electrospray analysis involves dissolving these isolated fatty acids in a solvent system that permits ESI, such as a methanol/ water system containing ammonium acetate (5 mM) to provide electrical conductivity to the spray solvent.

As discussed below, the saturated fatty acids are difficult to collisionally activate to yield structural informative ions, unlike the case for EI mass spectrometry. While molecular weight information is obtained, a definitive structure such as possible branching points are difficult, if not impossible, to assign from electrospray generated positive or negative ions. When high resolution mass analysis is employed to detect these fatty acid molecular species, the exact elemental composition of the ion can be readily calculated, since these are usually below m/z 500 and high resolution mass analyzers currently marketed readily achieve 5 ppm or less mass error measurement. When using ESI, definitive analysis of a fatty acid, even for saturated fatty acids, requires additional information such as chromatographic retention times relative to standard fatty acids to unambiguously assign the structure of these molecules. This is due to potential methyl branched alkyl chains and double bond regioisomers. Alternatively, making a derivative such as a methyl ester followed by GC-MS analysis remains a powerful strategy, as previously described.[1]

1.1   Carboxylate Anions [M − H]⁻

Most common fatty acids typically have a single, polar functional group, namely, the carboxylic acid moiety. In aqueous solution, the pKa of most fatty acids is between 4 and 5, making them highly ionized at neutral pH as the RCOO⁻ form and, not surprisingly, when electrosprayed in a solvent system buffered to pH 5 or above, yields a robust appearance of the carboxylate anion [M − H]⁻ emerging from the ion source. The collisional activation of fatty acids is exemplified by four different fatty acids that are typically encountered in biological extracts (Figure 1.1). The [M − H]⁻ ions derived from common fatty acids found in most biological samples are listed in Table 1.1 along with product ions observed following collisional activation. As can be seen from these examples, very few product ions are observed for saturated and monounsaturated fatty acids; however, there are product ions that

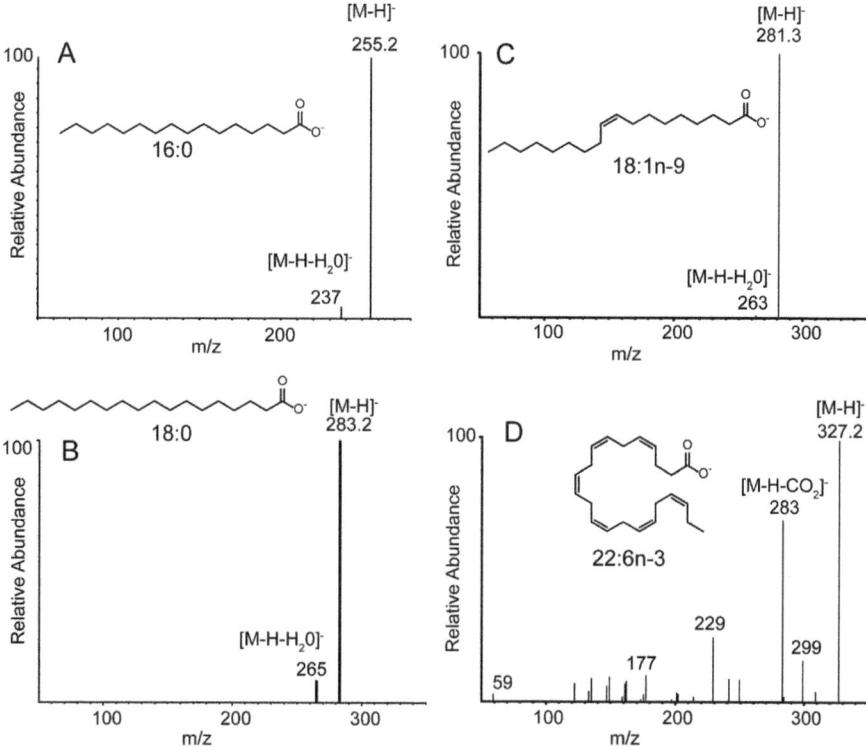

Figure 1.1 Electrospray ionization (negative ions) and tandem mass spectrometry of common fatty acids as their carboxylate anions. (A) Product ions obtained following collisional activation of palmitic acid (16 : 0), m/z 255; (B) product ions obtained following collisional activation of stearic acid (18 : 0) $[M - H]^-$ at m/z 383; (C) product ions obtained following collisional activation of oleic acid (18 : 1, n-9) $[M - H]^-$ at m/z 281; (D) product ions obtained following collisional activation of docosahexaenoic acid (22 : 6, n-3) $[M - H]^-$ at m/z 327. These MS/MS spectra were obtained using a tandem quadrupole mass spectrometer.

have been reported when alkali attachment cations are collisionally activated, as will be discussed below.

Collisional activation of the $[M - H]^-$ from saturated fatty acids typically only yields $[M - H - H_2O]^-$,[5] which is not structurally definitive. Yet this ion is interesting, since it must involve the loss of two protons from the alkyl chain of the fatty acid. Study of this ion using techniques such as resonance capture ionization has suggested a cyclic structure for this ion (Scheme 1.1),[6] but in theory it could proceed from a negative ion charge localized at a site along the alkyl chain (migration of H^+ to the carboxylate anion) followed by the loss of the small neutral water molecule. This does occur to some extent with saturated fatty acids, but is more favorable for polyunsaturated fatty acids when the charge site on the alkyl chain can be delocalized over many atoms.

Table 1.1 Negative molecular ions [M − H]⁻ from the electrospray ionization of common fatty acids and expected losses of H_2O and CO_2 after collisional activation in a tandem mass spectrometer.

Abbreviation[a]	n-x[b] nomenclature	Δ-nomenclature[c] Double bond positions	Common name as free acid	[M − H]⁻ m/z	Elemental composition	[M − H₂O]⁻ m/z	[M-CO₂]⁻ m/z
12 : 0			Lauric	199.170	$C_{12}H_{23}O_2$	181.159	
14 : 1	n-5	9	Myristoleic	225.186	$C_{14}H_{25}O_2$	207.175	
14 : 0			Myristic	227.201	$C_{14}H_{27}O_2$	209.190	
16 : 1	n-7	9	Palmitoleic	253.216	$C_{16}H_{29}O_2$	235.205	
16 : 0			Palmitic	255.232	$C_{16}H_{31}O_2$	237.221	
18 : 4	n-3	6,9,12,15	Stearidonic	275.202	$C_{18}H_{27}O_2$	257.191	231.212
18 : 3	n-6	6,9,12	Gamma linolenic	277.217	$C_{18}H_{29}O_2$	259.206	233.227
18 : 3	n-3	9,12,15	Alpha linolenic	277.217	$C_{18}H_{29}O_2$	259.206	233.227
18 : 2	n-6	9,12	Linoleic	279.232	$C_{18}H_{31}O_2$	261.221	
18 : 1	n-9	9	Oleic	281.248	$C_{18}H_{33}O_2$	263.237	
18 : 0			Stearic	283.264	$C_{18}H_{35}O_2$	265.253	
20 : 5	n-3	5,8,11,14,17	Eicosapentaenoic	301.217	$C_{20}H_{29}O_2$	283.206	257.227
20 : 4	n-6	5,8,11,14	Arachidonic	303.232	$C_{20}H_{31}O_2$	285.221	259.242
20 : 3	n-6	8,11,14	Dihomo-γ-linoleic	305.248	$C_{20}H_{33}O_2$	287.237	261.258
20 : 3	n-9	5,8,11	Meade	305.248	$C_{20}H_{33}O_2$	287.237	261.258
20 : 2	n-9	11,14	Eicosadienoic	307.264	$C_{20}H_{35}O_2$	289.253	
20 : 1	n-9	11	Eicosenoic	309.279	$C_{20}H_{37}O_2$	291.268	
20 : 0			Arachidic	311.295	$C_{20}H_{39}O_2$	293.284	
22 : 6	n-3	4,7,10,13,16,19	Docosahexaenoic	327.232	$C_{22}H_{31}O_2$	309.221	283.242
22 : 5	n-3	7,10,13,16,19	Docosapentaenoic	329.248	$C_{22}H_{33}O_2$	311.237	285.258
22 : 4	n-6	7,10,13,16	Adrenic	331.264	$C_{22}H_{35}O_2$	313.253	287.274
22 : 1	n-9	13	Erucic	337.311	$C_{22}H_{41}O_2$	319.300	
24 : 1	n-9	15	Nervonic	365.349	$C_{24}H_{45}O_2$	347.338	
24 : 0			Lignoceric	367.358	$C_{24}H_{47}O_2$	349.347	

[a]Total carbon atoms: total number of double bonds.
[b]Position of the double bond counting from the terminal methyl carbon atom. IUPAC favors the n-x nomenclature rather than "omega or ω-x" nomenclature.
[c]The Δˣ nomenclature indicates the position of all double bonds counting from the carboxylic acid moiety.

Structural characterization of saturated fatty acids has been extensively pursued by electron ionization mass spectrometry, where ion formation is much more energetic and decomposition processes that lead to structurally informative product ions are more favorable. In addition, chromatographic separation, in particular using capillary gas chromatography or high performance liquid chromatography, are powerful strategies to combine with molecular weight information (mass spectrometry) to assign a more definitive structure to the observed molecular anions through strategies such as co-elution with authentic standards. An example is the liquid chromatographic separation of a series of common fatty acids revealed by the observed molecular anions (Figure 1.2).

[M-H]-
Arrows
-H2O
[M-H2O]-

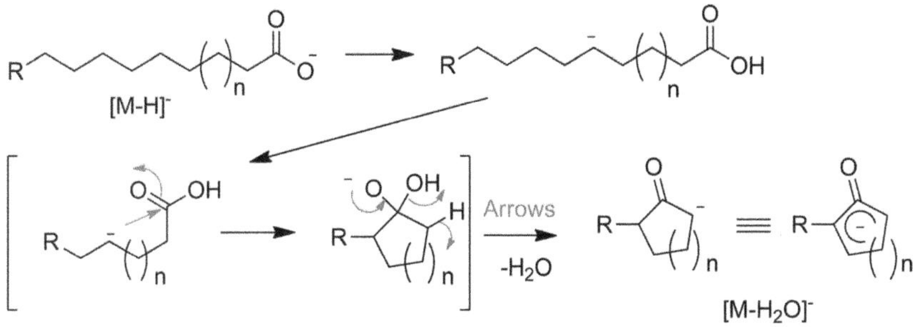

Scheme 1.1

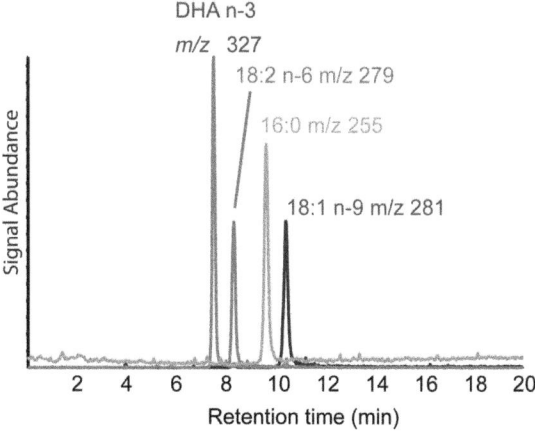

DHA n-3
m/z 327
18:2 n-6 m/z 279
16:0 m/z 255
18:1 n-9 m/z 281
Signal Abundance
2 4 6 8 10 12 14 16 18 20
Retention time (min)

Figure 1.2 Reversed phase HPLC separation of common fatty acids found in biological extracts, including palmitic (16 : 0), oleic (18 : 1, n-9), linoleic (18 : 2, n-6), and docosahexaenoic acid (22 : 6, n-3). Ions were detected as carboxylate anions and the elution of each fatty acid indicated as an extracted ion chromatogram for the corresponding carboxylate anions.

1.2   Alkali Metal Adduct Fatty Acid Cations

ESI can also be used to generate positive ions from fatty acids and this mode of ion formation has largely been employed by making alkali metal adducts such as $[M - H + Na_2]^+$, $[M - H + Li_2]^+$, and $[M - H + Ba]^+$. Such ions are interesting in that they have a complete closed shell of electrons and are thus quite stable. The formation of these adduct ions requires dissolving the free fatty acid in an appropriate solvent typically containing 1–3 mM alkali metal salt as the acetate or formate form of the corresponding cation. The yield of the attachment ion products can be less than quantitative in that multiple adduct species are often observed such as $[M + Na]^+$, $[M + Li]^+$, as well as the desired $[M - H + 2Li]^+$ for lithiated species (Table 1.2). In such cases the origin of the sodium ions is likely from the borosilicate glass surfaces used in vials to prepare the samples. There have not been detailed studies reported to optimize a single adduct of fatty acid from such alkali metal salts and multiple species of cations from a single fatty acid is a complication not seen in negative ion ESI where just a single $[M - H]^-$ is readily observed.

An advantage of the mass spectrometric analysis of such positive ions has been reports of product ions being formed following collisional activation, even at relatively low energies. For saturated fatty acids a series of alkali ions formed by fast atom bombardment ionization was reported a number of years ago following collisional activation of $[M - H + 2Li]$ and $[M - H + Ba]^+$ by charge site remote fragmentation mechanisms (CRF), also termed "remote site fragmentation".[7] This CRF process took place when precursor ions were accelerated to high velocities just prior to collision with a neutral gas molecule in a collision cell, initiating a high energy-decomposition reaction (Scheme 1.2). It now appears that low energy collisional activation of $[M - H + 2Li]^+$ ions generated by ESI in a tandem quadrupole instrument can also undergo CRF-driven decomposition.[8] Mechanisms for ion formation had been reported using isotope labeled fatty acids and include charge-driven as well as charge-remote rearrangements of carbon–carbon bonds. The general utility of these alkali attachment ions and interpretation of CID events for structural characterization of unknown fatty acids has not been widely reported, but this technique does offer some promise when applied to the analysis of unsaturated and polyunsaturated fatty acids.

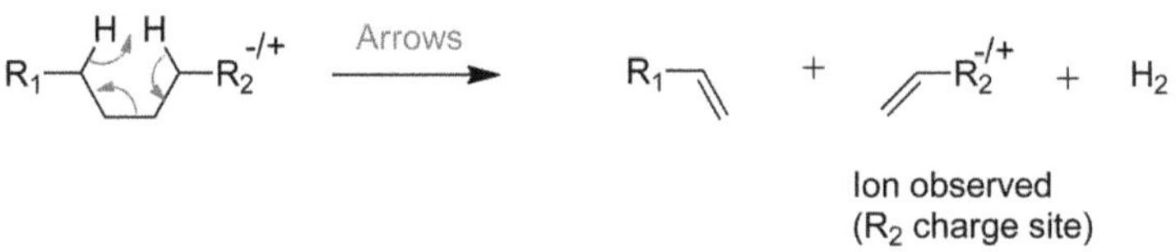

Scheme 1.2

Table 1.2   Positive ions from the electrospray ionization of common fatty acids as protonated, sodiated and lithiated species.

Abbreviation[a]	n-x[b] nomenclature	Common name as free acid	Elemental composition	[M + H]+ m/z	[M − H + 2Na]+ m/z	[M − H + 2Li]+ m/z
12 : 0		Lauric	$C_{12}H_{25}O_2$	201.185	245.148	213.201
14 : 1	n-5	Myristoleic	$C_{14}H_{27}O_2$	227.201	271.164	239.217
14 : 0		Myristic	$C_{14}H_{29}O_2$	229.217	273.180	241.233
16 : 1	n-7	Palmitoleic	$C_{16}H_{31}O_2$	255.232	299.195	267.248
16 : 0		Palmitic	$C_{16}H_{33}O_2$	257.248	301.211	269.264
18 : 4	n-3	Stearidonic	$C_{18}H_{29}O_2$	277.217	321.180	289.233
18 : 3	n-6	Gamma linolenic	$C_{18}H_{31}O_2$	279.232	323.195	291.248
18 : 3	n-3	Alpha linolenic	$C_{18}H_{31}O_2$	279.232	323.195	291.248
18 : 2	n-6	Linoleic	$C_{18}H_{33}O_2$	281.248	325.211	293.264
18 : 1	n-9	Oleic	$C_{18}H_{35}O_2$	283.264	327.227	295.280
18 : 0		Stearic	$C_{18}H_{37}O_2$	285.279	329.242	297.295
20 : 5	n-3	Eicosapentaenoic	$C_{20}H_{31}O_2$	303.232	347.195	315.248
20 : 4	n-6	Arachidonic	$C_{20}H_{33}O_2$	305.248	349.211	317.264
20 : 3	n-6	Dihomo-γ-linoleic	$C_{20}H_{35}O_2$	307.264	351.227	319.280
20 : 3	n-9	Meade	$C_{20}H_{35}O_2$	307.264	351.227	319.280
20 : 2	n-9	Eicosadienoic	$C_{20}H_{37}O_2$	309.279	353.242	321.295
20 : 1	n-9	Eicosenoic	$C_{20}H_{39}O_2$	311.295	355.258	323.311
20 : 0		Arachidic	$C_{20}H_{41}O_2$	313.311	357.274	325.327
22 : 6	n-3	Docosahexaenoic	$C_{22}H_{33}O_2$	329.248	373.211	341.264
22 : 5	n-3	Docosapentaenoic	$C_{22}H_{35}O_2$	331.264	375.227	343.280
22 : 4	n-6	Adrenic	$C_{22}H_{37}O_2$	333.279	377.242	345.295
22 : 1	n-9	Erucic	$C_{22}H_{43}O_2$	339.326	383.289	351.342
24 : 1	n-9	Nervonic	$C_{24}H_{47}O_2$	367.358	411.321	379.374
24 : 0		Lignoceric	$C_{24}H_{49}O_2$	369.373	413.336	381.389

[a]Total carbon atoms: total number of double bonds.
[b]Position of the double bond counting from the terminal methyl carbon atom. IUPAC favors the n-x nomenclature rather than "omega or ω-x" nomenclature.

1.3 Monounsaturated and Polyunsaturated Fatty Acids

The situation for analysis of unsaturated fatty acids is, on one hand, much better in terms of obtaining abundant product ions following collisional activation of molecular ion species, but on the other hand, it is made more difficult because of the introduction of stereochemical uncertainty such as double bond position and double bond geometry for unsaturated fatty acids. The addition of the π-orbitals to the molecular structure of these fatty acids facilitates carbon–carbon bond cleavage for either the carboxylate anion or the metalated carboxylate cation. In some cases product ion formation is relevant to the position of the double bonds and as the number of double bonds increases, additional pathways become open for ion decomposition events. The mechanisms for the reactions have only been partially studied but nonetheless, they have become definitive for structural identification of biologically derived fatty acids.

1.3.1 Monounsaturated Fatty Acid Negative Ions

The major product ions from a number of monounsaturated fatty acid molecular anions $[M - H]^-$ observed at low collision energy, is a loss of water $[M - H - H_2O]^-$; the exact origin or origins of the protons lost along with the carboxylate oxygen atom have not been fully elucidated, but most likely they are the result of formation of a cyclic anion (Scheme 1.1).

Additional product ions include cleavage of the carbon–carbon bond adjacent (α) or vinylic (β) to the double bond (Figure 1.3). A mechanism consistent with octadec-9-enoate (18 : 1 n − 9) and the unsaturated fatty acids presented in this figure could involve a change remote fragmentation.[5] The driving force for such a mechanism would involve breaking of the C–H bond allylic to the double bond by the electron withdrawing properties of the π-orbital with concerted formation of a stable conjugated double bond on one fragment, loss of hydrogen gas (H_2) and a stable olefin (Scheme 1.2). An alternative reaction for the abundant ion at m/z 127 (Figure 1.3A) is an "ene" reaction, seen quite frequently in lipid collisional spectra which is a proton abstraction reaction driven by a double bond, as shown in Scheme 1.3. This gas phase behavior is surprisingly common for many unsaturated fatty acids, either as carboxylate anions or metalated cations, and since the carbon–carbon bond formation does not involve those atoms bearing the charge site, the term "remote site fragmentation" is relevant.

1.3.2 Monounsaturated Fatty Acid Alkali Attachment Cations

The dilithiated octadec-9-enoate $[M - H + Li_2]^+$ cation can be used as an example of the behavior of monounsaturated fatty acids that have a closed shell of electrons (Figure 1.4A). These species are readily formed by ESI when alkali metal salts such as LiCl or Li-acetate are added to the electrospray

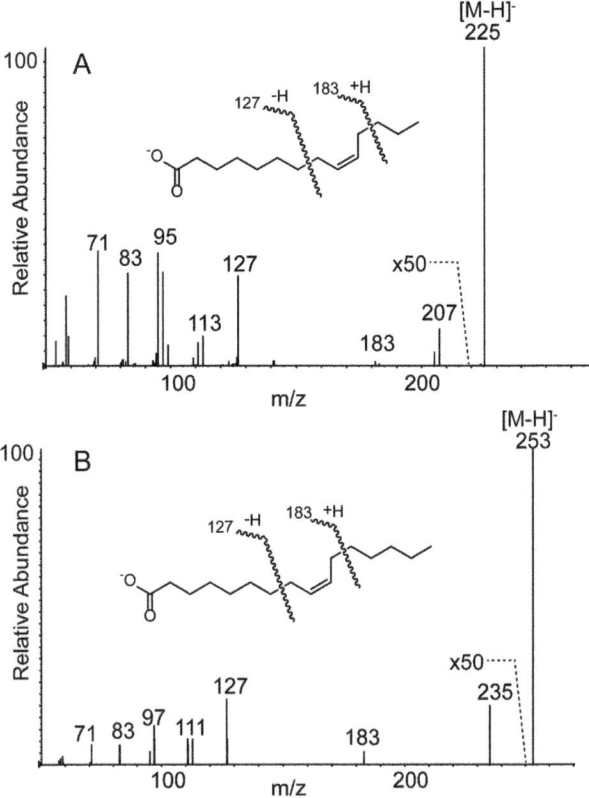

Figure 1.3 Electrospray ionization (negative ions) and tandem mass spectrometry of monounsaturated fatty acids. (A) Product ions obtained following collisional activation of the carboxylate anion of 9-tetradecenoic acid [M − H]⁻ at *m/z* 225. (B) Product ions obtained following collisional activation of 9,13-octadecadienoic acid [M − H]⁻ at *m/z* 253. These MS/MS spectra were obtained using a tandem quadrupole mass spectrometer.

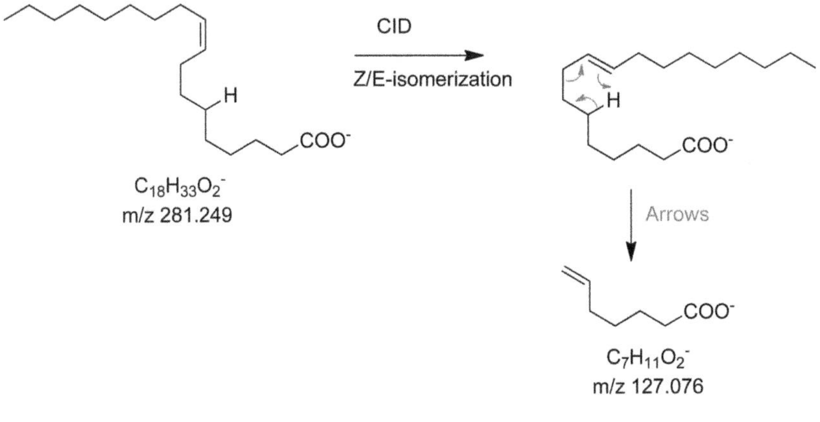

Scheme 1.3

solvent system. Unfortunately, the formation of this cationized carboxylate may not be quantitative and other cationic species are often observed, including $[M + H]^+$, $[M + Li]^+$, $[M - H + Li_2]^{+2}$, and even $[M - H + Na_2]^+$. The exact alkali metal salt, ESI conditions and even the ESI ion source employed, can alter the relative ratios of all of these ions. While this makes it less attractive to study these species, the advantage of making these adducts is that detailed studies have been reported concerning their behavior and thus they provide specific information that is useful in predicting subtle changes in structure with an unknown fatty acid.

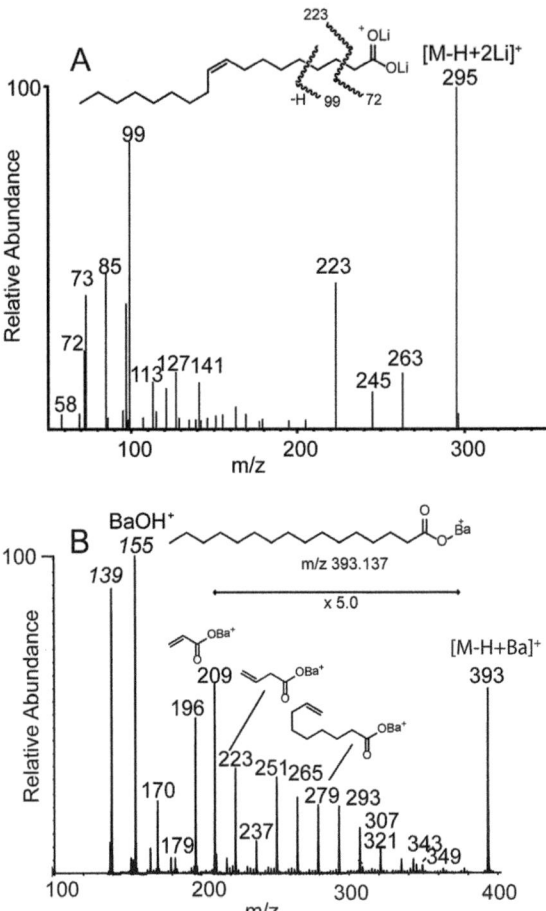

Figure 1.4 Electrospray ionization (positive ions) and tandem mass spectrometry of metalated carboxylate anions. (A) Product ions obtained following collisional activation of the dilithiated carboxylate anion of oleic acid $[M - H + 2Li]^+$ at m/z 295; (B) product ions obtained following collisional activation of the bariated carboxylate anion of palmitic acid $[M - H + Ba]^+$ at m/z 393. These MS/MS spectra were obtained using a tandem quadrupole mass spectrometer.

A common abundant product ion for the dilithiated cation from mono-unsaturated and polyunsaturated fatty acids with a first double bond at least seven carbons from the carboxyl group appears at an odd mass and corresponds to a charge remote fragmentation involving the double bond nearest the cationized carboxyl group. This arises from oleate (18 : 1 n-9) following cleavage of carbon 4-5 and a single proton rearrangement, presumably from carbon-4. A mechanism consistent with this ion formation has been suggested to be a charge remote elimination reaction of H_2 and formation of two double bonds after cleavage of carbon 4-5 (Scheme 1.4).[7]

Another consistent remote site mechanism that does not involve the loss of H_2 would be driven by proton abstraction by the double bond. This ene reaction drives cleavage of the carbon–carbon bond, 5-carbon atoms removed, leading to a cyclopentane neutral and a terminal olefin attached to the dilithiated carboxylate cations (Scheme 1.5).

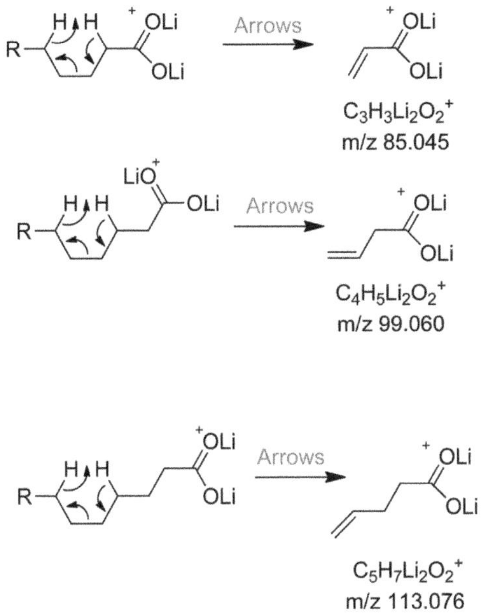

$C_3H_3Li_2O_2^+$
m/z 85.045

$C_4H_5Li_2O_2^+$
m/z 99.060

$C_5H_7Li_2O_2^+$
m/z 113.076

Scheme 1.4

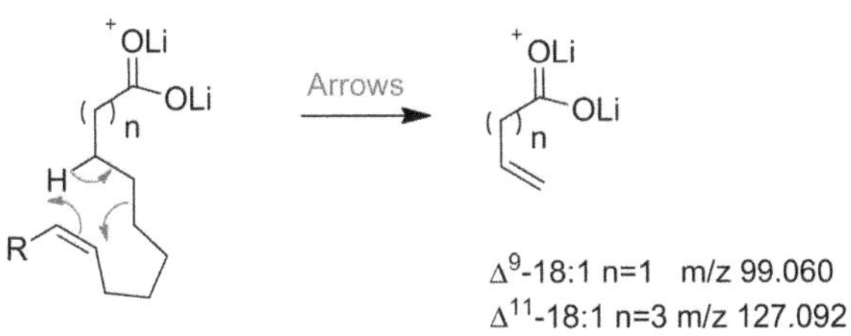

Δ^9-18:1 n=1   m/z 99.060
Δ^{11}-18:1 n=3 m/z 127.092

Scheme 1.5

A series of radical cation ions (even *m/z*) are observed at a low *m/z* that corresponds to homolytic bond cleavage near the metalated carboxylate cation. Virtually all $[M - H + 2Li]^+$ fatty acids (saturated, monounsaturated, and polyunsaturated) have these characteristic ions,[8] but their abundance varies in an interesting manner that can be used to deduce a proximal position of the first double bond when it is quite near the carboxylate moiety. For Δ^4-fatty acids (*e.g.* docosahexaenoate-lithium adduct) the most abundant ion appears at *m/z* 72, likely abundant because of the loss of a neutral radical species delocalized by the Δ^4 double bond shown in Scheme 1.6.

The Δ^5-series, such as is found in arachidonic acid and EPA, has the metalated carboxylate radical cation as the most abundant low mass ion (*m/z* 58), perhaps due to a stabilized loss of the highly delocalized radical illustrated in Scheme 1.6. These Δ^5-species still have an abundant *m/z* 72 radical cation that can be stabilized by the resonance structure illustrated (Scheme 1.6). The Δ^6 unsaturated fatty acids yield a radical cation of highest abundance observed at *m/z* 72 due to both delocalization the radical site in the neutral hydrocarbon chain that is lost, and delocalization of the radical by the carboxylate electrons. For the location of the nearest double bond more distal than Δ^6, remote site fragmentation appears to be the most abundant process occurring in the production of these low mass product ions (Scheme 1.4 and 1.5).

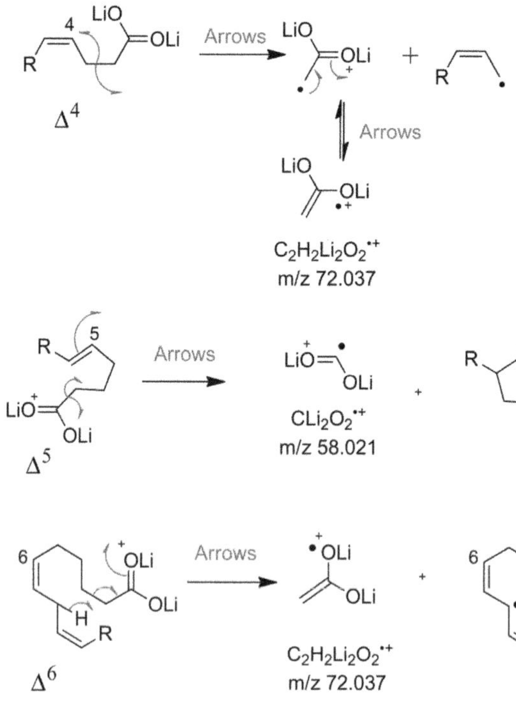

Scheme 1.6

These same ion series of low mass ions (Scheme 1.4) are also observed following CID of the barium ion adduct $[M - H + Ba]^+$ (Figure 1.4B) at m/z 196, 209, and 223, presumably following the same mechanism of decomposition of the molecular ion species at m/z 419 from bariated oleate.[9] The most abundant ion is seen at m/z 209 for this barium adduct that corresponds to charge remote cleavage of carbon bond 3-4, as illustrated for m/z 85 for lithiated carboxylate cations and may reflect unique weakening of the C–H bond at C-2 due to the affinity of Ba^{++} for the electrons at the carboxylate oxygen atoms (Scheme 1.4). An ion at m/z 196 is found quite abundant and likely an analog of the structure suggested in Scheme 1.6 for m/z 72, again due to the possible resonance structures of this ion.

1.3.3 Polyunsaturated Carboxylate Anions

Considerably more structurally relevant information is generated following collisional activation of polyunsaturated (as homoconjugated double bonds) fatty acid anions, as exemplified by the CID of arachidonic acid (AA) anion (Figure 1.5). In addition to the loss of H_2O (m/z 285) and CO_2 (m/z 259), the next most abundant product ion could, at first glance, correspond to cleavage of a double bond between carbon atoms 5 and 6 (m/z 205). This site of cleavage is consistent with the ion at m/z 212 from D_8-AA (which has all double bond methine hydrogen atoms labeled with deuterium) (Figure 1.5B), which is shifted to a +7 ion. However, the origin of this ion at m/z 205 must be more complex since there is also a significant abundance at +6 (m/z 211) from the D_8-AA. In addition, high resolution analysis of the ion at m/z 205 yields a doublet at m/z 205.196 ($C_{15}H_{24}$) and 205.123 ($C_{13}H_{17}O_2$).

Two possible pathways that would account for the carbon 5-6 double bond cleavage are presented in Scheme 1.7. One would be a charge remote double 1[3]-sigmatropic arrangement of the double bond to the $\Delta^{3,4}$ position then loss of neutral CO_2 and butadiene, yielding a carbanion at carbon-5 at m/z 205.196. This double bond rearrangement could also be charge driven by the carboxylate anion that could readily remove a proton from carbon-4, generating a conjugated carbanion site carbon-6 and shifting the double bond to Δ^4. This carbanion structure could undergo additional proton abstraction at carbon-13, then carbon-3, reverting to the same $\Delta^{3,4}$ carboxylate anion structure in Scheme 1.7. While the carbanion site would seem to be rather energetic, proton abstraction at carbon-13 would lead to a much more stable delocalized anion through conjugation with the two adjacent double bonds (Scheme 1.7).

The deuterium labeled arachidonate, as well as high resolution measurements of m/z 205, revealed formation of a second ion at m/z 205. The origin of this second product that is consistent with this data involves loss of the terminal 7 carbon atoms (along with both protons at C-14,15 in deuterium labeled in D_8-AA as well as the 19,19,20,20-D_4-AA),[5] but shifted to m/z 211 for D_8-AA, as in Figure 1.5B (Scheme 1.8). This mechanism involves an allylic proton transfer and vinylic bond cleavage described by Hsu and Turk and is

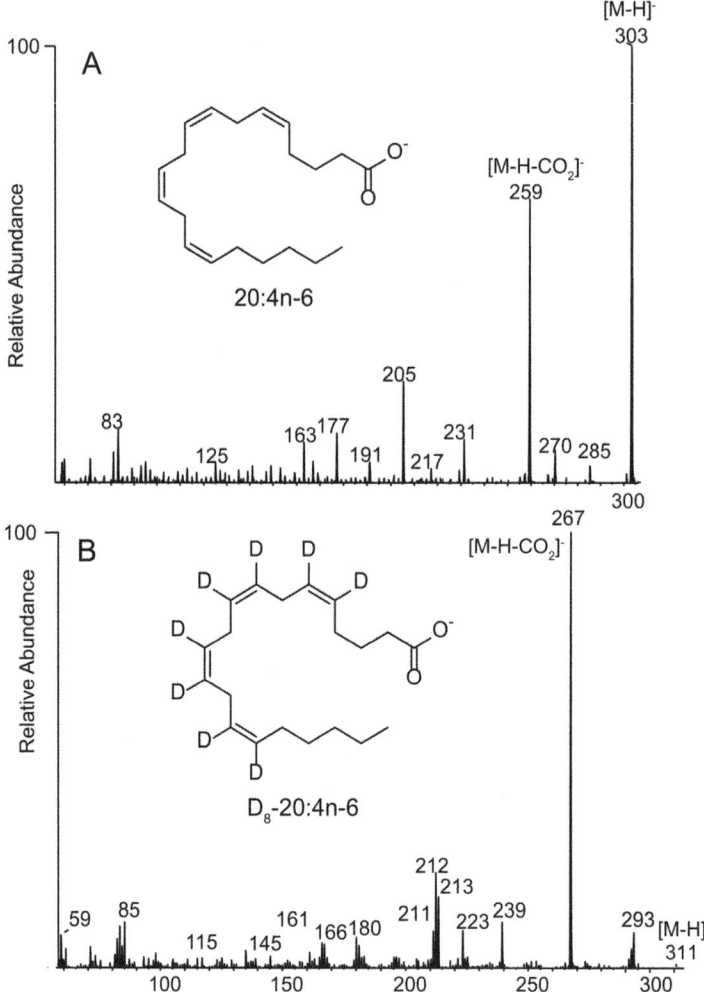

Figure 1.5 Electrospray ionization (negative ions) and collisional activation of the carboxylate anions of arachidonic acid and D$_8$-arachidonic acid. (A) Product ions obtained following collisional activation of arachidonic acid (20 : 4, n-6) [M − H]$^-$ at *m/z* 303; (B) product ions obtained following collisional activation of D$_8$-arachidonic acid [M − H]$^-$ at *m/z* 311. The location of the deuterium atoms in this labeled molecule is indicated in the structural inset. These MS/MS spectra were obtained using a tandem quadrupole mass spectrometer.

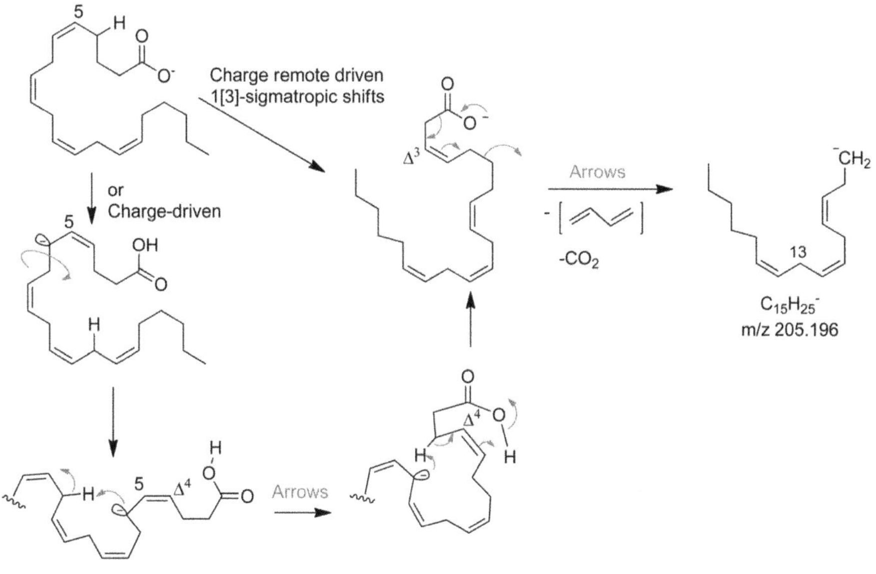

Scheme 1.7

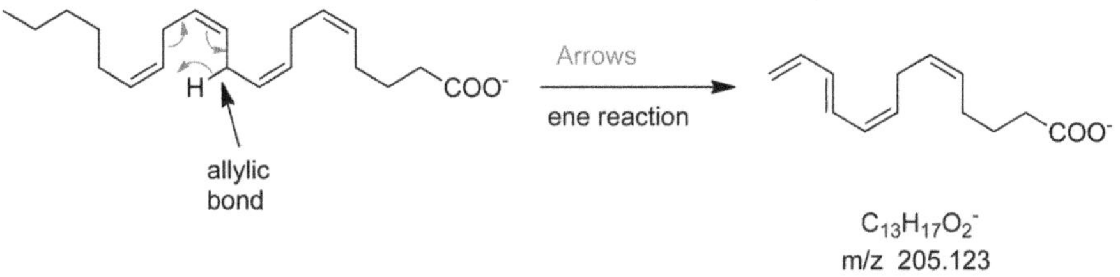

Scheme 1.8

another example of an ene proton transfer reaction.[8] The loss of this C_7H_{14} neutral species would leave a highly stable conjugated triene anion at m/z 205.123.

Docosahexaenoate anion behaves in a similar manner in terms of cleavage of susceptible double bonds, which are both allylic and vinylic. The product ion at m/z 229 (Figure 1.1) is likely formed by a similar mechanism as in Scheme 1.7 after shift of the double bond at Δ^4 to Δ^3 and charge driven loss of CO_2 and butadiene.

1.3.4 Positive Ions Polyunsaturated Fatty Acids

Metal ion attachment to the carboxylate moiety of fatty acids has been used to generate positive ions from polyunsaturated fatty acids. The advantage of this approach is to render a very stable closed shell ion (cationic) where this charge site does not easily leave the carboxyl moiety, even when ions receive additional energy. When collisions of this stable cation occur in a tandem quadrupole or ion trap-type mass spectrometer, excitation of the ion takes place, which is often relaxed by charge remote fragmentation mechanisms. Structurally useful reactions occur for the mono- and polyunsaturated fatty acid species as described for the Li$^+$ adducts by Hsu and Turk.[8] Since biosynthesis of polyunsaturated fatty acids (PUFA) leads to homoconjugated double bonds that are classified as to the double bond closest to the methyl terminus, indicated by "n- terminology" (*e.g.* n-6 or n-3 PUFAs). These positions are largely determined by two carbon atom elongases, such as EOVL1.[10] The position of the first double bond, counting from the carboxyl carbon atom, is designated by the "delta nomenclature" (*e.g.* Δ^5, Δ^9, Δ^{11}, *etc.*) that indicates the number of carbon atoms between the carboxyl group and the first double bond. The exact position of the first homoconjugated double bonds is controlled by fatty acyl CoA ester desaturases such as Δ^5-desaturase or Δ^6-desaturase.[10] These double bond positions become important targets for structure elucidation and CID mass spectrometry of these closed shell cations offered by alkali metal adducts.

Several examples of diverse PUFAs have been examined as Li$^+$ adducts and mechanism of product ion formation suggested (8). While these mechanisms have not been rigorously tested, they are consistent with high resolution measurements and limited availability of isotopically labeled PUFAs. The CID of three example PUFAs (arachidonic acid, Δ^5, ω-6; docosatetraenoic acid, Δ^7, ω-6; and docosahexaenoic acid, Δ^4, ω-3) illustrate major mechanisms of ion formation (Figure 1.6).

Allylic/vinylic carbon bond cleavages (A–V cleavage) with hydrogen transfer and formation of conjugated product ions or neutrals dominate the decomposition of these molecules after collisional activation (Scheme 1.9) and the observed ion depends upon which side (R$_1$ or R$_2$) the lithiated carboxyl group is located. The C–C bond cleaved in this scheme is both allylic to carbon *A* and vinylic to carbon *B*. Fragment X or Y can either be the resultant product ion since the charge remote fragmentation is driven by the stability of both products.

The ion of highest mass-to-charge ratio, which is also the most abundant in this region for each PUFA in Figure 1.6 correspond to this A–V cleavage at the n-terminus. This ion can classify the last double bond position in the PUFA and, therefore, the n-family type.

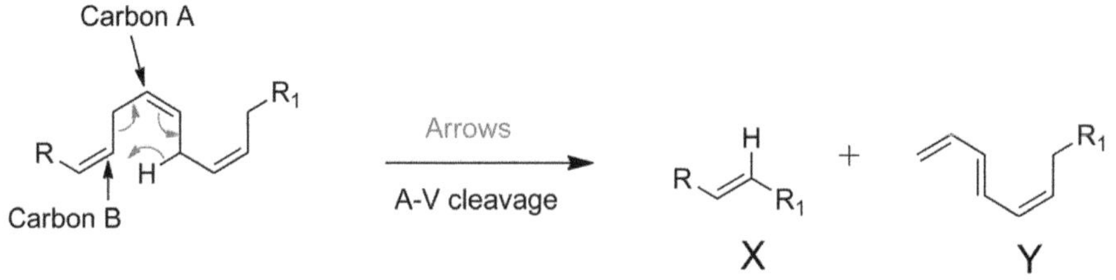

Scheme 1.9

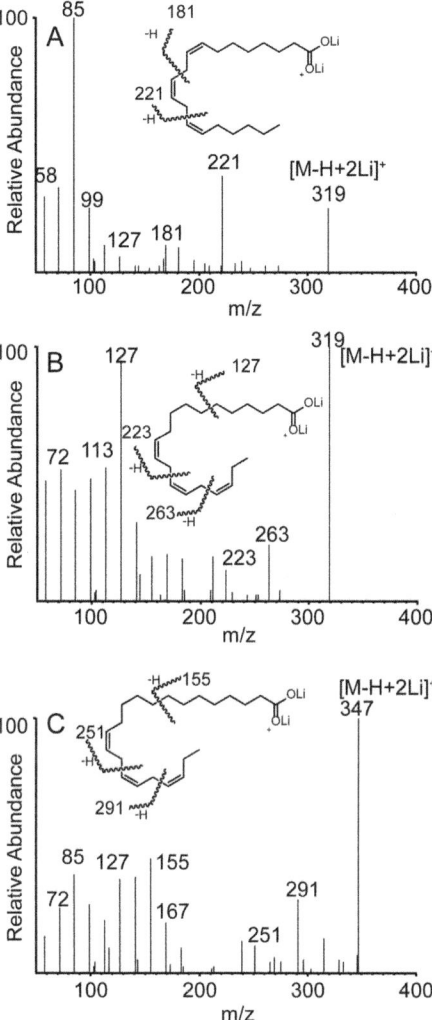

Figure 1.6 Electrospray ionization (positive ions) and tandem mass spectrometry of polyunsaturated fatty acids as their dilithiated salts. (A) Product ions obtained following collisional activation of dilithiated adducts of 8,11,14-eicosatrienoic acid $[M - H + 2Li]^+$ at m/z 319; (B) Product ions derived from the dilithiated adduct of 11,14,17-eicosatrienoic acid $[M - H + 2Li]^+$ at m/z 319; (C) Product ions obtained following collisional activation of the dilithiated adducts of 3,16,19-docosatrienoic acid $[M - H + 2Li]^+$ at m/z 347. This figure was redrawn from data presented in ref. 8.

From the highest mass n-series ion formed by the A–V mechanism, the other members in this series can be seen at sequential 40 Da intervals. For CID at m/z 341 from the $[M - H + 2Li]^+$ of DHA,[8] the n-series ions are observed at m/z 285, 245, 205, 165, 125, and 85. The mechanism of formation of m/z 285 is illustrated in Scheme 1.10 for this A–V series of charge remote fragmentations.

The other ion series that differs also by 40 Da (m/z 113, 153, 193, and 233) would arise from the same general mechanism, but initiated at a different allylic/vinylic site for the hydrogen transfer (Scheme 1.11). As discussed previously, in the saturated and monounsaturated fatty acids, the family members with the first double bond in a similar position relative to the carboxyl group can also be readily discerned. The mechanisms of formation of either the radical cations at low mass have been presented (Scheme 1.6) or the cations derived from the alternative A–V cleavage (m/z 233) are indicated in Scheme 1.11.

1.4 Hydroxy Fatty Acids

Many different types of hydroxy fatty acids are formed as intermediates of fatty acid biosynthesis and metabolism, as well as products of lipid peroxidation, to name just a few examples. Even the simplest monohydroxy fatty acid offers an additional channel for ion decomposition after collisional activation of either positive or negative ions in the tandem quadrupole mass spectrometer due to the lower activation energy imparted by the additional oxygen heteroatom.[11] As the complexity of hydroxy fatty acids increases with additional functional groups such as double bonds and additional carbinol moieties, the number of product ions increase as well as the amount of structural information that is readily gleaned from understanding the mechanism of ion formation. Most detailed mechanistic studies have been

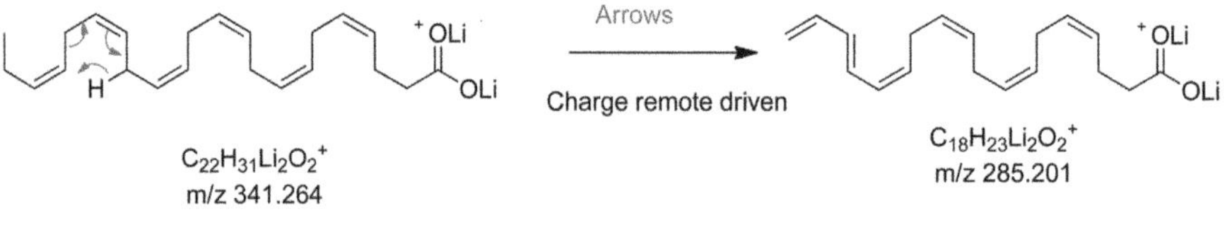

$C_{22}H_{31}Li_2O_2^+$
m/z 341.264

Arrows
Charge remote driven

$C_{18}H_{23}Li_2O_2^+$
m/z 285.201

Scheme 1.10

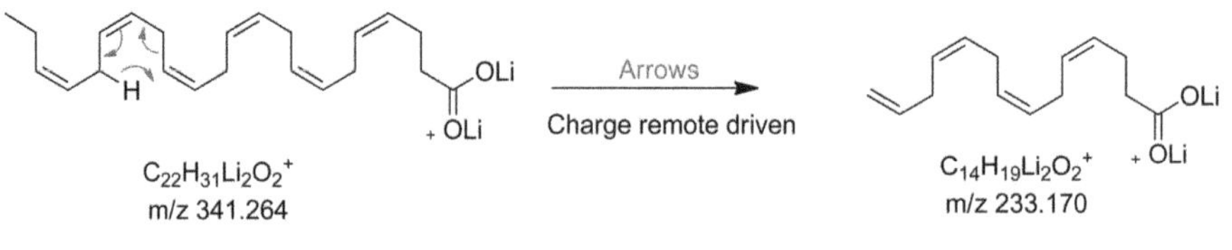

$C_{22}H_{31}Li_2O_2^+$
m/z 341.264

Arrows
Charge remote driven

$C_{14}H_{19}Li_2O_2^+$
m/z 233.170

Scheme 1.11

carried out following the collisional activation of negative ions (carboxylate anions) of these saturated species, but decomposition mechanisms for the carboxylate metalated cations have also been reported.[8]

1.4.1 Monohydroxy Fatty Acid Negative Ions

The singly hydroxylated fatty acid yields abundant fragment ions after collisional activation of $[M - H]^-$ and except for the 3-hydroxy fatty acids, the ions which are formed are surprisingly consistent, no matter at which position the hydroxyl group is present on the fatty alkyl chain. This renders structural information difficult to glean since the same mass losses are typically observed. This is illustrated for three positional isomers (Figure 1.7) 2-hydroxyhexadecanoate, 3-hydroxyhexadecanoate, and 15-hydroxypentadecanoate. The product ions for 2-hydroxyhexadecanoate and 15-hydroxypentadecanoate are surprisingly similar with a loss of 18, 46 and 74 Da from the molecular anion $[M - H]^-$. While the high mass ions formed following the collisional activation of 3-hydroxyhexadecanoate molecular anion yield high mass product ions corresponding to the losses of 18 and 46 (very low abundance) Da, the most abundant ion is observed at m/z 59, which is unique to the 3-hydroxy position. This product ion is likely due to a favorable cleavage between carbon-2 and carbon-3 as a result of a charge driven proton rearrangement (ene reaction) (Scheme 1.12) seen for carboxylic acids even under electron ionization conditions.[1]

This fragmentation mechanism has been shown to yield valuable structural information from α-alkyl-β-hydroxy fatty acids typically encountered as the mycolic acids unique to mycobacteria. A loss of a neutral aldehyde cleaving the α-alkyl group attached to the two carbon fragment of Scheme 1.12 was used to define the position of unsaturation and carbon-chain distribution in their branched-chain fatty acids using neutral loss and precursor ion scanning of a naturally occurring mixture of mycolic acids.[12]

A mechanism operating in the decomposition of these hydroxy fatty acids yielding the very common ions was studied in detail by Claeys and coworkers using ^{18}O-labeled carboxyl moieties and H/D-exchange of the proton on the carbinol oxygen atom.[13] The mechanism common to all these hydroxy fatty acids (Scheme 1.13) likely involves a charge-driven loss of CO_2 then loss of H_2 driven by the adjacent hydride ion (H^-) from carbon-3, attacking the proton attached to the hydroxyl group (no matter what position), yielding a stable alkoxide anion.

$C_2H_3O_2^-$
m/z 59.014

Scheme 1.12

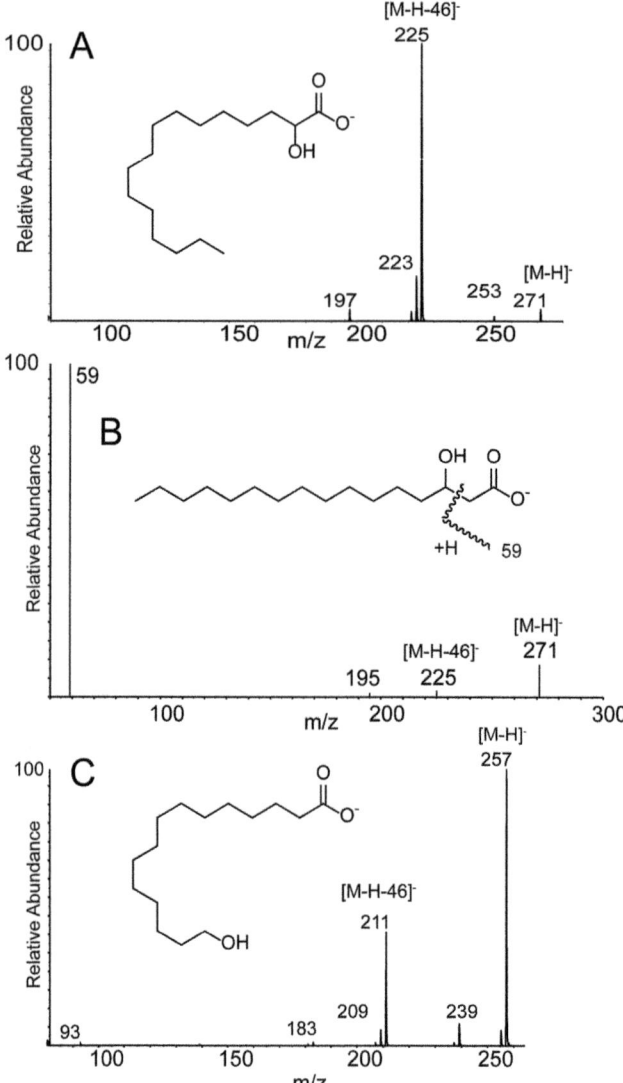

Figure 1.7 Electrospray ionization (negative ions) and tandem mass spectrometry of monohydroxy fatty acids. (A) Product ions obtained following collisional activation of 2-hydroxyhexadecanoic acid anions [M – H]⁻ at *m/z* 271; (B) production ions obtained following collisional activation of 3-hydroxyhexadecanic acid [M – H]⁻ at *m/z* 271; (C) collisional activation of 15-hydroxypentadecanoic acid [M – H]⁻ at *m/z* 257. These MS/MS spectra were obtained using a tandem quadrupole mass spectrometer.

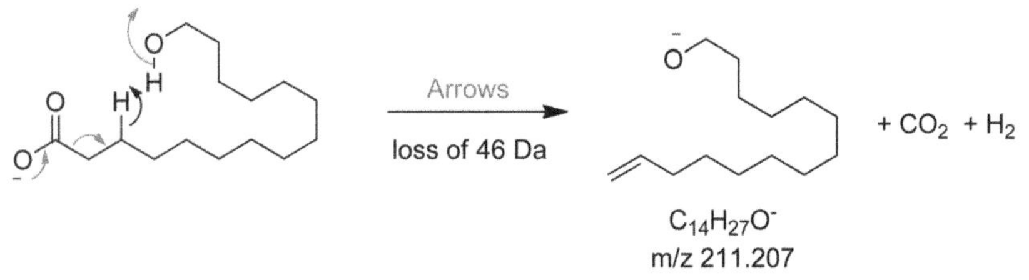

Scheme 1.13

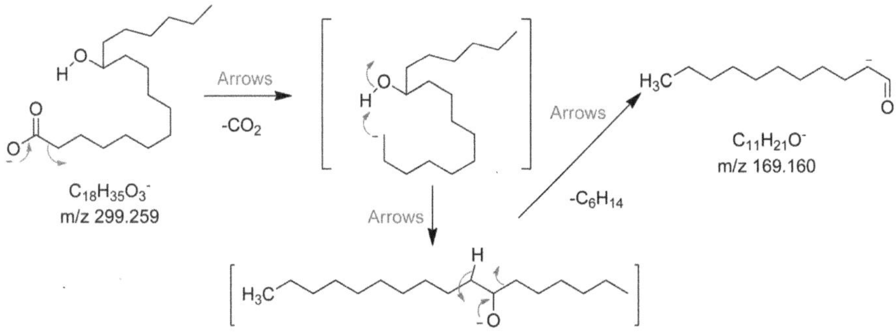

Scheme 1.14

Even if the carbinol site is quite remote from the carboxyl moiety (*e.g.* 15-hydroxypentadecanoic fatty acid, Figure 1.7C) this is quite an efficient process. When the hydroxyl group is closer to the carbinol group, such as for 12-hydroxyoctadecanoic acid,[13] ions can also be seen from a charge driven loss of a neutral hydrocarbon chain alpha to the nascent alkoxide anion at m/z 169. A mechanism which can account for the formation of the neutral hydrocarbon would involve formation of an alkoxide anion that could be stabilized as an aldehydic anion (Scheme 1.14).

This type of α-cleavage to the hydroxyl group is even more abundant for dihydroxy fatty acids, where the anion site can be on two separate carbinol moieties. These vicinal diols can be readily made from olefins by OsO_4 oxidation to locate double bonds in an unsaturated fatty acid.[14] The facile formation of these product ions is consistent with the mechanism shown in Scheme 1.14 and supported by stable isotope labeled molecules.[13]

1.4.2 Unsaturated Hydroxy Fatty Acids

The introduction of a double bond into the fatty acid acyl chain that also has a hydroxyl group has a significant impact on the ease of decomposition of carboxylate anions $[M - H]^-$ in the tandem quadrupole (low energy) mass spectrometer (Figure 1.8). Since biochemical as well as chemical oxidation of unsaturated fatty acids, and in particular polyunsaturated fatty acids, leads to many different hydroxylated fatty acids, a large number of lipids are found in this class both from enzymatic as well as nonenzymatic products of lipid peroxidation. Major fragment ions are observed, corresponding to cleavage adjacent to the carbinol C–C bond, but allylic to a double bond. Several general mechanisms have been suggested to operate in the collision induced decomposition of unsaturated hydroxy fatty acid molecular ions generated by electrospray ionization.[11,15–17] Both charge driven and charge remote mechanisms (Scheme 1.15) appear to operate and have been proposed to account for the majority of the most abundant product ions.

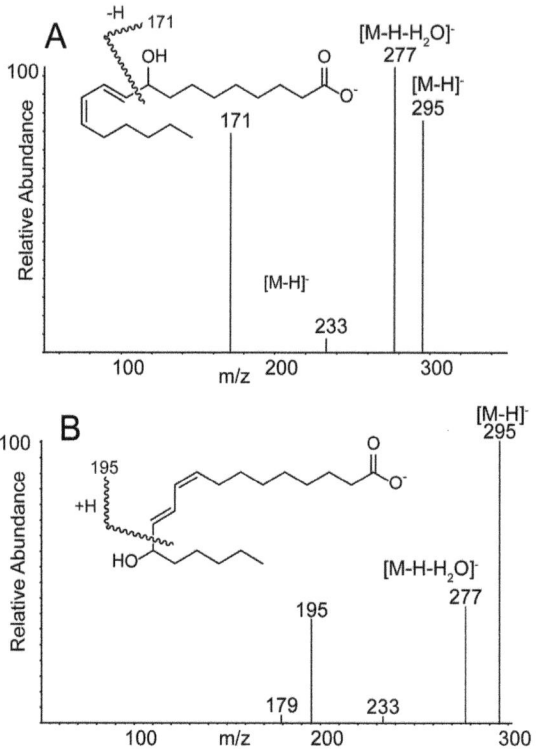

Figure 1.8   Electrospray ionization (negative ions) and tandem mass spectrometry of monohydroxy polyunsaturated fatty acids. (A) Product ions obtained following collisional activation of the carboxylate anion of 9-hydroxy-10,12-octadecadienoic acid [M − H]⁻ at *m/z* 295; (B) product ions obtained following collisional activation of 13-hydroxy-9,11-octadecadienoic acid [M − H]⁻ at *m/z* 295. These MS/MS spectra were obtained using a tandem quadrupole mass spectrometer.

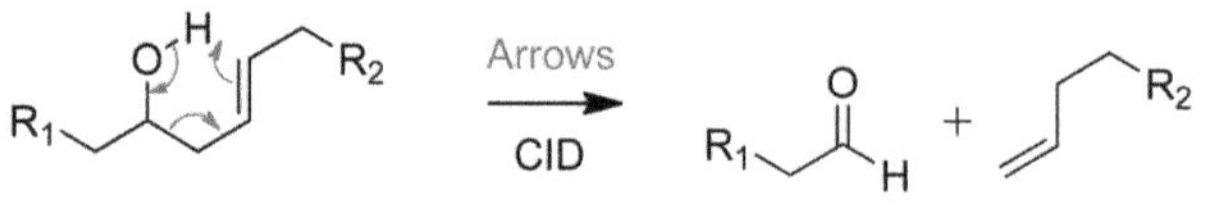

Scheme 1.15

1.4.2.1 Charge Remote Allylic Fragmentation (Ene Reaction)

Either R_1 or R_2 could retain the charge site (carboxylate anion or even the closed-shell alkali metal attachment positive ion) and many examples are available to illustrate this reaction, as well as isotope labeling and high resolution measurements to support this mechanism.[16] A variant of this highly favorable allylic reaction is cleavage of the carbon–carbon bond (vinylic to the alcoholic moiety) carbon atom (Scheme 1.16), but this decomposition pathway is likely preceded by double bond rearrangements. For conjugated dienyl fatty acids, which are common products of lipoxygenase reactions and peroxidation of polyunsaturated fatty acid, a 1[5]-sigmatropic shift likely precedes the carbon–carbon cleavage step which alters the position of the closest double bond to an allylic position, which is much more favorable to cleave, in that a very stable aldehyde and olefin result.

1.4.2.2 Charge Remote Allylic Fragmentation Preceded by Double Bond Migration

A major driving force for these decomposition reactions would appear to be related to the slightly weakened bond allylic to the double bond, which is rendered this way in part by the electronegativity of the oxygen atom at the immediately adjacent carbon atom. In specific cases, multiple double bond migration steps precede the C–C allylic cleavage step (*e.g.* 1[5]- sigmatropic shift). It would appear that increasing conjugation of the nascent olefin being formed from a diene to triene further enhances the probability for this fragmentation taking place.[15]

1.4.2.3 Charge Driven Allylic Fragmentation

A second general mechanism for carbon–carbon cleavage of unsaturated hydroxy fatty acid following collisional activation involves charge driven events. Such events are often observed in a negative ion mode, where the

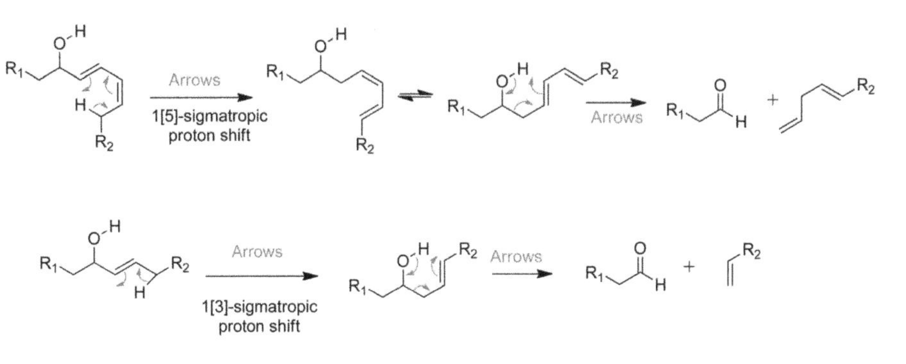

Scheme 1.16

rather stable carboxylate anion abstracts the proton of the alcohol functional group, leaving an alkoxide anion at that position. This more reactive charge localized species can readily undergo aldehyde formation (neutral species) on either side of the carbon–oxygen bond, but typically it is most abundant when a stabilized anion can form by charged delocalization by one or more double bonds in conjugation with the bond being cleaved and formation of the carbon-centered negative charge at this site (Scheme 1.17).

Again, the initial carboxylate anion can either be in R_1 or R_2 and the most abundant product ion would be the anion with electron density delocalized by an allylic double bond. A variant of this mechanism would involve an initial charge remote double bond migration, to render a double bond allylic to the carbon–oxygen alkoxide structure. Again, the bond allylic to the double bond is the one most readily cleaved, yielding the highest ion current, in the illustrated case (Scheme 1.18) with the R_2 portion of the starting unsaturated fatty alcohol.

Another interesting mechanism reported to occur for the CID of unsaturated alcohol fatty acids is the oxy-Cope type rearrangement (Scheme 1.19) found for specific polyunsaturated hydroxy fatty acids with a 3-hydroxy-1,5-diene structural unit embedded in the carboxylic acid structure.[15] The unique arrangement of molecular orbitals and charge localization appearing as an

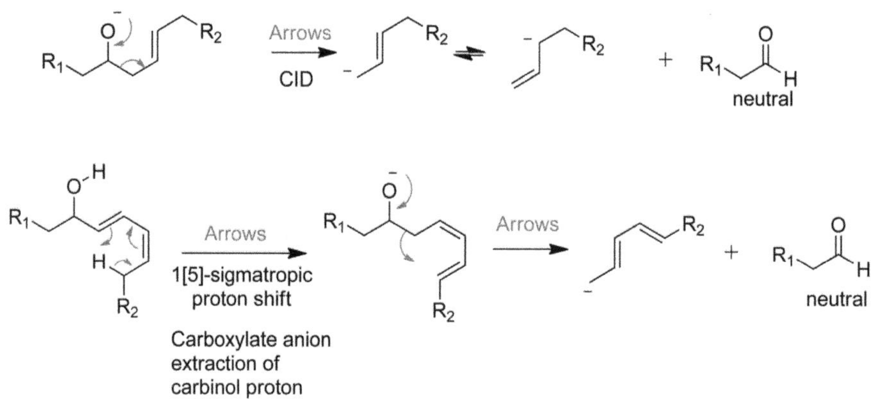

Scheme 1.17

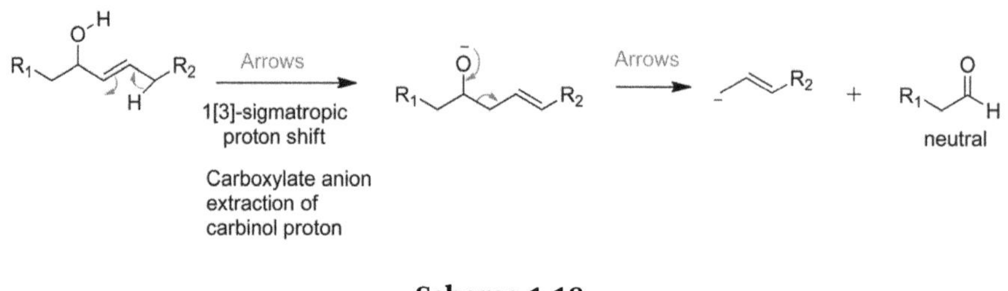

Scheme 1.18

alkoxide anion render this cyclic double bond rearrangement quite favorable.[18] The formation of such product ions have been described previously, and are remarkable in that the product ions are radical anions.[16] This means for non-nitrogen-containing hydroxy, unsaturated fatty acids, this ion would appear at an even mass-to-charge ratio.

Specific examples of each of these decomposition mechanisms can be observed for the 18-carbon hydroxy fatty acids in Figure 1.8. The abundant fragment ion observed following CID of 9-hydroxyoctadeca[10,12]dienoate at m/z 171 (Figure 1.8A), could be a result of the terminal aldehyde formation that would be a charge remote allylic fragmentation preceded by migration of both double bonds in a 1[5]-sigmatropic shift as described (Scheme 1.16). A charge-driven mechanism for decomposition of this hydroxy unsaturated fatty acid anion (*e.g.* Scheme 1.18) would lead to migration of the charge site to the methylene terminal hydrocarbon chain. This ion would occur at m/z 123, but is not observed (Figure 1.8A).

The isomeric fatty acid 13-hydroxyoctadeca[9,11]dienoate also has a rather simple collisional mass spectrum with a major product ion at m/z 195 (Figure 1.8B). This ion could result by both charge-driven and charge-remote mechanisms after 1[5]-sigmatropic shifts of the $\Delta^{9,11}$-double bond to the $\Delta^{8,11}$ positions which renders the carbon–carbon bond cleaved being allylic to the conjugated diene (charge-driven mechanism, Scheme 1.18) or alternatively the newly formed Δ^{10}-double bond being in a favorable position for the charge remote allylic fragmentation mechanism (Scheme 1.17).

These same reaction pathways can be evoked to rationalize the product ions observed following collisional activation of monohydroxy arachidonic acid isomers as negative ions (Figure 1.9). The most abundant ions, aside from the loss of small neutral molecules such as H_2O and CO_2, correspond to carbon–carbon cleavage adjacent to the hydroxyl group. It is important to follow the movement of the proton when one additional or one less mass unit appears in the observed fragment ion. For the CID of 5-HETE, 8-HETE, and 9-HETE the loss of the proton is a result of charge remote double bond migration prior to allylic fragmentation as specifically indicated in the suggested mechanism to rationalize the appearance of m/z 115 from 5-HETE (Scheme 1.20).

The ion at m/z 203 observed from 5-HETE (Figure 1.9A) would most likely be the result of a charged drive reaction, since the site of ionization is moved from the carboxyl moiety and transferred to the alkyl chain (Scheme 1.21).

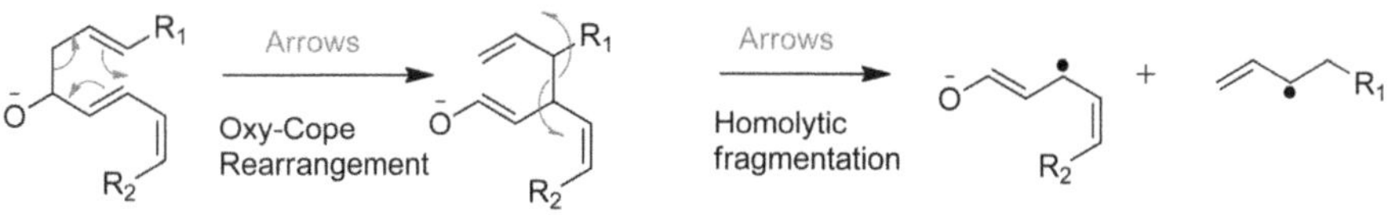

Scheme 1.19

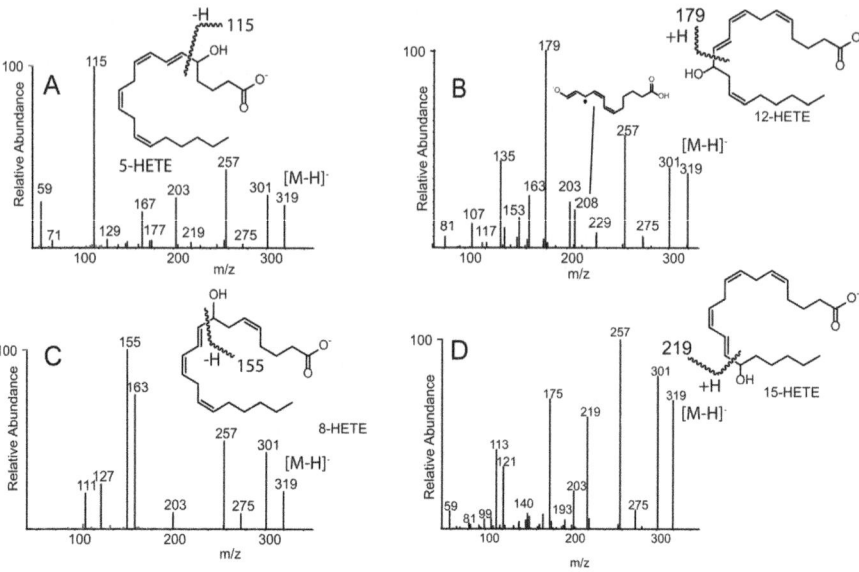

Figure 1.9 Electrospray ionization (negative ions) and tandem mass spectrometry of monohydroxy arachidonate metabolite isomers. (A) Product ions obtained following collisional activation of 5-hydroxy-6,8,11,14-eicosatetraenoic acid [M − H]⁻ at m/z 319; (B) Product ions obtained following collisional activation of 12-hydroxy-5,8,10,14-eicosatetra-enoic acid [M − H]⁻ at m/z 319; (C) Product ions obtained following collisional activation of 8-hydroxy-5,9,11,14-eicosatetraenoic acid [M − H]⁻ at m/z 319; (D) Product ions obtained following collisional activation of 15-hydroxy-5,8,11,13-eicosatetraenoic acid [M − H]⁻ at m/z 319. These MS/MS spectra were obtained using a tandem quadrupole mass spectrometer.

Scheme 1.20

The addition of a proton to the observed fragment ion following CID of the [M − H]⁻ can be illustrated by 12-HETE major product ion at m/z 179 (Figure 1.9B). This likely involves a charged driven mechanism where the carbinol proton at C-12 is transferred to the carboxylate anion prior to car-bon–carbon bond cleavage, but this was likely preceded by a 1[5]-sigmatropic proton shift to render a saturated carbon atom allylic to the hydroxyl group as well as making a conjugated triene group from carbon-5 to carbon-10 (Scheme 1.22). The ion observed at m/z 208 (even mass) suggests that it could be

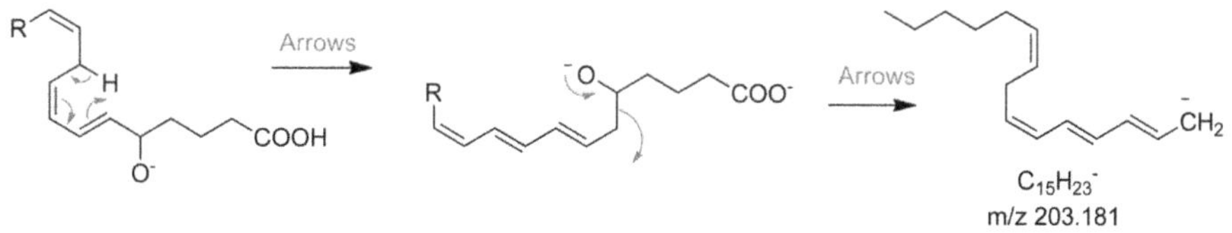

Scheme 1.21

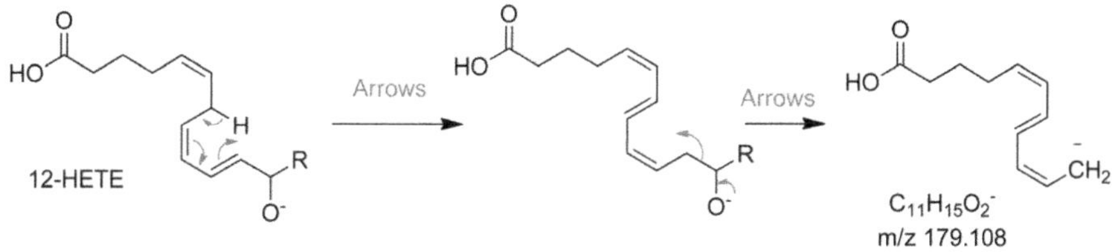

Scheme 1.22

a radical anion and a product of the oxy-Cope reaction indicated in (Figure 1.9B).

It is interesting to note that not all of the hydroxy arachidonate species yield this oxy-Cope reaction product in major abundance and this due to the fact that 5- and 15-HETE do not have the structural requirements necessary for the oxy-Cope reaction and that only 8, 9, and 12-HETEs have the possibility of further delocalization of the radical site over a number of carbon atoms because of the specific arrangement of the initial hydroxy group. However, only with the 12-HETE structure can the radical site be delocalized over 7-carbon atoms in the product ion. This is illustrated in Scheme 1.23 for a 1[3]-sigmatropic shift of an initially formed m/z 208 to further delocalize the radical site. This extended conjugation of the radical anion would be possible for the 8-HETE (Figure 1.9C) structure with the radical delocalized carbon atoms, but only a very small ion is observed at m/z 192. This suggests that there is additional competition for the anion charge site, resulting in shifting of the carboxyl group negative charge prior to radical cleavage. The delocalized radical would then be lost as an uncharged neutral species and therefore not observed. The 9-HETE structure would appear to fit the structural requirements for the initial oxy-Cope rearrangement, but the radical anion product could not be delocalized by extended conjugation as for the other structures.

1.4.3 Epoxy and Unsaturated Epoxy Fatty Acids

The addition of an epoxy group into the fatty acyl chain, especially with one or more double bonds, also present in the fatty acid, introduces sufficient reactivity after collisional activation to drive a number of bond cleavage pathways. These reactions have not been systematically studied, but based upon a structural variation of product ions for many different epoxides, mechanisms have been suggested to account for the major species observed. Often the ions rationalized are indicated by drawing lines to indicate cleavage across the 3-membered epoxide ring, but details are not discussed as to the mechanisms actually operating. A likely process involves collision induced carbon–oxygen bond cleavage of the already strained epoxide (indicated as bond a and b in Scheme 1.24), leaving a radical-centered oxygen atom at one site and another site with a carbon-centered radical. The carbon-centered radical could readily abstract hydrogen atoms from methylene groups remote from the original epoxide with the most likely being four or five carbon atoms removed and in this way separate the two free radicals from each other, terminating any reversible C–O bond cleavage/formation (Scheme 1.24 proceeds after radical cleavage of the b-bond).

The oxygen-centered radical could then pair with an electron from the original carbon–carbon bond of the epoxide, cleaving that bond to a neutral aldehyde moiety and reforming a carbon-radical at the adjacent epoxide carbon atom. This radical could then pair with the remote carbon radical to cyclize the formation of a neutral 5- or 6-membered ring (Scheme 1.25).

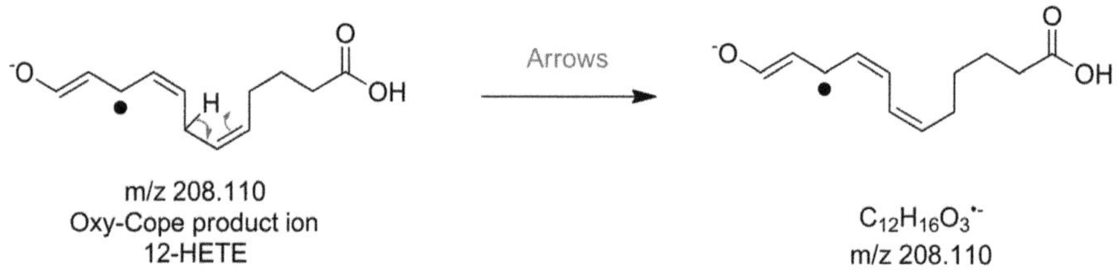
Arrows
OH
H
m/z 208.110
Oxy-Cope product ion
12-HETE
$C_{12}H_{16}O_3^{\bullet-}$
m/z 208.110

Scheme 1.23

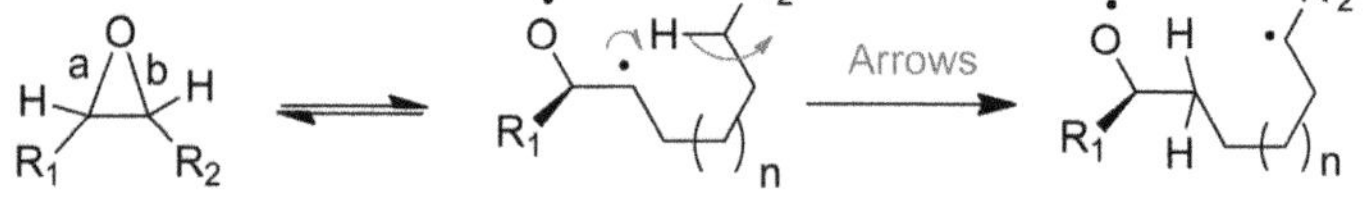
a
b
H
R1
R2
R'2
Arrows
n

Scheme 1.24

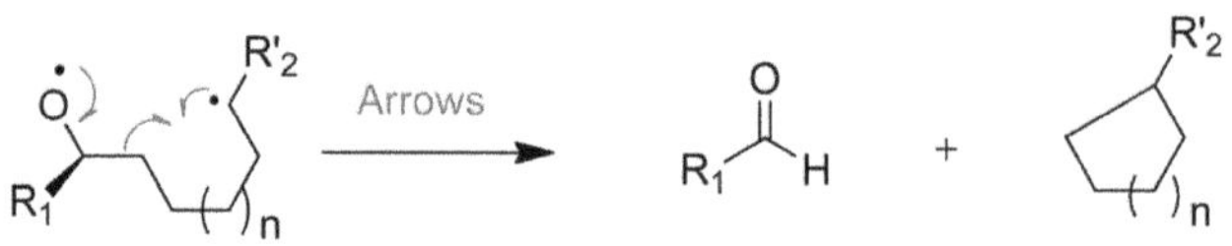
R'2
Arrows
R1
H
n
+

Scheme 1.25

The site of the remote carboxylate anion would determine the actual observed product ion. Initial cleavage of the a-bond (Scheme 1.24) would lead to a separate ion by the same mechanism, again in the case of Scheme 1.25 the carboxylate anion at R_1 or R_2 would determine the ion observed. The summary of these mechanisms is shown in Scheme 1.26 for a specific 9(10)-epoxyoctadeanoic acid. This type of rationalization is often depicted when the CID mass spectra of fatty acid epoxides are presented (Figure 1.10) and in fact the major product ions from collisional activation of 9(10)-epoxyoctadecanoate anion were *m/z* 171 and 155.[19] These authors have also published that *trans*-9(10)-epoxystearic acid behave identically, but abundance ratios differ from that of the *cis*-9(10)-epoxystearic acid for *m/z* 155/171.

The presence of one or more double bonds in the fatty acid chain provides additional decomposition pathways. The formation of epoxides by the enzymatic reaction catalyzed by various cytochrome P-450 is a common occurrence in many cells with polyunsaturated fatty acids, rendering formation of epoxides, homoconjugated to one or more double bonds, sometimes on both sides of the epoxide. A commonly observed fragment ion comes from an additional decomposition mechanism corresponding to

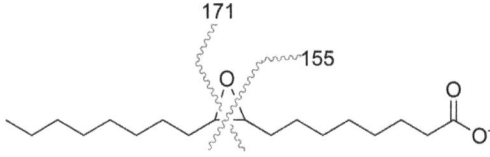

Scheme 1.26

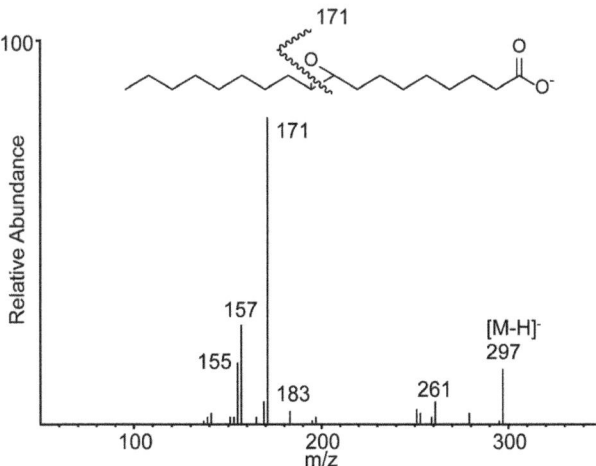

Figure 1.10 Electrospray ionization (negative ions) and tandem mass spectrometry of an epoxy fatty acid. The product ions obtained following collisional activation of the carboxylate anion of 9(10)-epoxy-octadecanoic acid [M − H]⁻ at *m/z* 297. This figure was redrawn from the data from ref. 19.

facile hydrogen atom rearrangement following homolytic cleavage of an epoxide carbon–oxygen bond, likely facilitated by the double bond that can participate in the hydrogen rearrangement (Scheme 1.27).

A good example of this mechanism is seen in the epoxides of linoleic acid termed leukotoxin A and leukotoxin B (Figure 1.11), which show the cleavage reactions across the epoxide ring described in Schemes 1.24–1.26, as well as the hydrogen rearrangements on the side of the epoxide homoconjugated to the double bond (Scheme 1.27). In the case of leukotoxin A, the ion m/z 183 loses a hydrogen atom during the rearrangement while for leukotoxin B a similar fragment ion at m/z 183 gains a hydrogen atom although the exact mass and elemental compositions of the same nominal m/z 183 ions are different (Scheme 1.28).

These same mechanisms of epoxide decompositions are observed for the collisional activation of the carboxylate anions of the P-450 metabolites of arachidonic acid which have been termed EETs (epoxyeicosatrienoic acids) and the CID mass spectra for the four regioisomers are presented in Figure 1.12. The cleavage reactions whose mechanisms are discussed above are indicated on the structural annotations of each mass spectrum.

An interesting ion observed following CID of 5(6)-EET appears at m/z 191 (Figure 1.12A) that does not contain the carboxylate anion and therefore arises following a charge driven reaction where the location of anionic site appears to be at carbon-7. The proximity of the 5(6)-epoxide to the carboxylate anion facilitates this latter group to attack the epoxide at carbon-5 followed by epoxide ring opening and alkoxide anion formation (Scheme 1.29). This can then form an aldehyde following scission of carbon-6, leaving the anion site at carbon-7 delocalized by the adjacent double bond. This ion further fragments to yield the hydrocarbon anions at m/z 163, 137, and 99.[20]

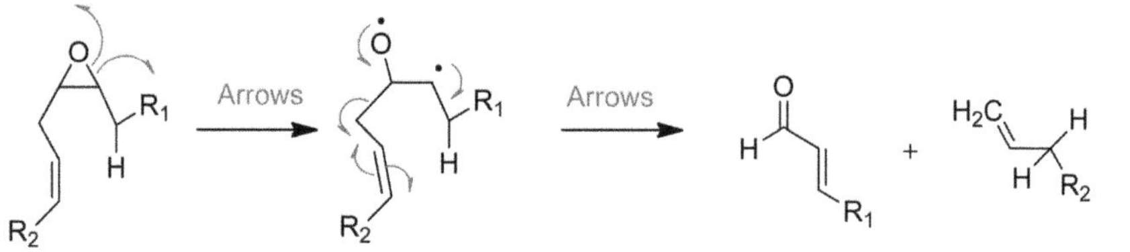

Scheme 1.27

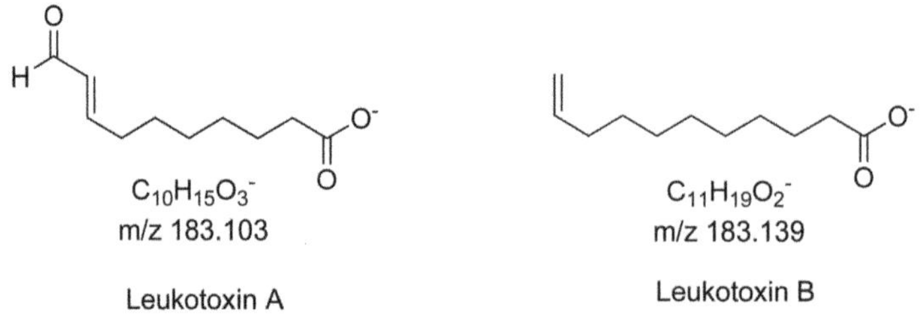

$C_{10}H_{15}O_3^-$
m/z 183.103

Leukotoxin A

$C_{11}H_{19}O_2^-$
m/z 183.139

Leukotoxin B

Scheme 1.28

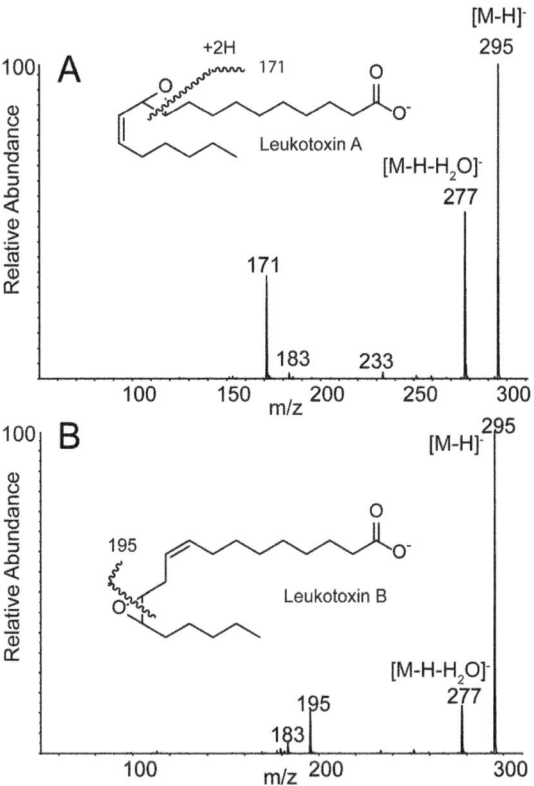

Figure 1.11 Electrospray ionization (negative ions) and tandem mass spectrometry of epoxy metabolites of linoleic acid termed leukotoxin A and leukotoxin B. (A) Product ions obtained following collisional activation of leukotoxin A, 9(10)epoxy-12-octadecaenoic acid [M − H]⁻ at *m/z* 295; (B) Product ions obtained following collisional activation of 12(13)epoxy-9-octadecaenoic acid [M − H]⁻ at *m/z* 295. This figure was redrawn from data stored on the LIPIDMAPS website (Lipidmaps.org), as tandem mass spectra of standards.

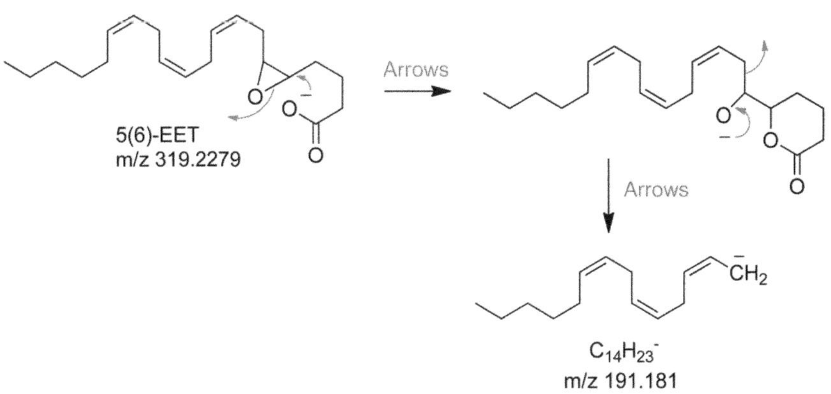

5(6)-EET
m/z 319.2279

Arrows →

Arrows ↓

$C_{14}H_{23}^-$
m/z 191.181

Scheme 1.29

1.4.4   Keto and Hydroperoxy Unsaturated Fatty Acids

Polyunsaturated fatty acids that contain a homoconjugated 1,4-diene moiety in the alkyl chain is a common occurrence in biology due to the mechanism of unsaturated fatty acid biosynthesis and the specific desaturases expressed in plant and animal cells. Such fatty acyl groups readily undergo radical abstraction of a methylene proton in biological systems leading to formation of a conjugated pentadienyl radical, which rapidly reacts with diatomic oxygen to form an oxygen-centered, conjugated hydroperoxy radical that eventually abstracts a hydrogen atom to form a conjugated lipid hydroperoxide species. These lipid hydroperoxides are surprisingly stable and can be isolated from biological matrices as such. Ubiquitous peroxidases and hydroperoxidases, however, often reduce these hydroperoxides to a conjugated diene fatty acid alcohol. Such conjugated alcohols can be enzymatically and chemically oxidized to unsaturated keto fatty acids.

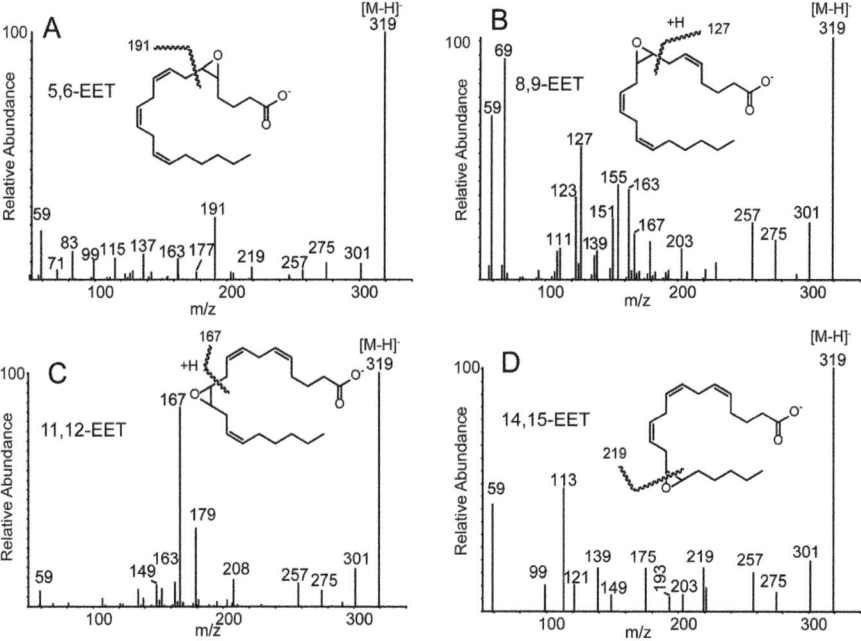

Figure 1.12   Electrospray ionization (negative ions) and tandem mass spectrometry of epoxyeicosatrienoic acid isomers derived from arachidonic acid. (A) Collisional activation of carboxylate anion from 5(6)-epoxy8,11,14-eicosatrienoic acid [M − H]⁻ at *m/z* 319; (B) Product ions obtained following collisional activation of 8(9)-epoxy-5,11,14-eicosatrienoic acid [M − H]⁻ at *m/z* 319; (C) Product ions obtained following collisional activation of 11(12)-epoxy-5,8,14-eicosatrienoic acid [M − H]⁻ at *m/z* 319; (D) Product ions obtained following collisional activation of 14(15)-epoxy-5,8,11-eicosatrienoic acid [M − H]⁻ at *m/z* 319. These MS/MS spectra were obtained using a tandem quadrupole mass spectrometer.

This section discusses the mass spectrometry of both keto fatty acids and hydroperoxy fatty acids when these functional groups are conjugated with double bonds since they undergo very similar collision induced decomposition reactions even though it would appear that the functional groups are quite different. The discussions here will center around the collisional activation of negative molecular ions $[M - H]^-$, but lipid hydroperoxides have been analyzed as positive $[M + NH_4]^+$ ions as well.[20]

The unsaturated hydroperoxy fatty acids behave virtually identically to the keto fatty acids and the only abundant ion which distinguishes these two classes of molecules is the molecular anion $[M - H]^-$ in which the hydroperoxy species is 18 amu higher in mass than the corresponding keto unsaturated fatty acid. Collisional activation of the $[M - H]^-$ from the unsaturated hydroperoxy fatty acids (Figure 1.13) results in facile loss of H_2O and all subsequent product ions suggest this dehydration step involves the loss of the hydrogen atom on the same carbon atom bearing the hydroperoxy moiety.[21] This would result in formation of a ketone at that position.

An alternative mechanism would involve the loss of a hydrogen atom from another site such as a saturated carbon atom 3 or 4 removed from the hydroperoxy oxygen atom, resulting in formation of a stable epoxide, furan, or pyran, respectively. In fact, loss of H_2O to form a keto-like structure is a dominant reaction based on isotope labeling studies,[22] but formation of the ring structure appears to be a reasonable option as well. The mechanism for the formation of the keto moiety (Scheme 1.30) has been proposed based upon isotope labeling studies.[22]

The similar behavior of unsaturated keto fatty acids is illustrated in Figure 1.14 for different, but isomeric, ketones derived from linoleic acid. These molecules have been studied in some detail by collisional activation aided by the availability of isotope labeled material (three deuterium atoms at the methine carbon atoms that make up the double bonds).

The formation of the abundant product ion from both 9-hydroperoxy-octadeca-10,12-dienoate (not shown) as well as 9-oxo-octadeca-10,12-dien-oate appear at *m/z* 197 and 185 (Figure 1.14A). These ions at first glance would appear quite unlikely since they involve cleavage between carbon atoms 10-11 (at double bond location) and in between the conjugated diene at carbon atoms 11-12 as well as proton transfer reactions. It was suggested that the mechanism of formation of these ions proceeded after an initial enolization of the oxyl group to form a vinyl alcohol based upon the isotope labeling studies.[23]

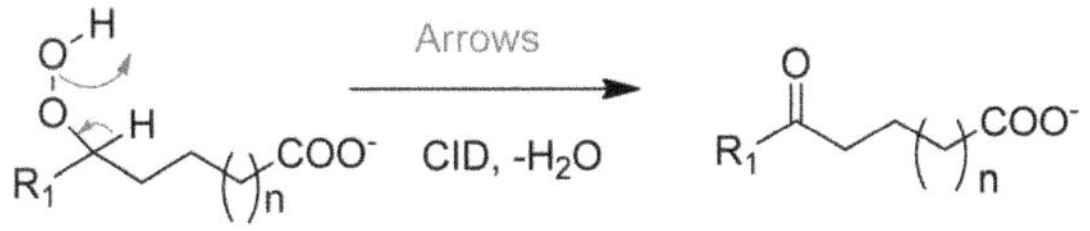

Scheme 1.30

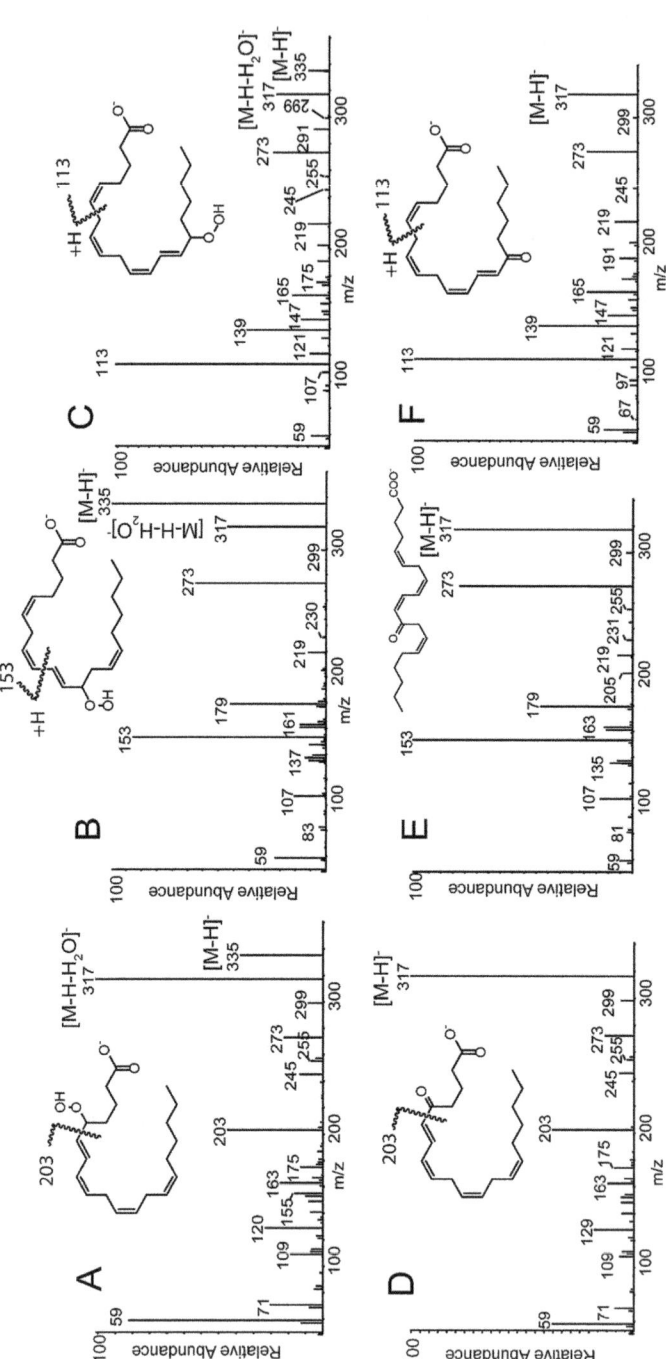

Figure 1.13 Electrospray ionization (negative ions) and tandem mass spectrometry of three regioisomeric hydroperoxides and corresponding ketones of arachidonic acid. (A) Product ions obtained following collisional activation of 5-hydroperoxy-6,8,11,14-eicosatetraenoic acid [M − H]⁻ at *m/z* 335; (B) product ions obtained following collisional activation of 12-hydroperoxy-5,8,10,14-eicosatetraenoic acid [M − H]⁻ at *m/z* 335; (C) product ions obtained following collisional activation of 15-hydroperoxy-5,8,11,13-eicosatetraenoic acid [M − H]⁻ at *m/z* 335; (D) product ions obtained following collisional activation of 5-oxo-6,8,11,14-eicosatetraenoic acid [M − H]⁻ at *m/z* 317; (E) Product ions obtained following collisional activation of 12-oxo-5,8,10,14-eicosatetraenoic acid [M − H]⁻ at *m/z* 317; (F) Product ions obtained following collisional activation of 15-oxo-5,8,11,13-eicosatetraenoic acid [M − H]⁻ at *m/z* 317. These MS/MS spectra were obtained using a tandem quadrupole mass spectrometer.

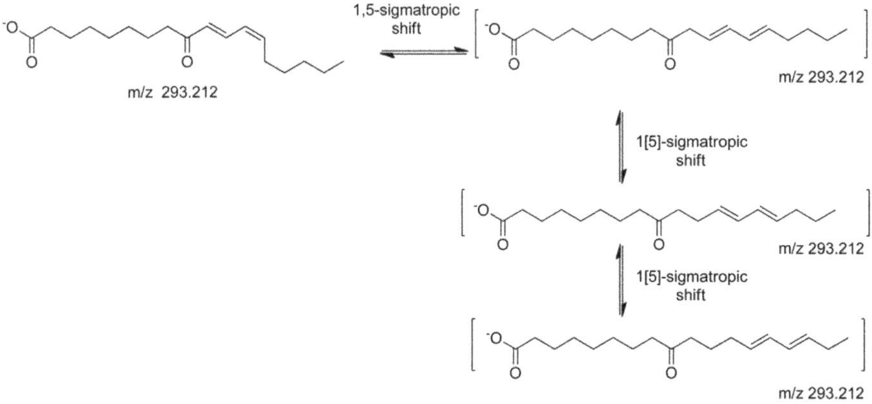

Scheme 1.31

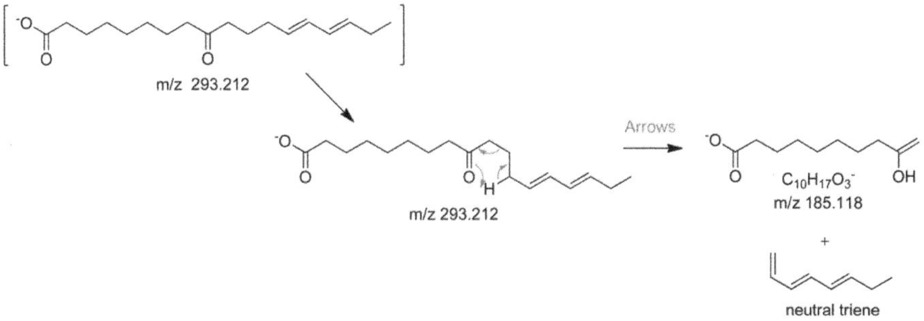

Scheme 1.32

A fragmentation mechanism for *m/z* 185 also consistent with the isotope labeled analogs and the sequential 1[5]-sigmatropic proton shifts of the double bonds, proceeds from the diene no longer conjugated to the keto moiety, that can readily decompose to very stable neutral species in product ion fragments by remote site fragmentation mechanisms (Scheme 1.31).

If one invokes reversibly of these double bond migrations, deuterium scrambling would result specifically for the [10,11,12-^{2}H$_3$]-9-oxo-octadeca-10,12-dienoic acid that reversibly rearrange following collisional activation to 9-oxo-octadeca-13,15-dienoate anion. Reversible 1[5]-sigmatropic shift would exchange protons for deuterium labeled atoms at positions between carbons 10 and 13. The mechanism leading to the scission of the carbon 10-11 bond (*m/z* 185) from the transient 9-oxo-octadeca-13,15-dienoate anion could proceed by an energetically favorably ene reaction often (Scheme 1.32). The carbon-12 proton (γ-proton to oxo moiety) would be quite favorably trans-ferred due to it being vinylic to the diene at carbon-13. The resulting ion would result in a vinyl alcohol carboxylate anion product and a neutral, conjugated triene. The same 9-oxo-octadeca-13,15-dienoate anion could undergo a sepa-rate proton transfer reaction leading to carbon 11-12 scission (Scheme 1.33).

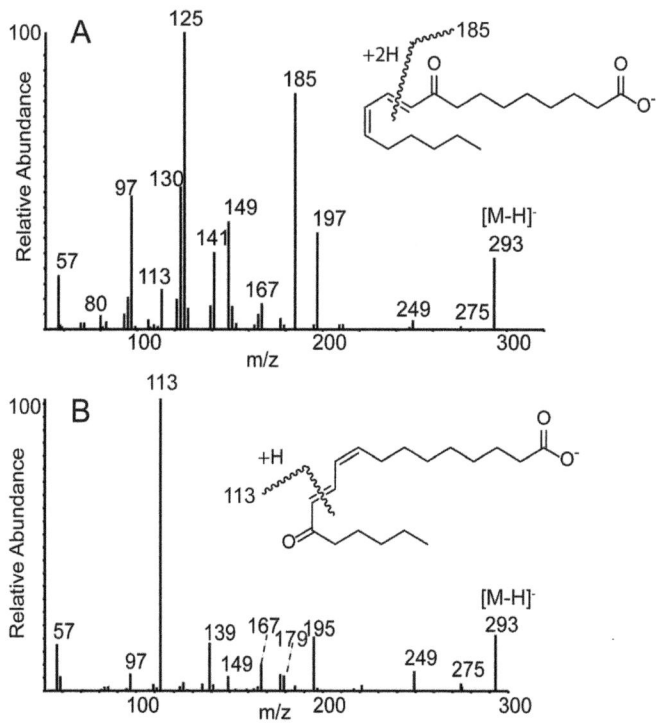

Scheme 1.33

Figure 1.14 Electrospray ionization (negative ions) and tandem mass spectrometry of keto fatty acids derived from linoleic acid. (A) Product ions obtained following collisional activation of 9-oxo-11,13-octadecadienoic acid [M − H]⁻ at *m/z* 293; (B) product ions obtained following collisional activation of 13-oxo-9,11-octadecadienoic acid [M − H]⁻ at *m/z* 293. These MS/MS spectra were obtained using a tandem quadrupole mass spectrometer.

Alternatively, the second product ion at *m/z* 197 would result after a further 1[5]-sigmatropic shift to form the 9-oxo-[14,16]-octadecadienoate anion that could undergo the previously discussed (Scheme 1.2) charge remote loss of H_2 forming *m/z* 197 as the dienone and formation of a 7-carbon, conjugated triene neutral hydrocarbon. These mechanisms are consistent with the observed ions in the deuterium labeled species that allow one to track proton transfer reactions when specific protons are deuterium atoms. However, the

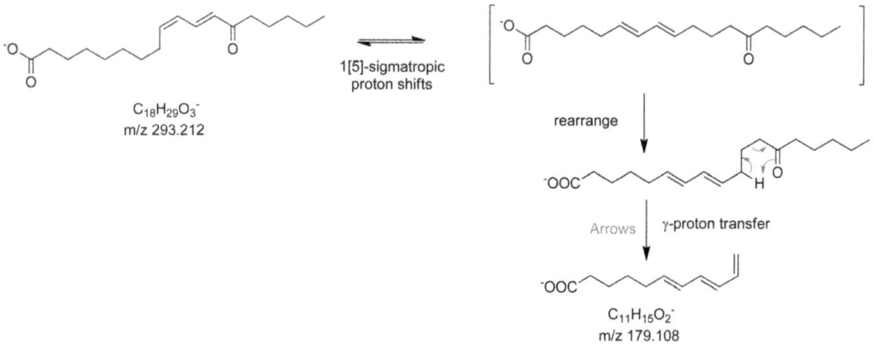

Scheme 1.34

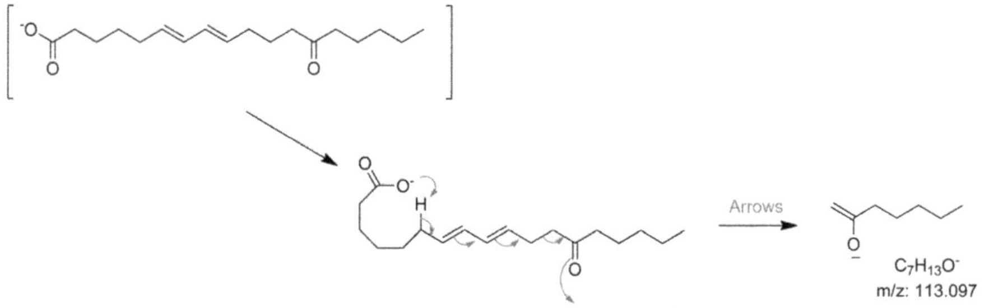

Scheme 1.35

difference in products formed are the neutral molecules because the charge site remains on the product identical to both mechanisms.

These same mechanisms appear to be operating to some extent for the regioisomer 13-oxo-[9,11]-octadecadienoate anion after collisional activation (Figure 1.14B). After two 1[5]-sigmatropic proton shifts, the 13-oxo-[6,8]-octadecadienoate anion could undergo the facile γ-proton transfer (ene reaction) (Scheme 1.34), but this mechanism would lead to a conjugated triene on the portion of the molecule containing the carboxylate anion and the ion observed at m/z 179 which is not that abundant.[21]

The most abundant product ion observed for 13-oxo-[9,11]-octadecadienoic acid could also be a result of the intermediate rearranged ion discussed above that moves the conjugated diene closer to the carboxylate anion. This charge driven product ion could then arise from cleavage of carbon atoms 11-12 and the alkoxide anion formation at m/z 113 (Scheme 1.35).

References

1. R. C. Murphy, *Mass Spectrometry of Lipids*, Plenum Press, New York, 1993.
2. J. B. Fenn, Ion formation from charged droplets: Roles of geometry, energy, and time, *J. Am. Soc. Mass Spectrom.*, 1993, **4**, 524–535.
3. J. Folch, M. Lees and G. H. Sloane Stanley, A simple method for the isolation and purification of total lipides from animal tissues, *J. Biol. Chem.*, 1957, **226**, 497–509.
4. E. G. Bligh and W. J. Dyer, A rapid method of total lipid extraction and purification, *Can. J. Biochem. Physiol.*, 1959, **37**, 911–917.
5. J. L. Kerwin, A. M. Wiens and L. H. Ericsson, Identification of fatty acids by electrospray mass spectrometry and tandem mass spectrometry, *J. Mass Spectrom.*, 1996, **31**, 184–192.
6. V. G. Voinov, H. Van den Heuvel and M. Claeys, Resonant electron capture mass spectrometry of free fatty acids: examination of ion structures using deuterium-labeled fatty acids and collisional activation, *J. Mass Spectrom.*, 2002, **37**, 313–321.
7. E. Davoli and M. L. Gross, Charge remote fragmentation of fatty acids cationized with alkaline earth metal ions, *J. Am. Soc. Mass Spectrom.*, 1990, **1**, 320–324.
8. F. F. Hsu and J. Turk, Distinction among isomeric unsaturated fatty acids as lithiated adducts by electrospray ionization mass spectrometry using low energy collisionally activated dissociation on a triple stage quadrupole instrument, *J. Am. Soc. Mass Spectrom.*, 1999, **10**, 600–612.
9. N. Zehethofer, D. M. Pinto and D. A. Volmer, Plasma free fatty acid profiling in a fish oil human intervention study using ultra-performance liquid chromatography/electrospray ionization tandem mass spectrometry, *Rapid Commun. Mass Spectrom.*, 2008, **22**, 2125–2133.
10. F. H. Chilton, R. C. Murphy, B. A. Wilson, S. Sergeant, H. Ainsworth, M. C. Seeds and R. A. Mathias, Diet-gene interactions and PUFA metabolism: a potential contributor to health disparities and human diseases, *Nutrients*, 2014, **6**, 1993–2022.
11. C. Cheng and M. L. Gross, Fragmentation mechanisms of oxofatty acids via high-energy collisional activation, *J. Am. Soc. Mass Spectrom.*, 1998, **9**, 620–627.
12. F. F. Hsu, K. Soehl, J. Turk and A. Haas, Characterization of mycolic acids from the pathogen Rhodococcus equi by tandem mass spectrometry with electrospray ionization, *Anal. Biochem.*, 2011, **409**, 112–122.
13. M. K. Moe, M. B. Strom, E. Jensen and M. Claeys, Negative electrospray ionization low-energy tandem mass spectrometry of hydroxylated fatty acids: a mechanistic study, *Rapid Commun. Mass Spectrom.*, 2004, **18**, 1731–1740.
14. T. J. Donohoe, R. M. Harris, S. Butterworth, J. N. Burrows, A. Cowley and J. S. Parker, New osmium-based reagent for the dihydroxylation of alkenes, *J. Org. Chem.*, 2006, **71**, 4481–4489.

15. P. Wheelan, J. A. Zirrolli and R. C. Murphy, Electrospray ionization and low energy tandem mass spectrometry of polyhydroxy unsaturated fatty acids, *J. Am. Soc. Mass Spectrom.*, 1996, 7, 140–149.

16. P. Wheelan, J. A. Zirrolli and R. C. Murphy, Low-energy fast atom bombardment tandem mass spectrometry of monohydroxy substituted unsaturated fatty acids, *Biol. Mass Spectrom.*, 1993, 22, 465–473.

17. E. H. Oliw, C. Su, T. Skogstrom and G. Benthin, Analysis of novel hydroperoxides and other metabolites of oleic, linoleic, and linolenic acids by liquid chromatography-mass spectrometry with ion trap MSn, *Lipids*, 1998, 33, 843–852.

18. D. A. Evans and A. M. Golob, [3,3]Sigmatropic rearrangements of 1,5-diene alkoxides. The powerful accelerating effects of the alkoxide substituent, *J. Am. Chem. Soc.*, 1975, 97, 4765–4766.

19. S. Goyal, S. Banerjee and S. Mazumdar, Oxygenation of monoenoic fatty acids by CYP175A1, an orphan cytochrome P450 from Thermus thermophilus HB27, *Biochemistry*, 2012, 51, 7880–7890.

20. K. Bernstrom, K. Kayganich and R. C. Murphy, Collisionally induced dissociation of epoxyeicosatrienoic acids and epoxyeicosatrienoic acid-phospholipid molecular species, *Anal. Biochem.*, 1991, 198, 203–211.

21. C. Schneider, P. Schreier and M. Herderich, Analysis of lipoxygenase-derived fatty acid hydroperoxides by electrospray ionization tandem mass spectrometry, *Lipids*, 1997, 32, 331–336.

22. D. K. Macmillan and R. C. Murphy, Analysis of lipid hydroperoxides and long-chain conjugated keto acids by negative ion electrospray mass spectrometry, *J. Am. Soc. Mass Spectrom.*, 1995, 6, 1190–1201.

23. E. H. Oliw, U. Garscha, T. Nilsson and M. Cristea, Payne rearrangement during analysis of epoxyalcohols of linoleic and alpha-linolenic acids by normal phase liquid chromatography with tandem mass spectrometry, *Anal. Biochem.*, 2006, 354, 111–126.

CHAPTER 2

Eicosanoid and Bioactive Lipid Mediators

Biosynthesis and release of small lipid molecules from cells is a common strategy that has evolved in multicellular organisms in order to signal nearby cells and coordinate biochemical events within cells in the immediate microenvironment. This biochemical coordination with the target cell is achieved by specific signaling proteins such as cell surface G-protein linked receptors that recognize these small lipid molecules. Many of these signal-inducing lipids are metabolites of arachidonic acid and have come to be called by the generic name "eicosanoids".[1]

Eicosanoid biosynthesis is initiated by enzymatic hydrolysis of esterified arachidonate (phospholipids), which provides the substrate for lip-oxygenases, cyclooxygenases, and specific P-450 enzymes. The bioactive products from 5-lipoxygenase are the various leukotrienes, while products from cyclooxygenase are the prostaglandins. P-450 derived metabolites include the various epoxyeicosatrienoic acids (EETs), and hydroxyeicosate-traenoic acids (HETEs).[2] In addition to these eicosanoids, other bioactive mediators are synthesized on demand, including ceramides, sphingosine-1-phosphate (S1P), ceramide-1-phosphate, lysophosphatidic acid (LPA) as well as metabolites of other fatty acids such as docosahexaenoic acid. All of these lipids operate by means of separate receptor recognition, including the S1P, LPA receptors and other specific G-protein coupled receptors.

Tandem mass spectrometry has played a central role not only in the initial structural identification of these lipids, but also in the quantitation of very low levels of all eicosanoids, eicosanoid metabolites, and other lipid mediators. In particular, the LC-MS/MS based analysis using stable isotope labeled internal standards has become the gold standard by which other analytical methods

New Developments in Mass Spectrometry No. 4
Tandem Mass Spectrometry of Lipids: Molecular Analysis of Complex Lipids
By Robert C Murphy
Published by the Royal Society of Chemistry, www.rsc.org

are compared. The structural identification of very low levels of these eicos-anoids is critically dependent upon mass spectrometric data, including that generated by electrospray ionization and tandem mass spectrometry.

This chapter will deal only with the ion chemistry of eicosanoids, their metabolites, and other lipid mediators, rather than the theory and practice of quantitative mass spectrometry that was previously discussed.[3] Metabolites of certain eicosanoids are included in this chapter, since they are the molecules that can be used to measure prostaglandin or leukotriene production *in vivo*. All eicosanoids are rapidly metabolized once they leave the biosynthetic cell, but some investigators mistakenly consider measurement of primary eicos-anoids in the blood as the measure of total body synthesis. Indeed it is possible to measure low levels of primary prostaglandins and leukotrienes in the blood, but these levels are merely those which escape local metabolism at the site of synthesis and this capability to measure eicosanoids in plasma has become possible due to the extraordinarily high sensitivity of modern mass spectrometers. There are also specific instances where primary prostaglan-dins are observed in blood, but this is merely an artifact of needle puncture or shear stress that cause blood cells, as they are drawn into a syringe, to initiate biosynthesis of prostaglandins, and in particular thromboxanes. These levels do not indicate the eicosanoids circulating in the blood, but rather an arti-factual initiation of eicosanoid biosynthesis during blood sampling. Urinary metabolites on the other hand integrate the production of prostaglandins and leukotrienes in the entire organism since the metabolites are collected and excreted in the urine. Nonetheless, it should be realized that only a small fraction of the total synthesis of these eicosanoids is reflected by the metab-olite appearing in urine and one has to consider the fact it is likely that less than 5–10% of the total body production of the prostaglandin appears in the urine and substantially less of that for certain leukotrienes.

Most eicosanoid lipid mediators are hydroxy fatty acids containing multiple double bonds that render favorable collision induced decomposition reactions after one or more double bond rearrangements. Simple losses of one or more hydroxyl group as H_2O and loss of CO_2 are typically observed, but these losses are not unique to the eicosanoid structure. Those product ions which involve cleavage of the 20-carbon chain of arachidonic acid are more related to the eicosanoid structure and often used for selected ion transitions to detect and quantitate these lipids in biological extracts. The more abundant product ions that are generated from collisional activated $[M - H]^-$ derived from a pri-mary eicosanoid or eicosanoid metabolite will be discussed. The ion chemistry of the most abundant eicosanoids has been reviewed in some detail.[4]

2.1 Prostaglandins

2.1.1 Prostaglandin E_2 (PGE$_2$) and Prostaglandin E_2 Metabolite (PGEM)

The tandem mass spectrometry of PGE$_2$ (Figure 2.1) has been studied using stable isotope analogs and high resolution analysis of product ions.[4,5] Elec-trospray ionization generates abundant $[M - H]^-$ for PGE$_2$ at m/z 351

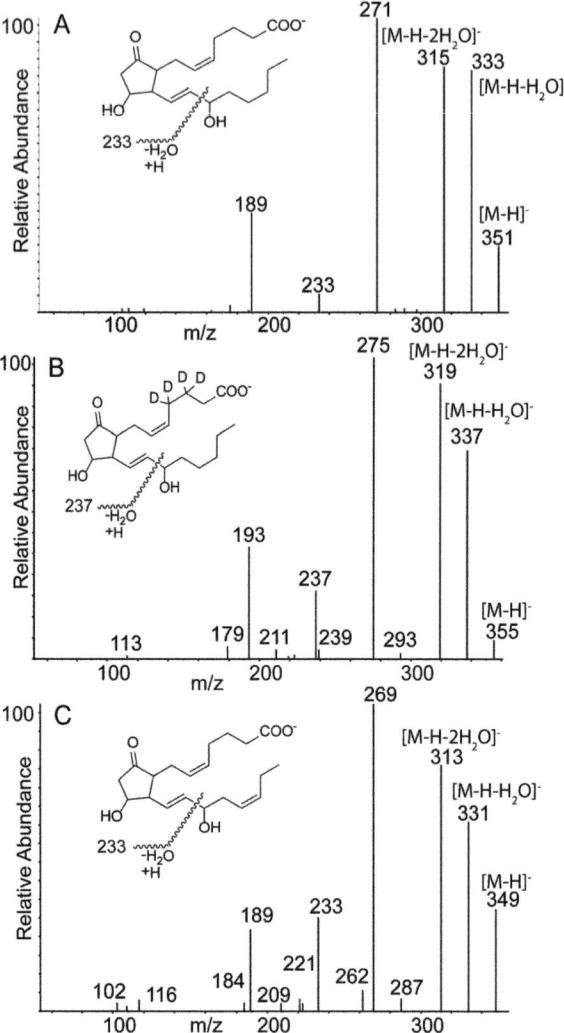

Figure 2.1 Electrospray ionization (negative ions) and tandem mass spectrometry of prostaglandin E. (A) Product ions following collisional activation of PGE$_2$ [M − H]$^-$ at *m/z* 351; (B) product ions obtained following collisional activation of [3,3,4,4-D$_4$]PGE$_2$ [M − H]$^-$ at *m/z* 355; (C) product ions obtained following collisional activation of PGE$_3$ [M − H]$^-$ at *m/z* 349. These MS/MS spectra were obtained using a tandem quadrupole mass spectrometer.

(Figure 2.1A), m/z 355 for d_4-PGE$_2$ (Figure 2.1B), and m/z 349 for PGE$_3$ (Figure 2.1C). Collisional activation of each $[M - H]^-$ readily leads to loss of H_2O (m/z 333), likely forming an α,β-unsaturated ketone within the cyclopentane ring (Scheme 2.1).

The abundant product ion at m/z 315 corresponds to the loss of two molecules of water, but likely also engages double bond migration in order to form a conjugated system. This dehydrated PGE$_2$ product ion would then facilitate loss of CO_2, which is for most fatty acids, a minor product ion, suggesting participation of the double bonds in the dehydrated PGE$_2$ in the formation of the abundant m/z 271. A possible mechanism for m/z 271 from m/z 315 could involve the double bond at carbon-5 undergoing two sigmatropic shifts to become conjugated with the ketone at C-9. Loss of CO_2 could then be a charged driven process with formation of the 6-membered ring and oxygen centered product anion (Scheme 2.2). This mechanism is consistent with d_4-PGE$_2$ and PGE$_3$ (Figure 2.1) and high resolution mass analysis, m/z 271.207 ($C_{19}H_{27}O^-$).

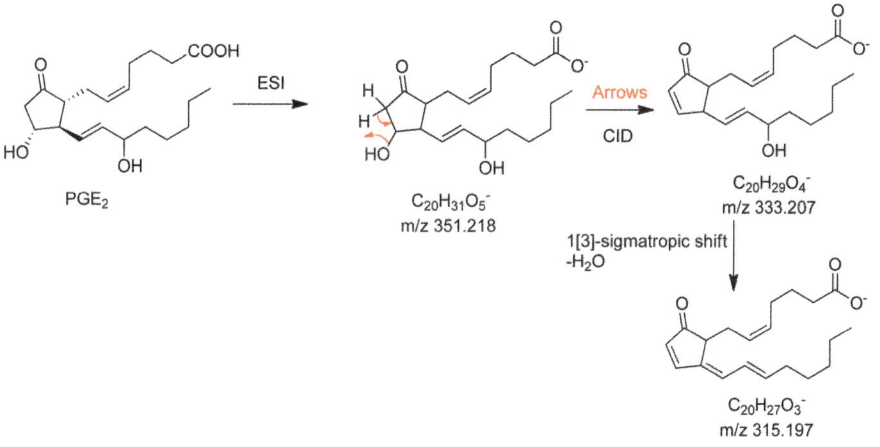

Scheme 2.1

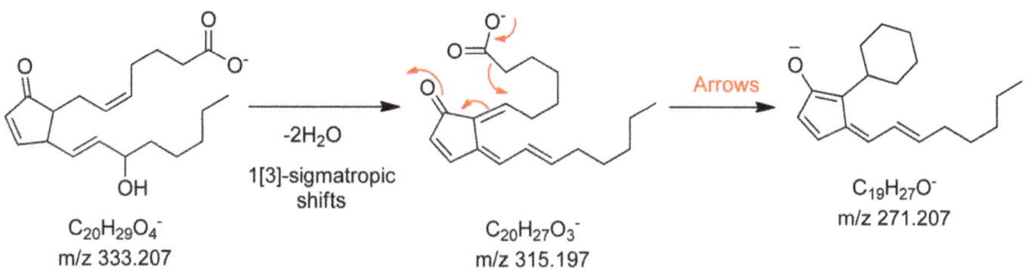

Scheme 2.2

The ion at m/z 233 is found both in PGE$_2$ and PGE$_3$, suggesting this ion is derived from the loss of the methyl terminus of the PGE structure. A charge driven cleavage of the C-14/C-15 bond from a dehydrated m/z 333 has been suggested to form a high delocalized anion (Scheme 2.3),[4] but could also be a charge remote ene-reaction of the conjugated diene (Scheme 2.4). This isomeric structure for m/z 233 would then facilitate the loss of CO$_2$ to form the highly conjugated enolate anion at m/z 189. The 2,2,3,3-d$_4$-PGE$_2$ (Figure 2.1B) has all of these ions shifted by 4 Da, suggesting that none of the protons at these positions are involved in product ion formation, not even a loss of H$_2$O.

The urinary metabolite of PGE$_2$, tetranor-PGE$_2$ often referred to as PGEM, is an extensively metabolized product and results from 4 cycles of β-oxidation from the carboxyl terminus, action of the 15-hydroxy prostaglandin dehydrogenase/15,14-reductase and ω-oxidation (Figure 2.2A). This dicarboxylic

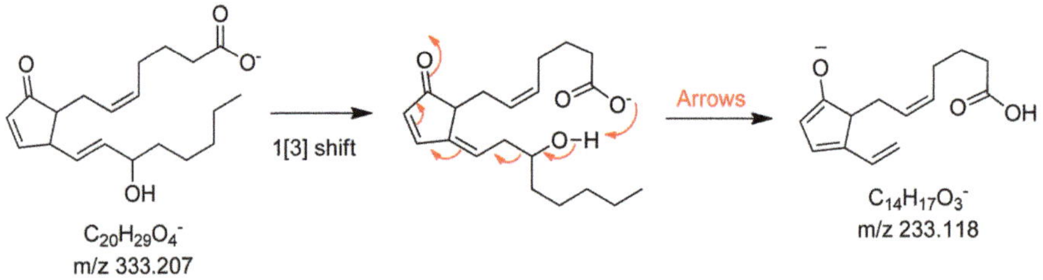

Scheme 2.3

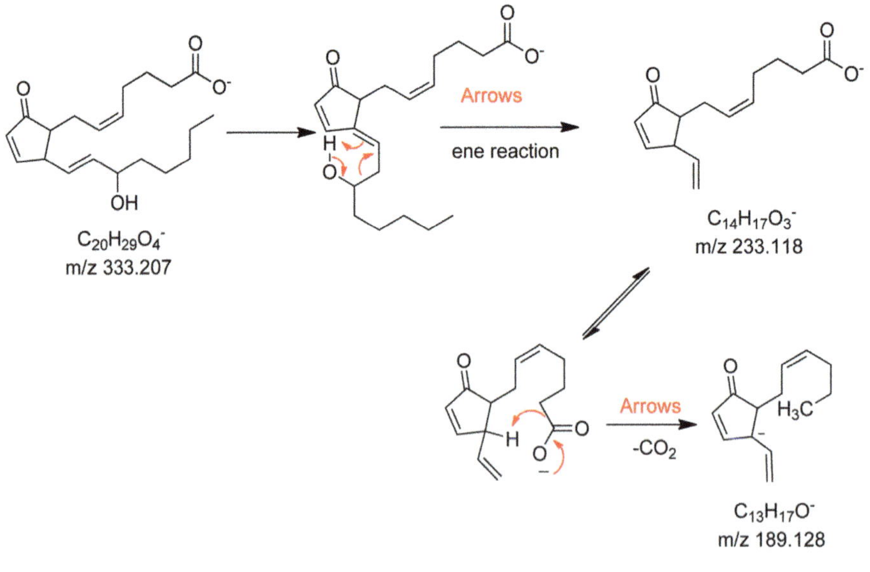

Scheme 2.4

acid has an altered collision induced product ion mass spectrum from that of PGE$_2$. The losses of one (*m/z* 309) and two (*m/z* 291) molecules of water are evident, but *m/z* 291 is significantly less abundant, when compared to the ion corresponding to the loss of two water molecules in native PGE$_2$ (Scheme 2.5). The formation of *m/z* 291 may involve some formation of a ketene like structure from the loss of H$_2$O from the carboxyl group closest to the keto-structure since there is little loss of HOD (*m/z* 296) evident in the D$_6$-PGEM (Figure 2.2B) where the three methylene groups adjacent to the terminal carboxyl group, contains two deuterium atoms each.

The analysis of PGEM can be difficult due to the short retention time on reversed phase HPLC. An alternative strategy is to synthesize the bis(me-thoxime) derivative,[6] which is more lipophilic, imparting a longer retention time on reversed phase columns (Figure 2.2C). Collisional activation of the resulting [M − H]$^-$, now at *m/z* 385, results in the abundant ion at *m/z* 336 which is a radical anion due to the loss of the methoxy group (31 Da), as a radical, and one neutral water molecule (Figure 2.2C) likely with the structure indicated in Scheme 2.6 with the radical site on the nitrogen atom delocalized by resonance.

2.1.2 Prostaglandin D$_2$ (PGD$_2$)

The tandem mass spectrometry of the electrospray generated [M − H]$^-$ from PGD$_2$ is almost identical to that of PGE$_2$. The abundance of *m/z* 233 and 189 is considerably higher, but changes in electrospray conditions and instrument variations make this difference in abundance impossible to unambiguously

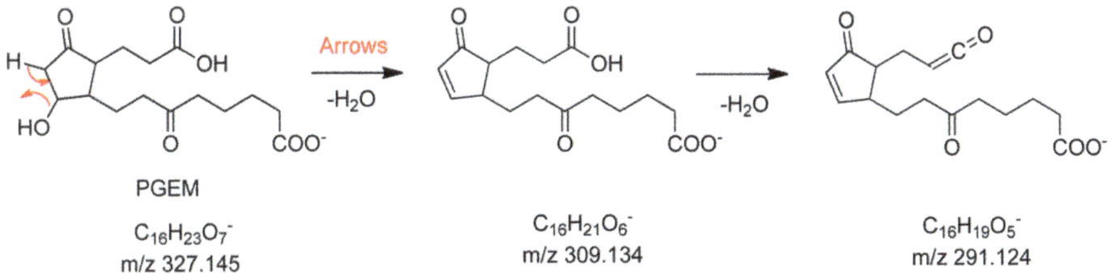

PGEM
C$_{16}$H$_{23}$O$_7^-$
m/z 327.145

C$_{16}$H$_{21}$O$_6^-$
m/z 309.134

C$_{16}$H$_{19}$O$_5^-$
m/z 291.124

Scheme 2.5

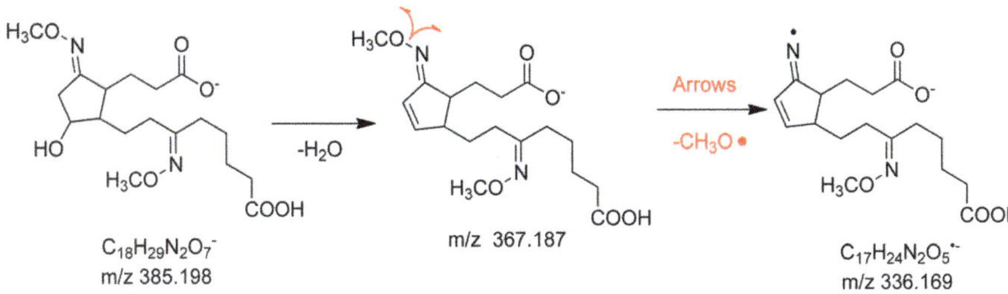

C$_{18}$H$_{29}$N$_2$O$_7^-$
m/z 385.198

m/z 367.187

C$_{17}$H$_{24}$N$_2$O$_5^{\cdot-}$
m/z 336.169

Scheme 2.6

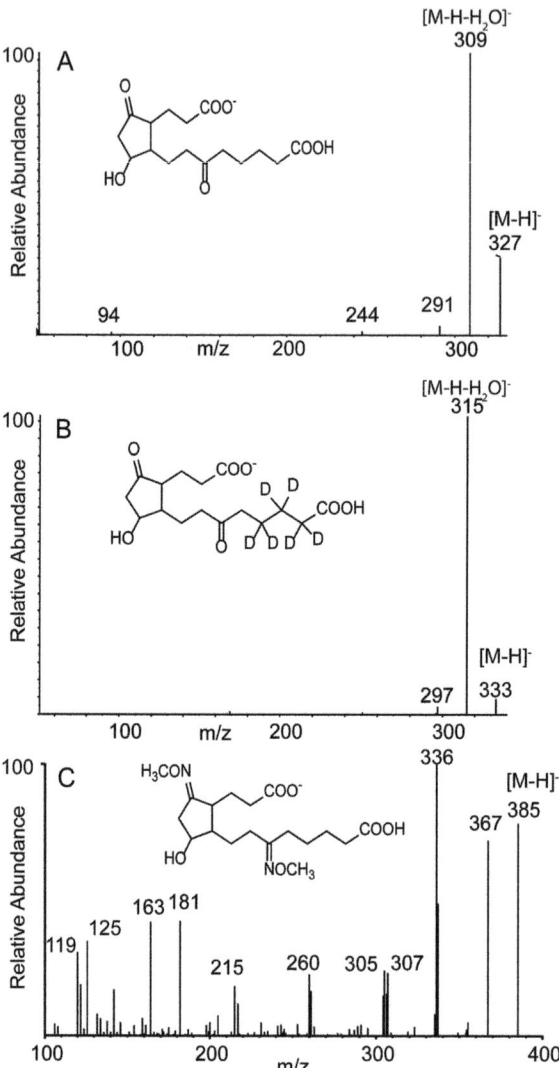

Figure 2.2 Electrospray ionization (negative ions) and tandem mass spectrometry of the urinary metabolite of PGE$_2$. (A) Product ions obtained following collisional activation of the tetranor-prostaglandin E metabolite [M − H]$^-$ at m/z 327; (B) product ions obtained following collisional activation of [2,2,3,3,4,4-D$_6$]-tetranor-PGE metabolite [M − H]$^-$ at m/z 333; (C) product ions obtained following collisional activation of the bis (methoxime) derivative of the tetranor-PGE$_2$ metabolite [M − H]$^-$ at m/z 385. These MS/MS spectra were obtained using a tandem quadrupole mass spectrometer.

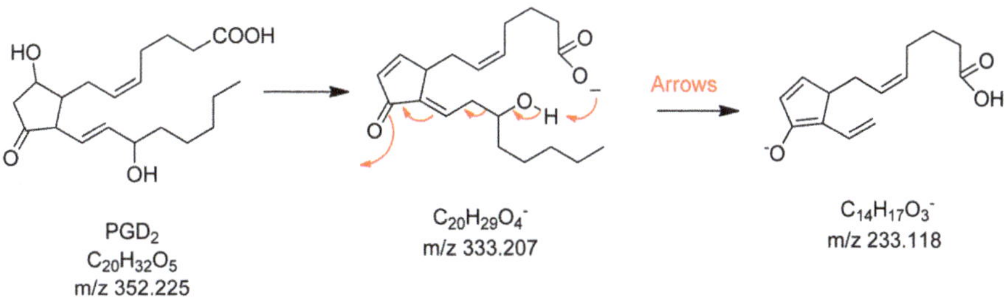

HO ~~~~COOH

PGD$_2$
C$_{20}$H$_{32}$O$_5$
m/z 352.225

C$_{20}$H$_{29}$O$_4^-$
m/z 333.207

Arrows

C$_{14}$H$_{17}$O$_3^-$
m/z 233.118

Scheme 2.7

characterize PGE$_2$ relative to PGD$_2$. These two isomeric eicosanoids are typically identified in biological extracts, by a slightly longer retention time for PGD$_2$ relative to the PGE$_2$ in reversed phase HPLC. The mechanism of formation of m/z 233 is likely very similar to that shown in Scheme 2.4 for PGE$_2$, but the isomeric ion at m/z 233 may be slightly more facile to form (Scheme 2.7).

Chemically, PGD$_2$ is more unstable than PGE$_2$ and readily dehydrates in solution to form quite reactive PGJ$_2$ metabolites that are α,β-unsaturated ketones. These electrophilic eicosanoids often form covalent adducts with nucleophiles in biological systems and analysis of these indicates interesting chemical events taking place within the cell. These dehydrated PGD$_2$ products or the structurally related isoprostane PGD$_2$ family members, are analyzed often as glutathione adducts eliminated in urine.[7] Mass spectrometric behavior of these cysteinyl peptide lipid adducts has not been extensively studied.

2.1.3 Prostaglandin F$_{2\alpha}$ (PGF$_{2\alpha}$)

The trihydroxy prostaglandin PGF$_{2\alpha}$ yields numerous product ions following collisional activation of the [M − H]$^-$ formed by electrospray ionization (Figure 2.3). The higher mass ions arise from the loss of one to three water molecules, along with the loss of 44 Da in combinations with these small neutral losses.

The most abundant high mass ion, measured at m/z 309.207 (C$_{18}$H$_{29}$O$_4$) corresponds to the loss of 44 Da, which would correspond to the loss of C$_2$H$_4$O rather than CO$_2$. Such an ion was suggested to arise from a charge remote fragmentation when the positive ion from bariated PGF$_{2\alpha}$ was collisionally activated at high energy.[8] When the carboxylate anion is activated at much lower energies (tandem quadrupole instrument), charged driven formation of an oxygen-centered anion at either C-9 or C-11 is more likely, followed by the loss of C$_2$H$_4$O from the cyclopentane ring (Scheme 2.8).

The delocalization of the carbon-centered anion by the adjacent double bond would stabilize the structure. This ion at m/z 309 could readily lose a water molecule from the C-15 hydroxyl group further stabilizing the anionic

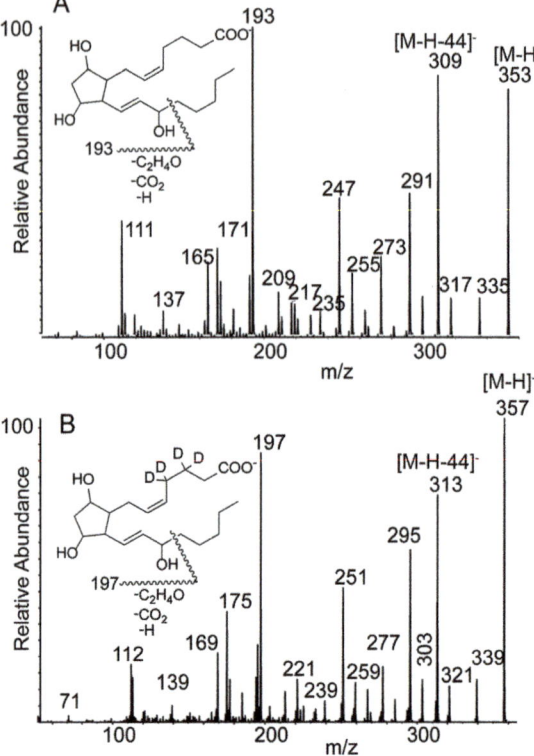

Figure 2.3 Electrospray ionization (negative ions) and tandem mass spectrometry of prostaglandin $F_{2\alpha}$. (A) Product ions obtained following collisional activation of $PGF_{2\alpha}$ $[M - H]^-$ at m/z 353; (B) product ions obtained following collisional activation of $[3,3,4,4\text{-}D_4]\text{-}PGF_{2\alpha}$ $[M - H]^-$ at m/z 357. These MS/MS spectra were obtained using a tandem quadrupole mass spectrometer.

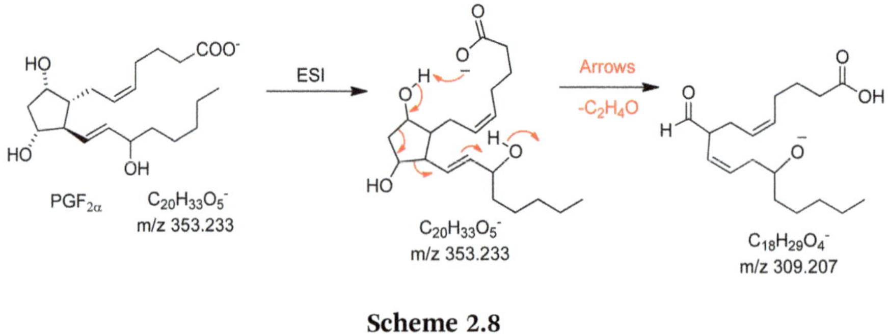

Scheme 2.8

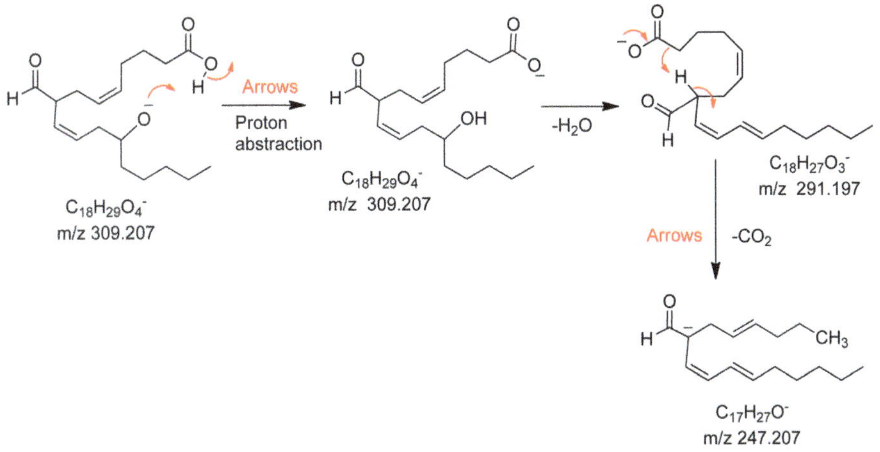

Scheme 2.9

site by extending conjugation (Scheme 2.9). The loss of an additional 44 Da, which now corresponds to the loss of CO_2, suggests that the charge site could move back to the carboxyl moiety because of the flexibility of this molecule and perhaps abstraction of a proton to further delocalize the resulting carbon-centered anion. These mechanisms are all consistent with high resolution data and limited deuterium isotope labeled D_4-PGF$_{2\alpha}$ (Figure 2.3B).

The most abundant product ion formed following CID of PGF$_{2\alpha}$ is observed at m/z 193.123 ($C_{12}H_{17}O_2$), which retains all deuterium atoms at carbons 3 and 4 (Figure 2.3B). The published structures of this ion are not entirely consistent with this high resolution and deuterium labeled data,[4] but one structure consistent with the high resolution measurements and deuterium labeled analogs is suggested in Scheme 2.10. The intermediate suggested in Scheme 2.8 could undergo a charge driven α-cleavage, loss of CO_2 and concerted proton rearrangement to form a carbon-centered anion stabilized by resonance (Scheme 2.10).

The biosynthesis of prostaglandins, including PGF$_{2\alpha}$ is the result of the action of prostaglandin H-synthase 1 or 2 (cyclooxygenase 1 or 2) which controls the precise stereochemistry of the prostaglandin products which are derived from free arachidonic acid.[1] However, free radical oxidation of the arachidonic acid while it is still esterified to phospholipids leads to the formation of a family of stable products that, once hydrolyzed from the phospholipid, are found to be isomeric to the prostaglandins.[9] These compounds have come to be called isoprostanes and they are not products of enzymatic biosynthesis. Examples are the F$_2$-isoprostanes isomeric to prostaglandin F$_{2\alpha}$ (Scheme 2.11).

Four basic regioisomeric structures are found and each family structure is a mixture of 16 different stereochemical isomers, leading to a total of 64 possible different F$_2$-isoprostanes. The four family members can be uniquely

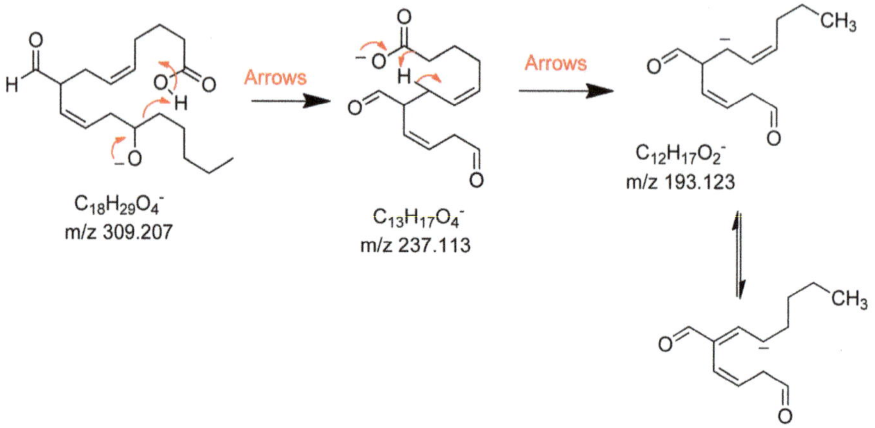

C₁₈H₂₉O₄⁻
m/z 309.207

C₁₃H₁₇O₄⁻
m/z 237.113

C₁₂H₁₇O₂⁻
m/z 193.123

Scheme 2.10

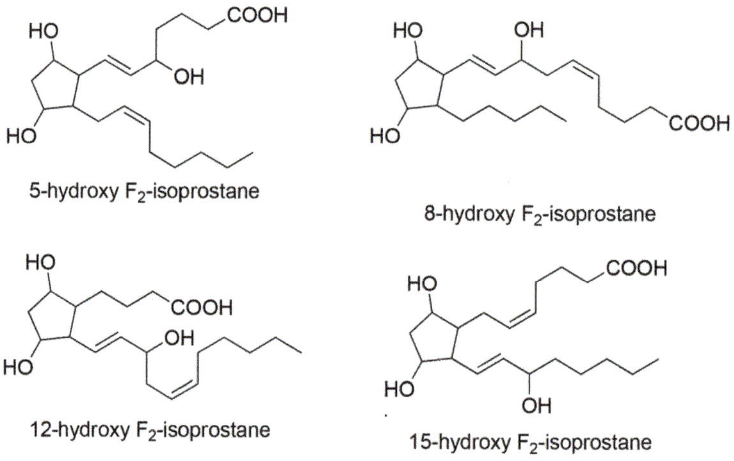

5-hydroxy F₂-isoprostane

8-hydroxy F₂-isoprostane

12-hydroxy F₂-isoprostane

15-hydroxy F₂-isoprostane

Scheme 2.11

identified by tandem mass spectrometry and the formation of several abundant ions have been found to be unique for each family.[10,11] These ions are largely driven by the location of the hydroxyl group, not in the cyclopentane ring, but on the side chain, which will be the basis of how they are designated here (Scheme 2.11). Each F_2-isoprostane has very similar neutral losses at high mass, such as loss of C_2H_4O (44 Da) and one or more water molecules.

The 5-hydroxy family member yields an abundant ion at m/z 115, likely due to a charge remote ene-reaction that could take place after a 1[3]-sigmatropic shift (Scheme 2.12). The prior loss of the C_2H_4O could form a product ion that is stabilized as an α,β-unsaturated ketone from which m/z 115 is derived, but this precursor structure is not illustrated.

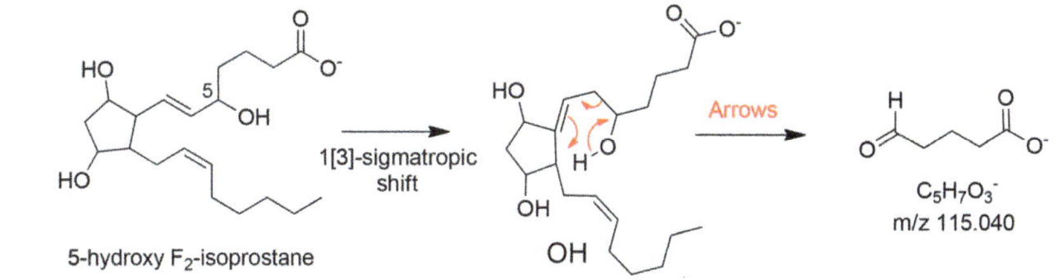

Scheme 2.12

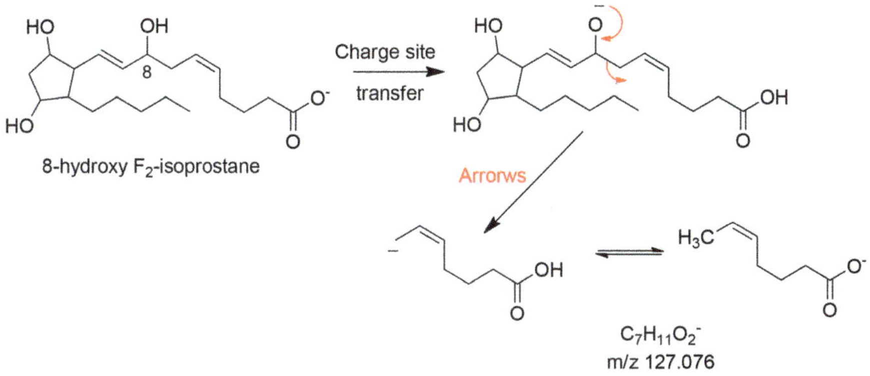

Scheme 2.13

The 8-hydroxy F_2-isoprostane yields an abundant ion at m/z 127, likely after the carboxylate anion removes a proton from the carbon-8 hydroxyl group. A resulting charge-driven formation of an aldehyde would leave a carbon-centered anionic site in conjugation with the first double bond (Scheme 2.13).

The 12-hydroxy F_2-isoprostane forms several unique ions including m/z 151 and 179.[10,11] The ion at m/z 179 is likely a result of initial charge remote cleavage of the β-hydroxy olefin moiety that can readily undergo an ene-reaction based cleavage to form a chain shortened aldehyde that after dehydration of either one of the cyclopentane alcohols could then lose CO_2 and result in a carbon-centered anion at C-2. This would be rather energetically unfavorable, but this C-2 anion could remove a proton from carbon-9 to form a carbon-centered anion at carbon-9 stabilized by adjacent double bonds. The ion at m/z 151 could also result from a similar mechanistic path to yield a product ion that would be the loss of CO_2 and cleavage of C3-4 bond with formation of ethylene (Scheme 2.14).

There are other isoprostanes, including E_2-, D_2-, J_2-, and A_2-isoprostanes, that have been reported following free radical reactions within cells.[9] These isoprostanes are similarly complex mixtures in terms of being a mixture of stereoisomers, but detailed mass spectrometry has not been carried out to establish specific ions relevant to their regioisomeric structures. Nonetheless,

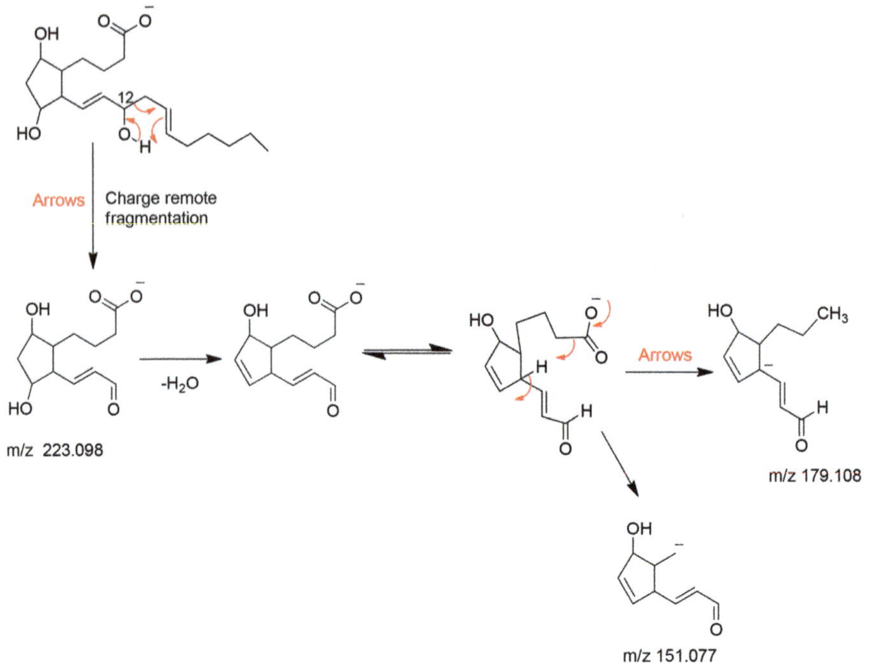

Scheme 2.14

LC-MS/MS is used to quantitatively assess the production of these unique products of phospholipid peroxidation and markers of free radical biology.[12]

2.1.4 Prostacyclin and 6-Keto-PGF$_{1\alpha}$

Prostacyclin, also called prostaglandin I$_2$, is a very bioactive eicosanoid in the cardiovascular system, and is also chemically unstable. This bicyclic structure is rapidly hydrolyzed in water to the biochemically inactive product 6-keto-PGF$_{1\alpha}$ (Scheme 2.15). Yet it is this inactive form and metabolites, including 2,3-dinor-PGF$_{1\alpha}$, that can be measured in biological fluids by tandem mass spectrometry after electrospray ionization. The collisional activation of 6-keto-PGF$_{1\alpha}$ (Figure 2.4) yields a distinctive set of product ions that suggest a strong influence of the keto moiety in the induced fragmentation.

The same behavior of loss of 44 Da as C$_2$H$_4$O and 1-3 molecules of water is characteristic of the higher mass product ions and is the result of a mechanism similarly described to that seen for PGF$_{2\alpha}$ (Scheme 2.8) possibly with a prior dehydration of the C-15 alcohol (Scheme 2.16). The 6-keto group can be attacked by a carbanion that may form at C-14 after a 1[3]-sigmatropic proton shift and formation of a cyclic carbinol that could readily dehydrate to conjugated cyclohexadiene intermediate (Scheme 2.16). The final step

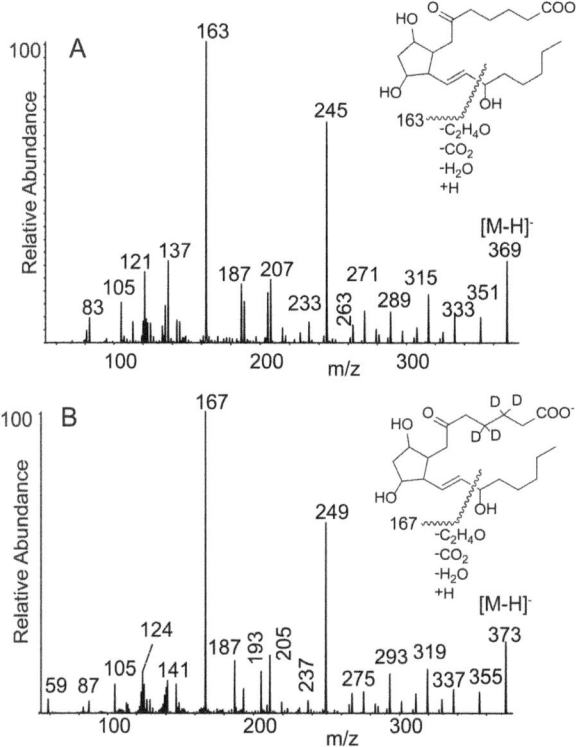

Figure 2.4   Electrospray ionization (negative ions) and tandem mass spectrometry of the hydrolysis product of prostacyclin, 6-keto-prostaglandin $F_{1\alpha}$. (A) Product ions obtained following collisional activation of 6-keto-PGF$_{1\alpha}$ [M − H]$^-$ at m/z 369; (B) product ions obtained following collisional activation of [3,3,4,4-D$_4$]-6-keto-PGF$_{1\alpha}$ [M − H]$^-$ at m/z 373. These MS/MS spectra were obtained using a tandem quadrupole mass spectrometer.

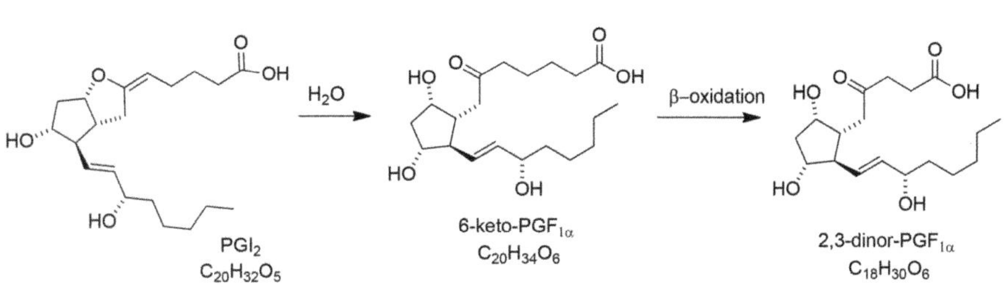

Scheme 2.15

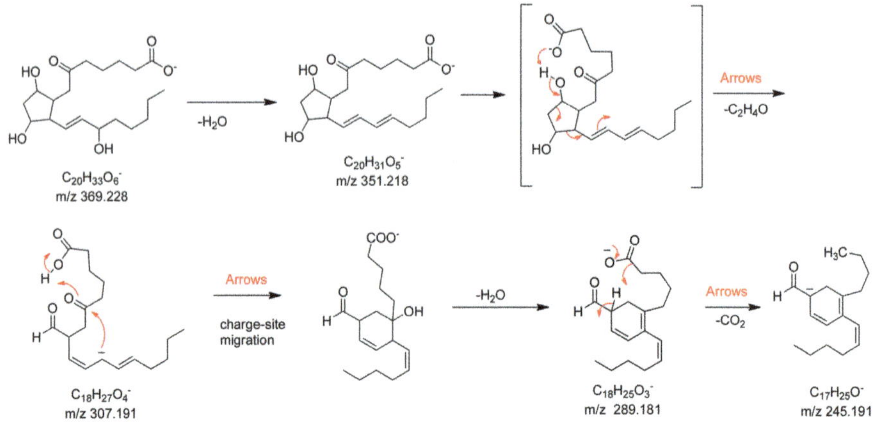

Scheme 2.16

indicated is the loss of CO_2, which drives formation of a highly stabilized anion ($C_{17}H_{28}O$). The length and flexibility of the carboxyl moiety attached to the side chain facilitates these changes in anionic sites for this molecule, as well as most of the eicosanoids. While the mechanism depicted suggests one series of step-wise reactions, more likely the loss of water, loss of C_2H_4O, a second loss of H_2O and final loss of water occurs in various orders. This mechanism is consistent with high resolution and deuterium labeling studies.

The most abundant ion measured at m/z 163.112 has the elemental composition $C_{11}H_{15}O$. The proton from carbon atoms at C-3 and C-4 remains in this ion based on the shift of the corresponding 3,3,4,4-tetradeutero-6-keto-PGF$_{1\alpha}$ product ion observed at m/z 167 (+4). This product ion is likely a result of an ene-reaction seen often in prostaglandins due to the 13,14-double bond shifted to carbon atom positions 12 and 13 followed by the loss of carbon-15 chain as a neutral aldehyde (Scheme 2.17). After transformation of the cyclopentane by the loss of C_4H_4O and loss of an additional molecule of water, a highly delocalized product ion is formed, driven by the loss of CO_2 from the carboxylate anion and charge site moved to carbon-8 by removal of the proton allylic to the aldehyde and delocalized by the conjugated diene (Scheme 2.17).

A urinary metabolite of prostacyclin is 2,3-dinor-6-keto-prostaglandin-F$_{1\alpha}$.[13] The tandem mass spectrometry of this eicosanoid [M − H]$^-$ is quite similar to that of the metabolic precursor, 6-keto-PGF$_{1\alpha}$ (Figure 2.4), with major ions shifted 28 Da lower in mass. This is consistent with the mechanisms presented for the CID behavior of 6-keto-PGF$_{1\alpha}$ which does not involve cleavage of bonds between carbon atoms 2 and 6, and retention of all deuterium atoms at carbons 3 and 4. An example is the ion observed at m/z 217, which could be formed by the mechanism outlined in Scheme 2.18, seen previously for the major product ion from 6-keto-PGF$_{1\alpha}$ at m/z 245.

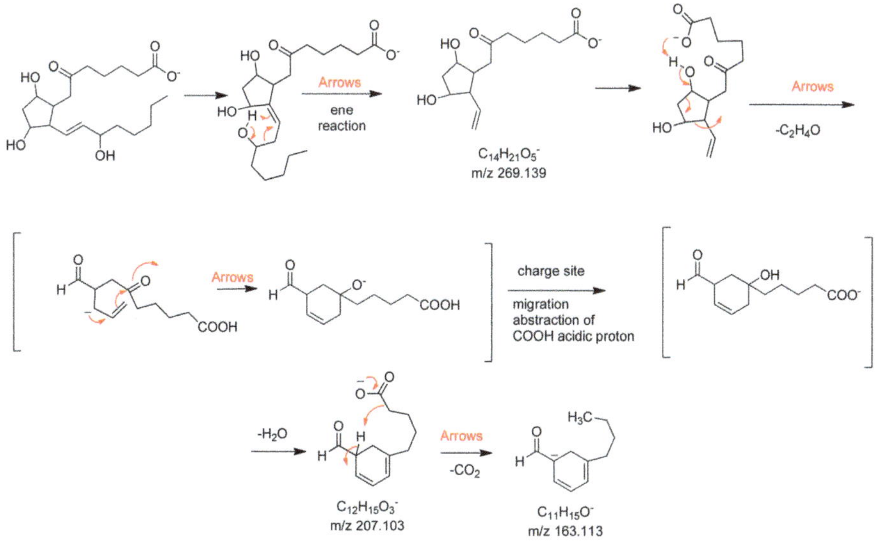

Scheme 2.17

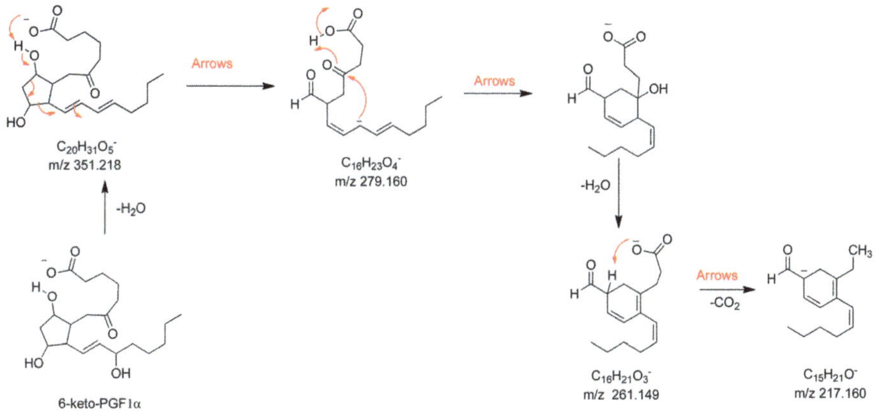

Scheme 2.18

The very abundant product ion at m/z 135 is likely derived by a mechanism similar to that suggested for m/z 163 from 6-keto-PGF$_{1\alpha}$ with the formation of a carbon centered anion delocalized by a conjugated diene (Scheme 2.19).

2.1.5   Thromboxane A$_2$, Thromboxane B$_2$, and Metabolites

The biologically active thromboxane is thromboxane A$_2$ (TxA$_2$), which is a product of arachidonate metabolism by cyclooxygenase and thromboxane synthase. However, TxA$_2$ has an extraordinarily short-half life and cannot be

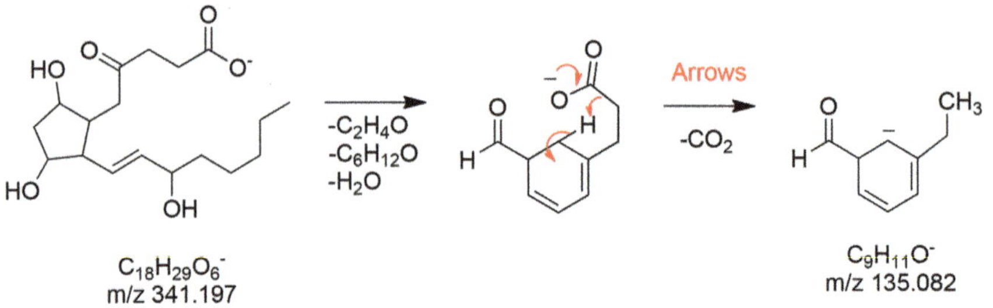
HO
HO
OH
O
O
O-
-C2H4O
-C6H12O
-H2O
Arrows
-CO2
H
H
CH3
C18H29O6-
m/z 341.197
C9H11O-
m/z 135.082

Scheme 2.19

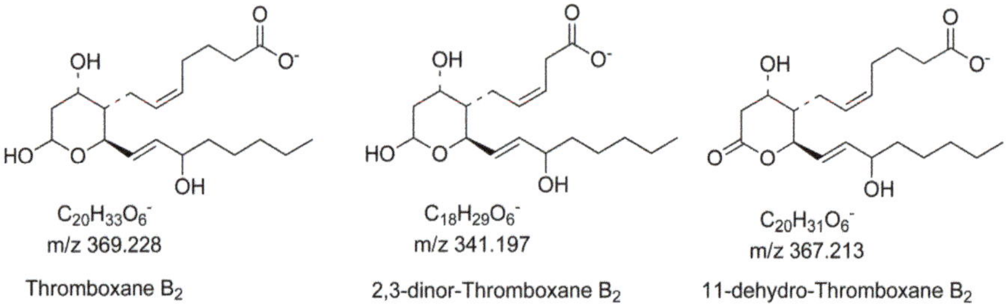
C20H33O6-
m/z 369.228
Thromboxane B2
C18H29O6-
m/z 341.197
2,3-dinor-Thromboxane B2
C20H31O6-
m/z 367.213
11-dehydro-Thromboxane B2

Scheme 2.20

isolated from biological systems as such. Rather, the hydrolysis of TxA_2 yields thromboxane B_2 (TxB_2), which is quite stable. It does undergo hemiacetal formation to form a 6-membered ring with the side chains, as the dominate form and the minor form being the ring opened aldehyde (Scheme 2.20). These two forms can be separated by reversed phase chromatography and, since hemiacetal formation and hydrolysis occurs during chromatographic separation, reversed phase HPLC analysis typically yields a major peak with a "saddle" in between a second and typically less abundant ring open form.[14]

Collisional activation of $[M - H]^-$ of TxB_2 (m/z 369) yields two abundant and quite characteristic major product ions at m/z 195 and 169 (Figure 2.6A). These same fragment ions were observed following high energy collision induced decomposition of $[M - H]^-$ of TxB_2 generated by fast atom bombardment mass spectrometry. The mechanism of formation of these ions has been suggested to be driven by charge remote processes.[8,15] Charge driven mechanisms are perhaps more likely to occur at the lower energies of the tandem quadrupole mass spectrometry, but these mechanisms can also lead to the same ions, as suggested in Schemes 2.21 and 2.22. In this first mechanism the carboxylate anion removes a proton from the hemiacetal hydroxyl group at carbon-11 and this initiates the formation of an α,β-unsaturated aldehyde neutral ($C_9H_{16}O$) from the terminal TxB_2 side chain and a concerted loss of H_2O, bringing the charge site back to the highly stable carboxylate anion.

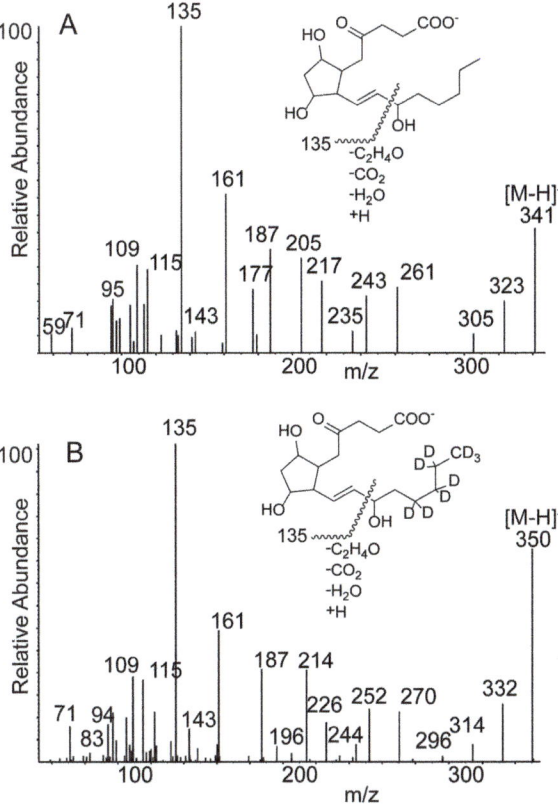

Figure 2.5 Electrospray ionization (negative ions) and tandem mass spectrometry of a urinary metabolite of prostacyclin. (A) Product ions obtained following collisional activation of 2,3-dinor-6-keto-PGF$_{1\alpha}$ [M – H]$^-$ at m/z 341; (B) product ions obtained following collisional activation of [17,17,18,18,19,19,20,20,20-D$_9$]-2,3-dinor-6-keto-PGF$_{1\alpha}$ [M – H]$^-$ at m/z 350. These MS/MS spectra were obtained using a tandem quadrupole mass spectrometer.

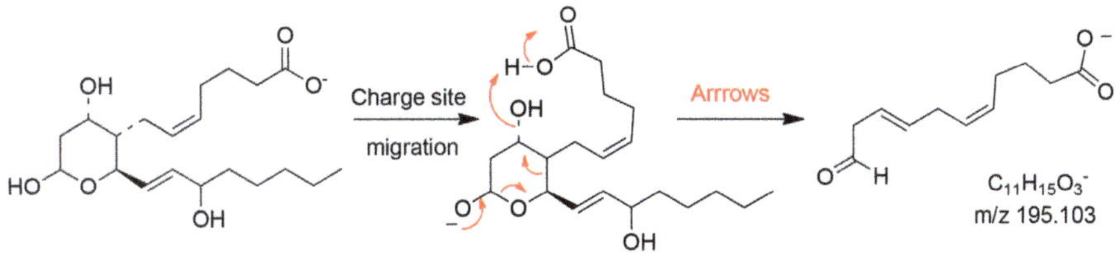

Scheme 2.21

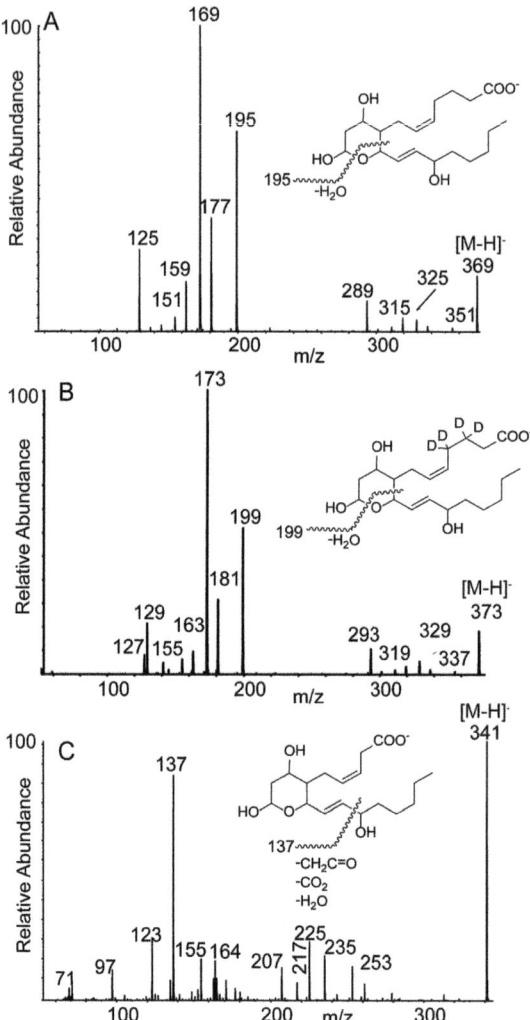

Figure 2.6 Electrospray ionization (negative ions) and tandem mass spectrometry of thromboxane and thromboxane urinary metabolite. (A) Product ions obtained following collisional activation of thromboxane B_2 (TxB_2) $[M - H]^-$ at m/z 369; (B) product ions obtained following collisional activation of $[3,3,4,4-D_4]$-TxB_2 $[M - H]^-$ at m/z 373; (C) product ions obtained following collisional activation of 2,3-dinor-TxB_2 $[M - H]^-$ at m/z 341. These MS/MS spectra were obtained using a tandem quadrupole mass spectrometer.

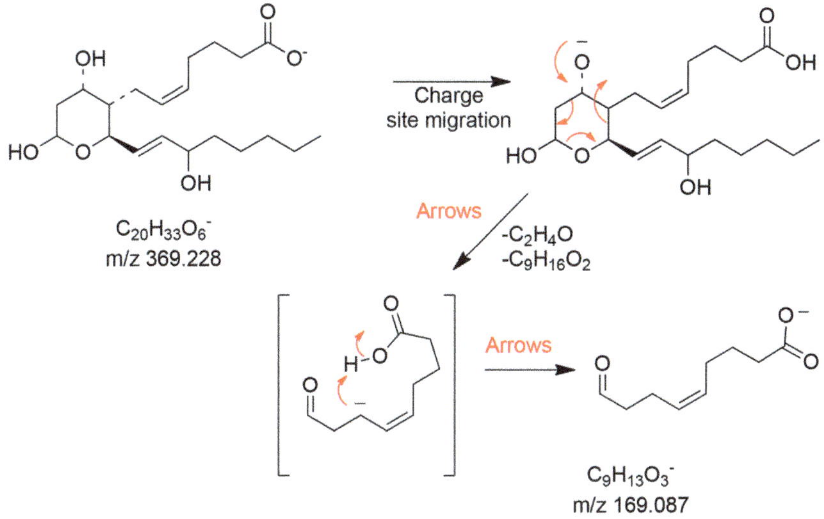

Scheme 2.22

The most abundant product ion derived from TxB$_2$ is observed at m/z 169 and can also be generated by a charge remote mechanism,[4] as well as a charge driven mechanism (Scheme 2.22). In the charge driven mechanism, a charge site migrates from the carboxyl group to the oxygen atom at carbon-9, which is quite readily accessible by the carboxylate anion as was found for other prostaglandins. Subsequent rearrangement of the alkoxide anion then leads to the loss of C$_2$H$_4$O as we have seen for PGF$_{2\alpha}$, and the loss of the same α,β-unsaturated aldehyde from the methyl terminus of TxB$_2$. Both of these major ions are shifted +4 Da in the 3,3,4,4-tetradeutero TxB$_2$ analog (Figure 2.6B). One or both of these ions are often used for selected reaction monitoring, as derived from the [M − H]$^-$, as a basis to quantitate the level of TxB$_2$ in biological systems.

The collisional activation of the [M − H]$^-$ from 2,3-dinor-TxB$_2$ is not surprisingly quite similar to that of the TxB$_2$ (Figure 2.5C). The major product ions at m/z 169 and 195 observed for TxB$_2$ are shifted 28 Da lower in mass (m/z 141 and 167) due to the shortening of the carboxyl terminus by two carbon atoms. These two carbon atoms do not alter in any significant way the mechanisms described for the formation of these ions, but their abundance is quite low. The more abundant product ions are m/z 155 and 137, which do not appear to be abundant in the tandem mass spectrometry of TxB$_2$ and are likely the result of reduced accessibility of the carboxylate anion charge site to facilitate migration of protons from remote positions. The ion at m/z 155 is likely a unique product of an initial ene-reaction and loss of the C-15 aldehyde, as seen and described previously (Scheme 2.4). These resultant product ions can then undergo a concerted loss of CO$_2$ and ketene from the "ring-open" form of 2,3-dinor-TxB$_2$ (Scheme 2.23). The loss of H$_2$O from this ion would lead to the most abundant product ion from this metabolite, observed at m/z 137.

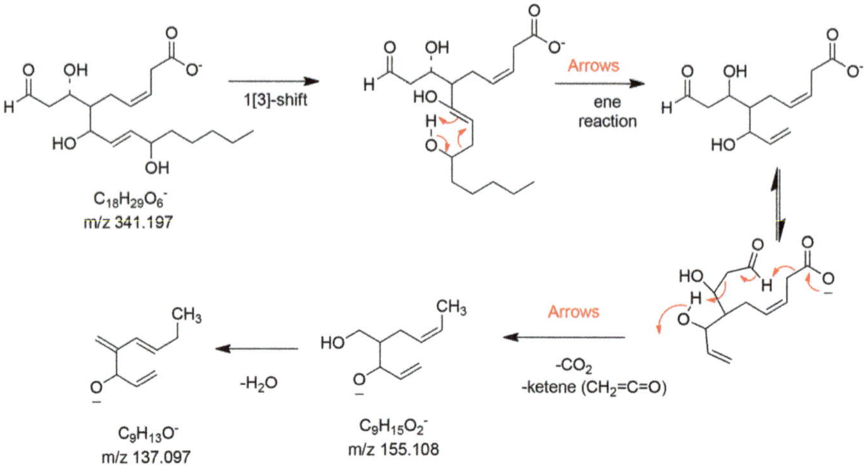

Scheme 2.23

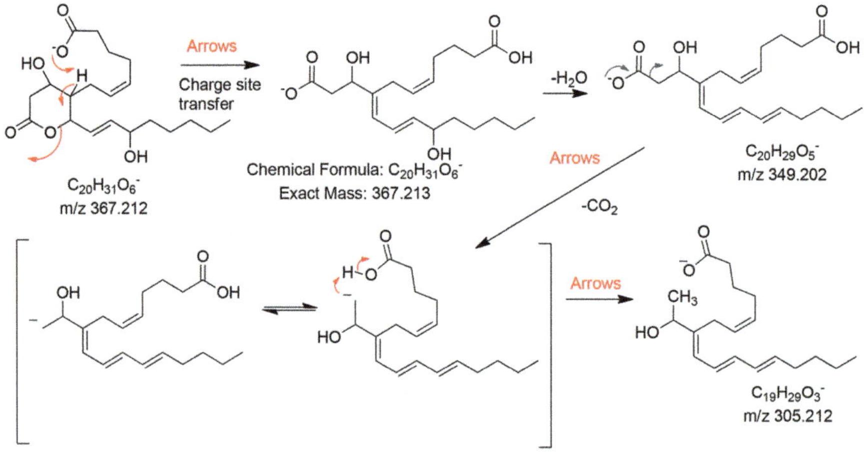

Scheme 2.24

The second common metabolite of TxB_2 is 11-dehydro-TxB_2 (Figure 2.7), which has had the ring-open aldehyde oxidized to a carboxylic acid and then lactonize to form a 6-membered ring structure indicated in Scheme 2.20. The ions observed from collisional activation of this dehydro form are quite significantly different from either TxB_2 or 2,3-dinor-TxB_2, due to this structural change from a hemiacetal to a lactone (Figure 2.7). Three abundant ions are observed at m/z 305, 243, and 161.

The initial charge site for the 11-dehydro-TxB_2 is undoubtedly the carboxylate anion and this can undergo a charge driven opening of the lactone ring, as previously suggested.[4] The ring-opened form can then lose water from the C-15 hydroxyl group to extend conjugation and lose CO_2 to form m/z 305 (Scheme 2.24).

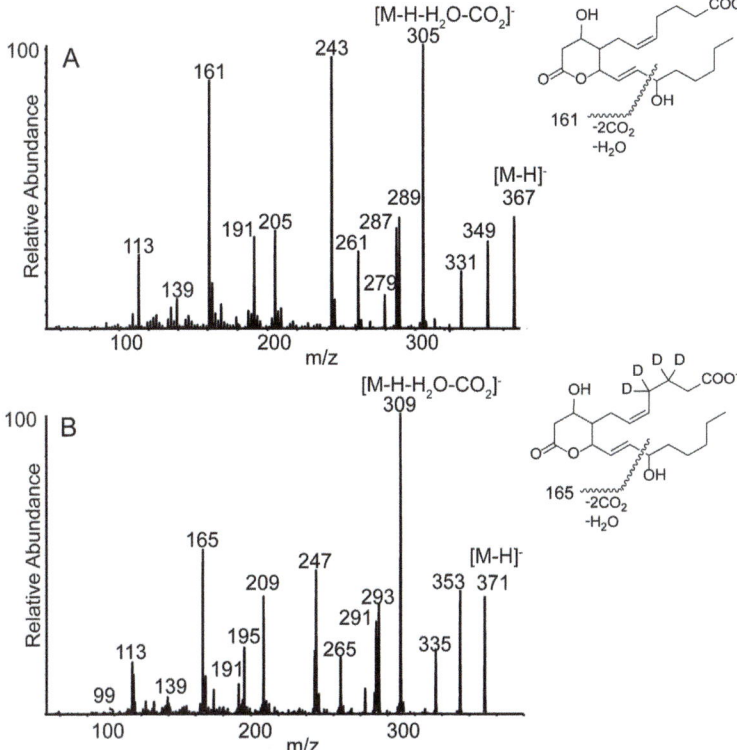

Figure 2.7 Electrospray ionization (negative ions) and tandem mass spectrometry of a urinary metabolite of thromboxane B_2. (A) Product ions obtained following collisional activation of 11-dehydro-TxB$_2$ [M − H]$^-$ at m/z 367; (B) production ions obtained following collisional activation of [3,3,4,4-D$_4$]-11-dehydro-TxB$_2$ [M − H]$^-$ at m/z 371. These MS/MS spectra were obtained using a tandem quadrupole mass spectrometer.

Scheme 2.25

The two very abundant ions at m/z 161 and 243 are quite interesting in that, by their exact mass, they are hydrocarbon ions containing no oxygen atoms. This requires loss of an additional water and CO_2 from m/z 305 to yield m/z 243 (Scheme 2.25).[16]

The mechanism of formation of the other hydrocarbon ion at m/z 161 could be quite similar to that seen in Scheme 2.25, except preceded by an ene-reaction to remove the methyl-terminus chain as a neutral aldehyde, which is

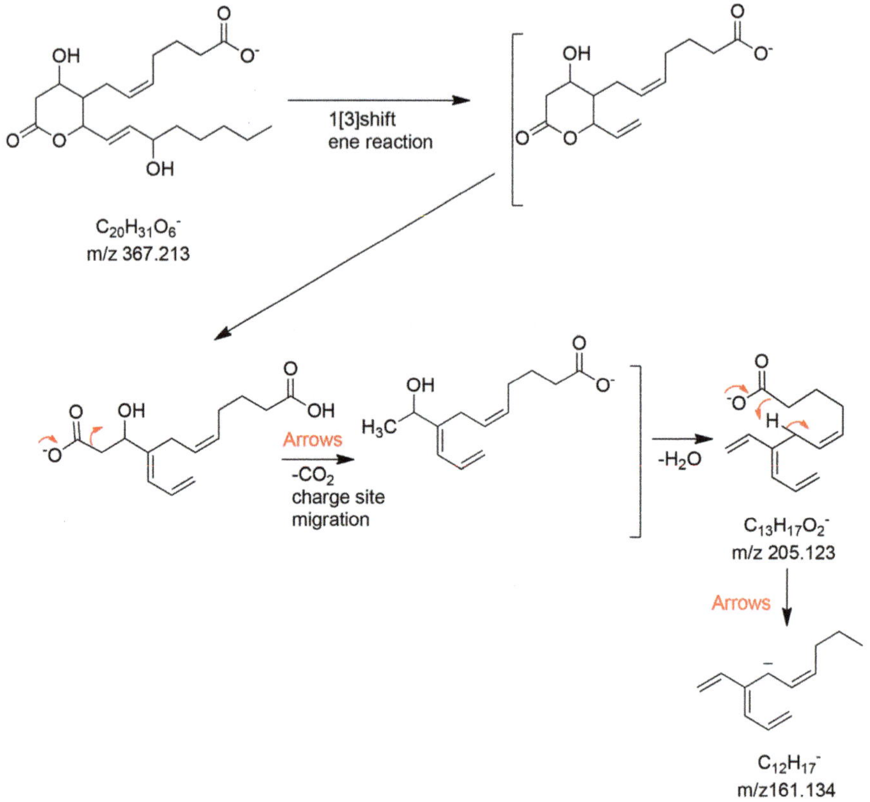

Scheme 2.26

a common reaction pathway for these eicosanoids. This chain shortened form then undergoes ring opening of the lactone and a charge site transfer with loss of CO_2 and water (Scheme 2.26).

2.2 Leukotrienes

The eicosanoids termed leukotrienes are derived by the action of the enzyme 5-lipoxygenase, which converts arachidonic acid to a chemically unstable, yet critical intermediate termed leukotriene A_4 (LTA$_4$). This conjugated triene epoxide (Scheme 2.27) is converted into biologically active lipid mediators leukotriene B_4 (LTB$_4$) by LTA$_4$-hydrolase and into leukotriene C_4 (LTC$_4$) by LTC$_4$-synthase which conjugates the peptide glutathione to LTA$_4$ by way of formation of a thioether bond. As a result of the action of unique peptidase cleavage reactions of LTC$_4$, the cysteinyl glycine adduct, leukotriene D_4 (LTD$_4$), and cysteinyl adduct leukotriene E_4 (LTE$_4$) are formed. These leukotrienes are chemically characterized as conjugated triene fatty acids

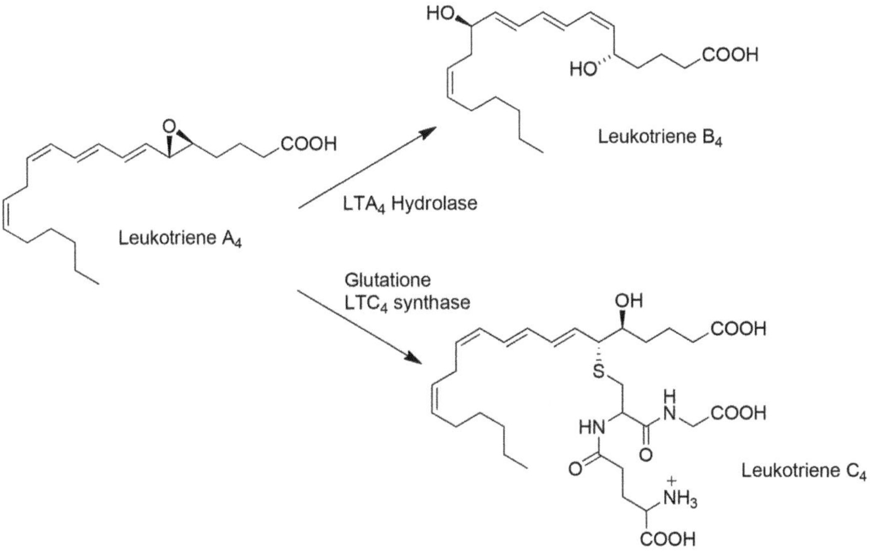

Scheme 2.27

with one or more hydroxyl substituents, as well as thioether bonds. All are biologically quite active.[17]

2.2.1 Leukotriene B$_4$ and Metabolites

The conjugated triene in LTB$_4$ is flanked on either side by two hydroxyl moieties that have a major influence on the product ions obtained after collisional activation of the $[M - H]^-$ at m/z 335 (Figure 2.8). The characteristic ions for LTB$_4$ are found at m/z 203 and 195 and the mechanism of formation of these ions has been studied in some detail using various isotope labeled analogs.[16] The most abundant product ion corresponds to cleavage of the carbon–carbon bond vinylic to the conjugated trienes, which at first glance would appear unlikely. However, the proposed mechanism of formation of this ion at m/z 195 ($C_{11}H_{15}O_3{}^-$) involves double bond rearrangement prior to this bond cleavage. An initial Diels–Alder type cyclization of the $[M - H]^-$ triene would form a cyclohexadiene with charge migration to the 12-alkoxide anion. This rearrangement enables a charge driven elimination of the methyl terminus as a neutral aldehyde and transfer of the charge site to the stable carboxylate anion with only minor twisting of the conjugated triene π-bond system into bringing the C-11 in close proximity to the carboxyl proton (Scheme 2.28).

The other abundant product ion derived from LTB$_4$ is at m/z 203.181 ($C_{15}H_{23}$). This ion is challenging to rationalize, since it contains no oxygen atoms from high resolution measurements and retains all four double bonds.

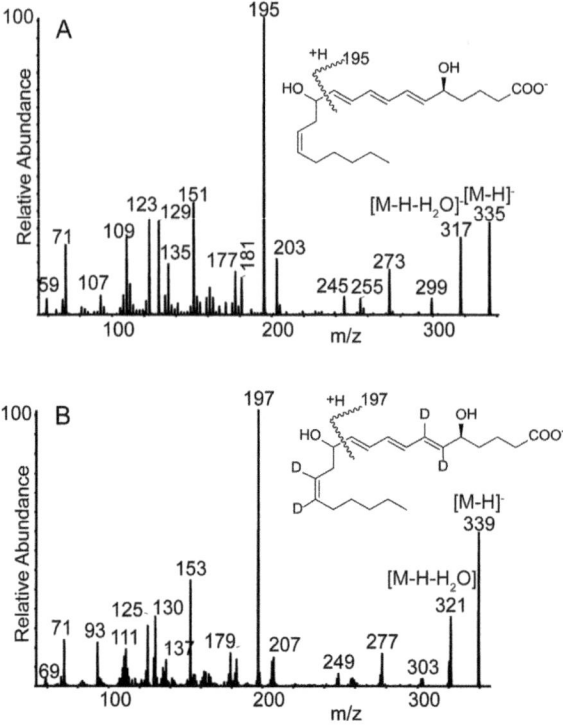

Figure 2.8 Electrospray ionization (negative ions) and tandem mass spectrometry of leukotriene B$_4$ (LTB$_4$). (A) Product ions obtained following collisional activation of LTB$_4$ [M − H]$^-$ at *m/z* 335; (B) Product ions obtained following collisional activation of [6,7,14,15-D$_4$]LTB$_4$ [M − H]$^-$ at *m/z* 339. These MS/MS spectra were obtained using a tandem quadrupole mass spectrometer.

Various stable isotope labeled variants and structural analogs support this ion being formed from the same Diels–Alder cyclic intermediate discussed above that can undergo a charge site remote loss of water and concomitant opening of the 6-membered ring to form a vinyl alcohol. Upon tautomerization to the keto form at carbon-5, there can be a charge-driven loss of ketene by removing the hydrogen atom at carbon-4 and by cleavage of carbon-5-6 adjacent to the keto moiety (Scheme 2.29).

The ion at *m/z* 129 could also be derived from the intermediate oxo-tetraene after two charge remote reactions that form and easily transfer a proton at carbon-8, that is allylic to the cyclohexadiene moiety. Following the loss of a neutral alkylbenzene, the ion at *m/z* 129 can result with the charge site remaining at the carboxylate anion (Scheme 2.30).

The common metabolites of LTB$_4$ are 20-hydroxy-LTB$_4$ and 20-carboxy-LTB$_4$ (Figure 2.9). Their tandem mass spectrometric behavior has been studied and the same major fragmentation mechanisms observed for LTB$_4$ are found for these metabolites.[16]

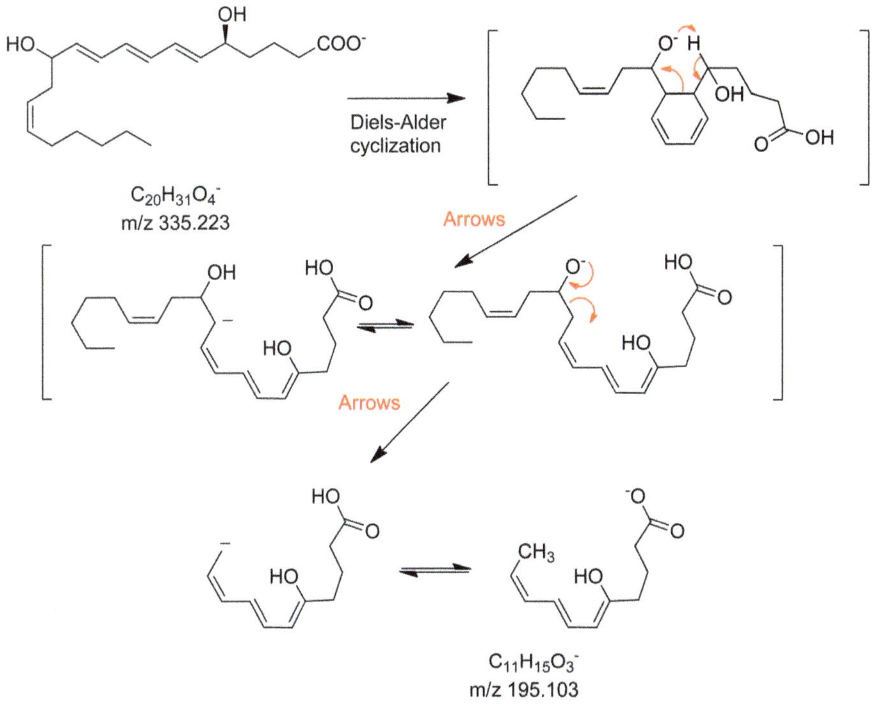

Scheme 2.28

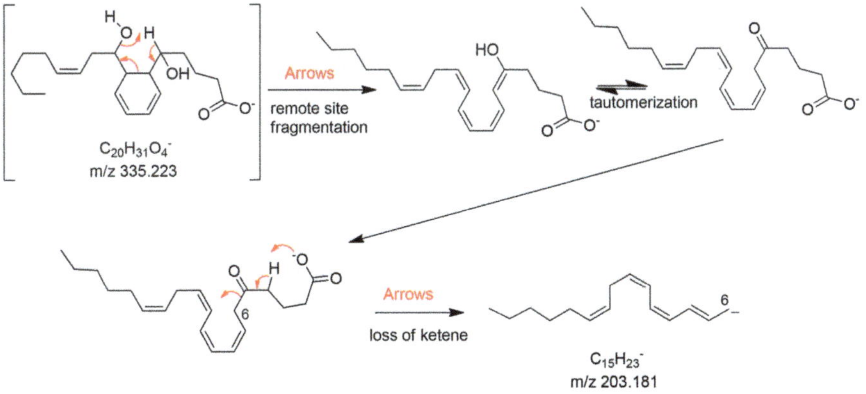

Scheme 2.29

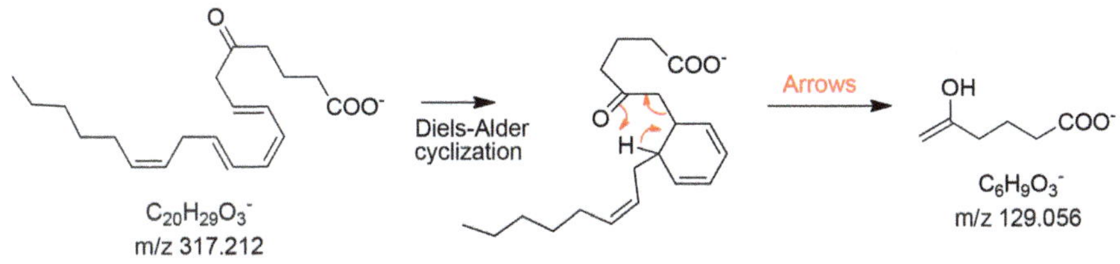

Scheme 2.30

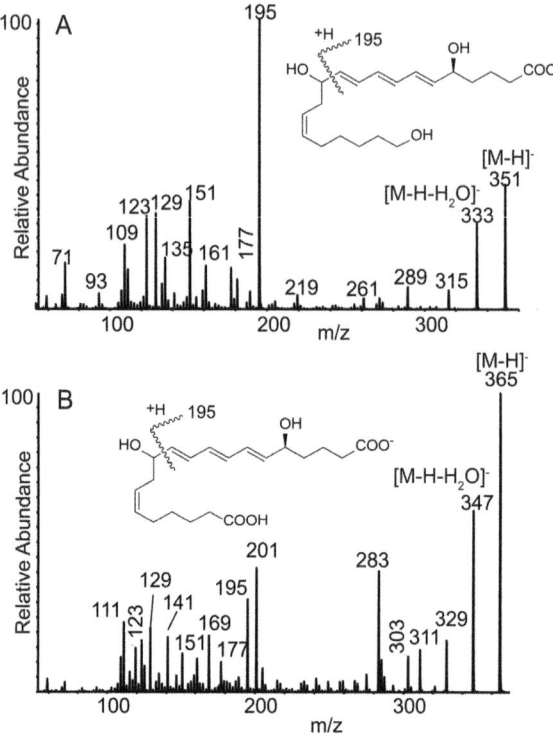

Figure 2.9 Electrospray ionization (negative ions) and tandem mass spectrometry of leukotriene B_4 metabolites. (A) Product ions obtained following collisional activation of 20-hydroxy-LTB$_4$ [M − H]$^-$ at m/z 351; (B) Product ions obtained following collisional activation of 20-carboxy-LTB$_4$ [M − H]$^-$ at m/z 365. These MS/MS spectra were obtained using a tandem quadrupole mass spectrometer.

2.2.2   Leukotriene C_4, D_4, and E_4

Electrospray ionization of these cysteinyl leukotrienes, LTC$_4$, LTD$_4$, and LTE$_4$ yield abundant positive (Figure 2.7), as well as negative, molecular ion species (Figure 2.8). The abundance of both polarities of ions is likely a result of the fact that these molecules contain both carboxyl, as well as amino functionalities that could easily lose a proton or accept a proton to form ionic species driven by the electrospray polarity.

The positive ion tandem mass spectrometry of LTC$_4$, D$_4$, and E$_4$ yields remarkably different product ions, suggesting the expected dominance of the peptide portion of the molecule to localize the positive charge (Figure 2.10). As expected, the relative abundances of the related product ions differ substantially as a result of structural differences. The most abundant product ion following collisional activation of the [M + H]$^+$ of LTC$_4$ appears at m/z 308.091, the same exact mass (elemental composition) of glutathione

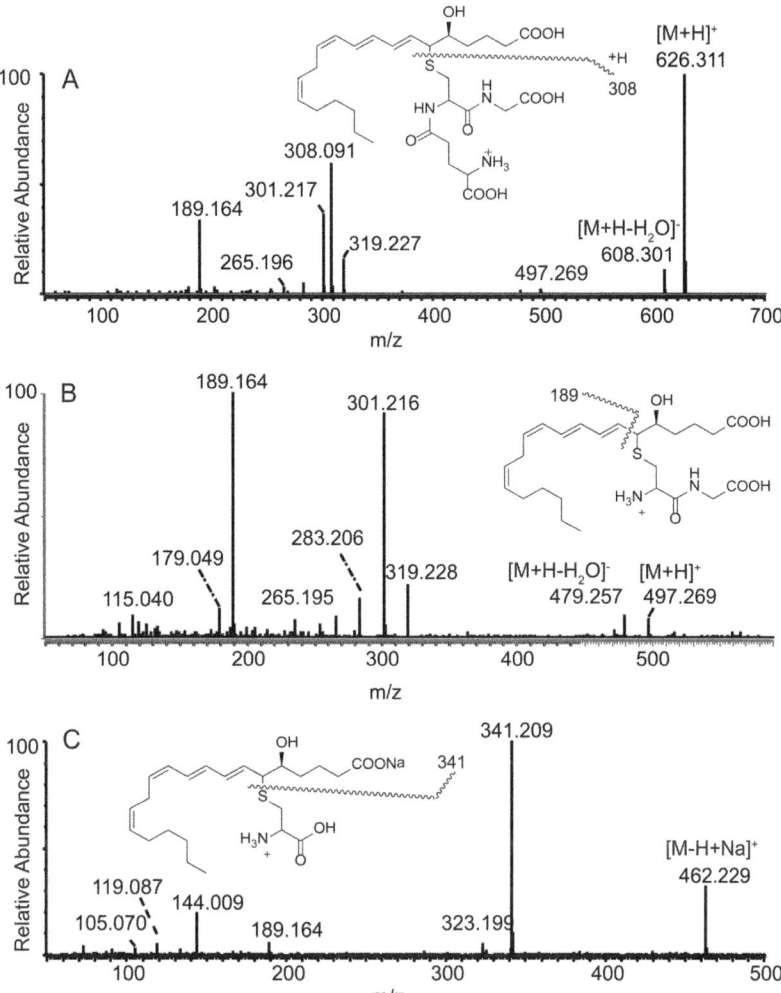

Figure 2.10 Electrospray ionization (positive ions) and tandem mass spectrometry of cysteinyl leukotrienes. (A) Product ions obtained following collisional activation of LTC$_4$ [M + H]$^+$ at *m/z* 626; (B) product ions obtained following collisional activation of LTD$_4$ [M + H]$^+$ at *m/z* 497; (C) product ions obtained following collisional activation of the sodium salt of LTE$_4$ [M − H + Na]$^+$ at *m/z* 341. These MS/MS spectra were obtained using a tandem quadrupole mass spectrometer.

(C$_{10}$H$_{18}$N$_3$O$_6$S). The corresponding ion from the [M + H]$^+$ of LTD$_4$ appears at *m/z* 179 (Figure 2.7B). The formation of these two abundant product ions is likely a straightforward charge remote proton transfer from the carbon-5 atom, resulting in loss of a neutral tetraene fatty acid and glutathione for LTC$_4$ and cysteinyl-glycine for LTD$_4$ (Scheme 2.31).

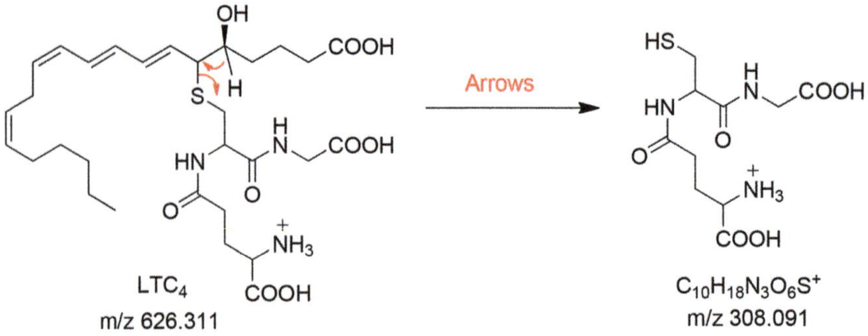

Scheme 2.31

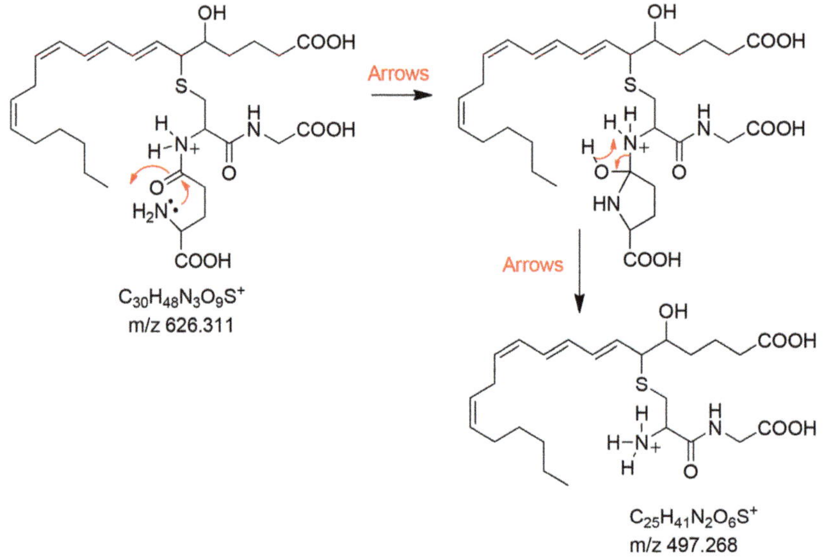

Scheme 2.32

The ion at m/z 497 (Figure 2.10A) is a typical "Y-type" ion that appears in peptide mass spectrometry and is a result of γ-glutamyl amino residue of LTC$_4$ attacking the peptide carbonyl when the site of proton attachment is at the peptide bond of the cysteinyl γ-glutamic acid moiety (Scheme 2.32).[18]

While the ion observed at m/z 319 is not particularly abundant in LTC$_4$, it is quite abundant in LTD$_4$. An ion at m/z 319 is not observed in the sodium adduct tandem mass spectrum of LTE$_4$, but the same ion species is observed 22 Da higher at m/z 341 (see below). The structure of this product ion corresponds to the 20-carbon lipid portion of the molecule common to all cysteinyl leukotrienes. The mechanism of formation of this ion has been suggested to be a charge-driven process involving the C-5 alcohol oxygen atom displacing the sulfur atom, which removes a proton from any of the

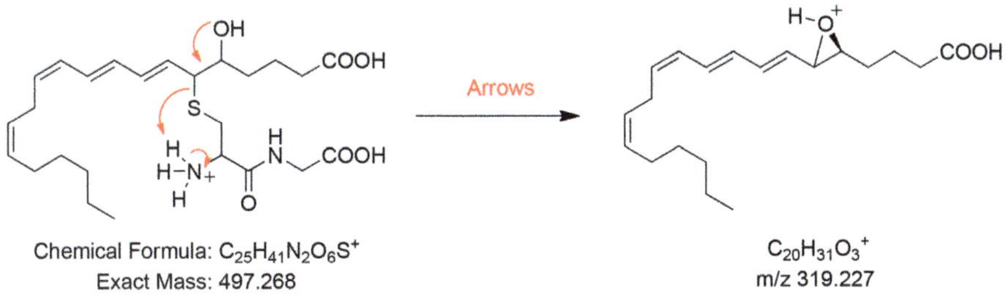

Chemical Formula: $C_{25}H_{41}N_2O_6S^+$
Exact Mass: 497.268

$C_{20}H_{31}O_3^+$
m/z 319.227

Scheme 2.33

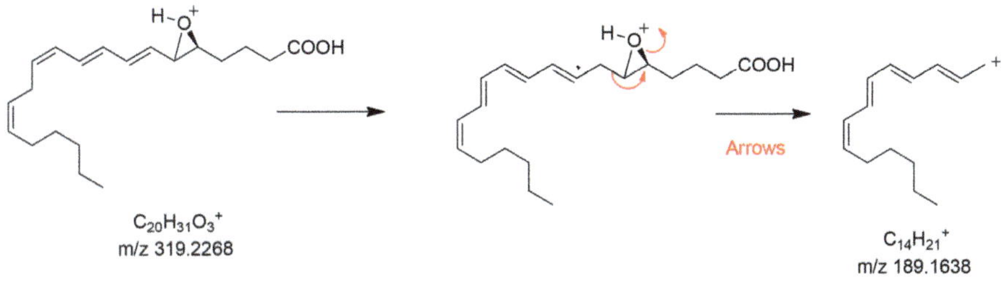

$C_{20}H_{31}O_3^+$
m/z 319.2268

$C_{14}H_{21}^+$
m/z 189.1638

Scheme 2.34

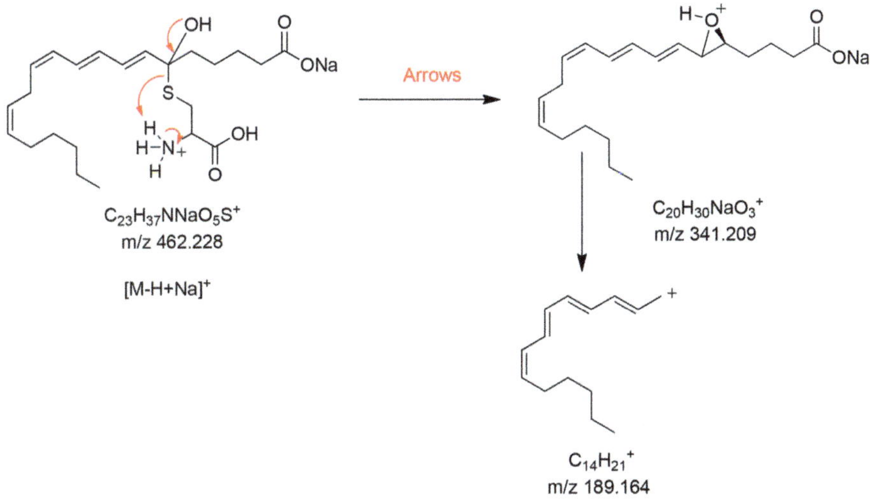

$C_{23}H_{37}NNaO_5S^+$
m/z 462.228

[M-H+Na]$^+$

$C_{20}H_{30}NaO_3^+$
m/z 341.209

$C_{14}H_{21}^+$
m/z 189.164

Scheme 2.35

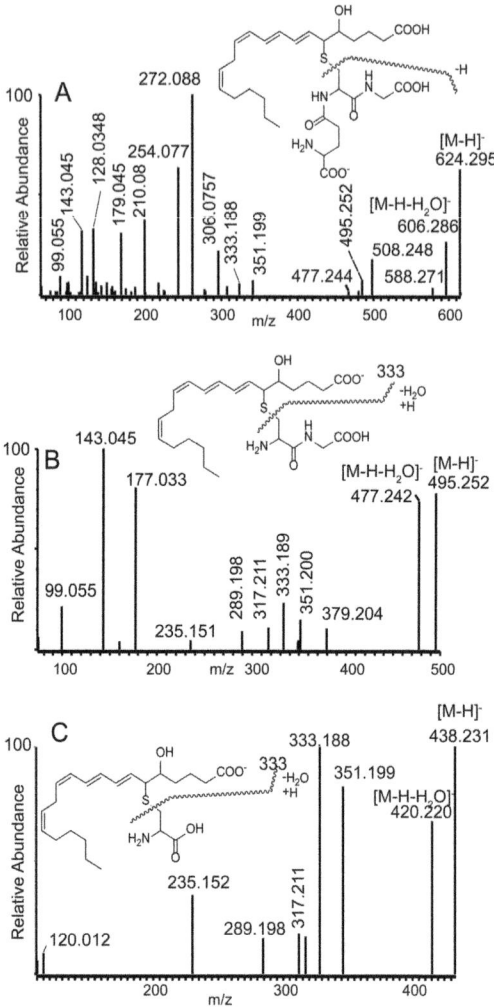

Figure 2.11   Electrospray ionization (negative ions) and tandem mass spectrometry of cysteinyl leukotrienes. (A) Product ions obtained following collisional activation of LTC$_4$ [M − H]$^-$ at m/z 624; (B) product ions obtained following collisional activation of LTD$_4$ [M − H]$^-$ at m/z 495; (C) product ions obtained following collisional activation of LTE$_4$ [M − H]$^-$ at m/z 438. These MS/MS spectra were obtained using a quadrupole time-of-flight mass spectrometer.

potential sites of protonation of the amino groups in the peptide side chains (Scheme 2.33).[19] The abundance of this ion-type in LTD$_4$ and LTE$_4$ is due to the high population of the proton attachment site at the cysteinyl terminal amino group, which is in an excellent position to transfer the charging proton to the emerging thioate anion group during the transition state (Scheme 2.32).

This ion at m/z 319 has been shown to be a precursor of m/z 301 by the loss of water by MS³ experiments.[19] The ion at m/z 189 can also be a product of m/z 319 by charge remote conjugation of all double bonds that would facilitate loss of the carbon-5-6 protonated epoxide by a charge driven formation of a highly stabilized carbocation $(C_{14}H_{21})$ (Scheme 2.34).

The collision induced dissociation of the sodium adduct of LTE_4 is presented in Figure 2.10C, which is somewhat different from the CID of $[M + H]^+$ of LTE_4 (see Figure 1C in ref. 19). Aside from the molecular ion at m/z 462 $[M - H + Na]^+$, the most abundant product ion is seen at m/z 342. This is the ion discussed above as a product ion of LTD_4 $[M + H]^+$ at m/z 319. In this case, the carboxyl group of the arachidonate derived lipid chain is a sodium salt (Scheme 2.35), which also loses water to form m/z 189.

The negative ion tandem mass spectra of LTC_4 and LTD_4 are quite similar (Figure 2.11A and B), but that of LTE_4 is significantly different (Figure 2.11C). LTC_4 and LTD_4 product ions are dominated by the peptide portion of the molecule (m/z 272 and 143, respectively). While for LTE_4, the most abundant ions appear at m/z 351 and 333 corresponding to the lipid portion of the molecule.[19] The formation of m/z 272 (LTC_4) and 143 (LTD_4) might reflect a dominance of charge remote fragmentation of the carbon–sulfur bond to form a thio-leaving group when the charge is quite distal to either the γ-glutamyl or glycine moiety (Scheme 2.36).

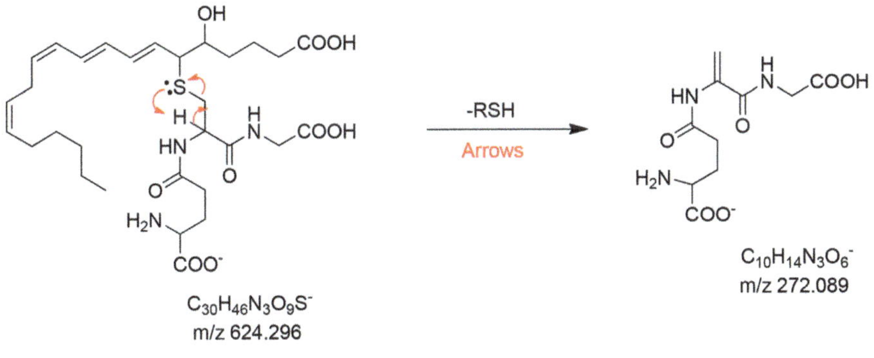

Scheme 2.36

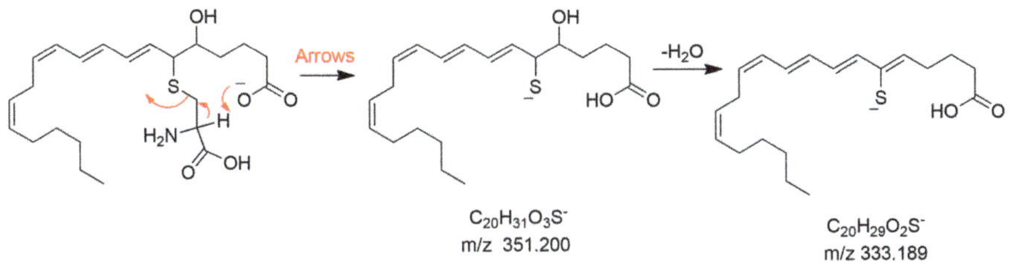

Scheme 2.37

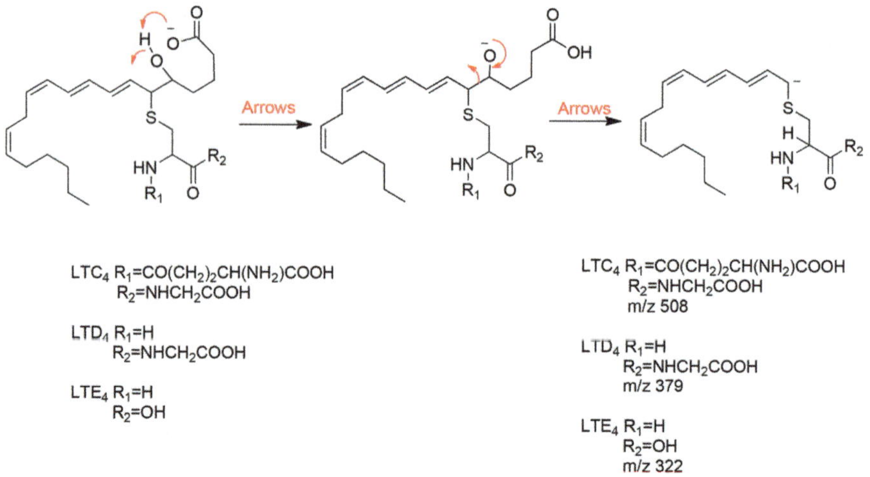

LTC$_4$ R$_1$=CO(CH$_2$)$_2$CH(NH$_2$)COOH
R$_2$=NHCH$_2$COOH

LTD$_4$ R$_1$=H
R$_2$=NHCH$_2$COOH

LTE$_4$ R$_1$=H
R$_2$=OH

LTC$_4$ R$_1$=CO(CH$_2$)$_2$CH(NH$_2$)COOH
R$_2$=NHCH$_2$COOH
m/z 508

LTD$_4$ R$_1$=H
R$_2$=NHCH$_2$COOH
m/z 379

LTE$_4$ R$_1$=H
R$_2$=OH
m/z 322

Scheme 2.38

Attack of a proton on the peptide backbone, when the lipid carboxyl group is ionized, facilitates a charge-driven reaction at the peptide portion of the molecule, resulting in the formation of a thiolate anion at m/z 351 after the carbon–sulfur bond is cleaved (Scheme 2.37). This rather abundant ion from LTE$_4$ could then lose water from the C-5 hydroxyl group to further stabilize the ion observed at m/z 333. These same ions are observed in the product ion spectra of LTC$_4$ and LTD$_4$.

The higher mass product ions from LTC$_4$ and LTD$_4$ appear at m/z 508 and 379, while only to some extent at m/z 322 from LTE$_4$. These ions have been suggested to be derived following a charge-driven reaction, originating from the formation of the alkoxide anion at carbon-5 followed by loss of a neutral aldehyde and a delocalized anion at carbon-6 (Scheme 2.38).[19]

References

1. W. L. Smith and R. C. Murphy, The Eicosanoids: Cyclooxygenase, lipoxygenase, and epoxygenase pathways, in *Biochemistry of Lipids, Lipoproteins and Membranes*, ed. D. E. Vance and J. E. Vance, Elsevier Science, Oxford, UK, 2008.
2. W. L. Smith, P. Borgeat, M. Hamberg, L. J. Roberts, A. Willis, S. Yamamoto, P. W. Ramwell, J. Rokach, B. Samuelsson and E. J. Corey, Nomenclature, *Methods Enzymol.*, 1990, **187**, 1–9.
3. R. C. Murphy, *Mass Spectrometry of Lipids*, Plenum Press, New York, 1993.
4. R. C. Murphy, R. M. Barkley, K. Zemski Berry, J. A. Hankin, K. Harrison, C. Johnson, J. Krank, A. McAnoy, C. Uhlson and S. Zarini, Electrospray ionization and tandem mass spectrometry of eicosanoids, *Anal. Biochem.*, 2005, **346**, 1–42.

5. J. A. Hankin, P. Wheelan and R. C. Murphy, Identification of novel metabolites of prostaglandin E_2 formed by isolated rat hepatocytes, *Arch. Biochem. Biophys.*, 1997, **340**, 317–330.

6. L. J. Murphey, M. K. Williams, S. C. Sanchez, L. M. Byrne, I. Csiki, J. A. Oates, D. H. Johnson and J. D. Morrow, Quantification of the major urinary metabolite of PGE2 by a liquid chromatographic/mass spectrometric assay: determination of cyclooxygenase-specific PGE2 synthesis in healthy humans and those with lung cancer, *Anal. Biochem.*, 2004, **334**, 266–275.

7. E. M. Brunoldi, G. Zanoni, G. Vidari, S. Sasi, M. L. Freeman, G. L. Milne and J. D. Morrow, Cyclopentenone prostaglandin, 15-deoxy-Δ12,14-PGJ$_2$, is metabolized by HepG2 cells via conjugation with glutathione, *Chem. Res. Toxicol.*, 2007, **20**, 1528–1535.

8. J. A. Zirrolli, E. Davoli, L. Bettazzoli, M. L. Gross and R. C. Murphy, Fast atom bombardment and collision-induced dissociation of prostaglandins and thromboxanes: Some examples of charge remote fragmentation, *J. Am. Soc. Mass Spectrom.*, 1990, **1**, 325–335.

9. J. D. Morrow, J. A. Awad, H. J. Boss, I. A. Blair and L. J. Roberts, Non-cyclooxygenase-derived prostanoids (F2-isoprostanes) are formed in situ on phospholipids, *Proc. Natl. Acad. Sci. U. S. A.*, 1992, **89**, 10721–10725.

10. R. J. Waugh and R. C. Murphy, Mass spectrometric analysis of four regioisomers of F2-isoprostanes formed by free radical oxidation of arachidonic acid, *J. Am. Soc. Mass Spectrom.*, 1996, **7**, 490–499.

11. J. A. Lawson, H. Li, J. Rokach, M. Adiyaman, S. W. Hwang, S. P. Khanapure and G. A. FitzGerald, Identification of two major F_2 isoprostanes, 8,12-iso- and 5-epi-8, 12-iso-isoprostane $F_{2\alpha}$-VI, in human urine, *J. Biol. Chem.*, 1998, **273**, 29295–29301.

12. C. Vigor, J. Bertrand-Michel, E. Pinot, C. Oger, J. Vercauteren, F. P. Le, J. M. Galano, J. C. Lee and T. Durand, Non-enzymatic lipid oxidation products in biological systems: Assessment of the metabolites from polyunsaturated fatty acids, *J. Chromatogr. B: Anal. Technol. Biomed. Life Sci.*, 2014, **964**, 65–78.

13. D. Kuklev, J. A. Hankin, C. L. Uhlson, Y. H. M. R. C. Hong and W. L. Smith, Major urinary metabolites of 6-keto-prostaglandin $F_{2\alpha}$ in mice, *J. Lipid Res.*, 2013, **54**, 1906–1914.

14. H. John and W. Schlegel, Reversed-phase high-performance liquid chromatographic method for the determination of the 11-hydroxythromboxane B2 anomers equilibrium, *J. Chromatogr. B: Biomed. Sci. Appl.*, 1997, **698**, 9–15.

15. C. Cheng and M. L. Gross, Applications and mechanisms of charge-remote fragmentation, *Mass Spectrom. Rev.*, 2000, **19**, 398–420.

16. A. C. Gucinski, E. D. Dodds, W. Li and V. H. Wysocki, Understanding and exploiting Peptide fragment ion intensities using experimental and informatic approaches, *Methods Mol. Biol.*, 2010, **604**, 73–94.

17. P. Wheelan, J. A. Zirrolli and R. C. Murphy, Negative ion electrospray tandem mass spectrometric structural characterization of leukotriene B_4(LTB$_4$) and LTB$_4$-derived metabolites, *J. Am. Soc. Mass Spectrom.*, 1996, **7**, 129–139.
18. R. C. Murphy and M. Gijon, Biosynthesis and metabolism of leukotrienes, *Biochem. J.*, 2007, **405**, 379–395.
19. J. M. Hevko and R. C. Murphy, Electrospray ionization and tandem mass spectrometry of cysteinyl eicosanoids: leukotriene C_4 and FOG$_7$, *J. Am. Soc. Mass Spectrom.*, 2001, **12**, 763–771.

CHAPTER 3

Fatty Acyl Esters and Amides

A wide variety of esters and amide lipids are found in biological systems where the fatty acyl moiety is derivatized by a rather straightforward condensation reaction with various alcohols or amines. The structures of these derivatives can be fairly simple, but in some case, such as the case of the fatty acyl thioesters of coenzyme A, very complex lipids can result. Mass spectrometry continues to play a central role in structural characterization of these naturally occurring fatty acid derivatives and with the development of electrospray ionization, the very complex and most polar esters/amides can be readily analyzed as to their molecular weight and elemental composition when using high resolution mass analysis.

The lipid ester/amide family is arguably the largest class of all lipids since this classification includes fatty acid esters of glycerol (triglycerides and glycerophospholipids) as well as sphingolipids which are amides resulting from the condensation of a long-chain base with a fatty acid. These specific esters and amides, organized by Lipid Maps into separate categories, will be discussed in subsequent chapters.[1]

3.1 Amides

3.1.1 Primary Fatty Acid Amides

A number of different primary amides are found in biological systems and serve important yet quite diverse biological functions. The primary amides of unsaturated and monounsaturated fatty acids are known to have significant activities and are thought to regulate many biochemical events within cells.[2] These neutral molecules are present in biological systems and

New Developments in Mass Spectrometry No. 4
Tandem Mass Spectrometry of Lipids: Molecular Analysis of Complex Lipids
By Robert C Murphy
© Robert C Murphy 2015
Published by the Royal Society of Chemistry, www.rsc.org

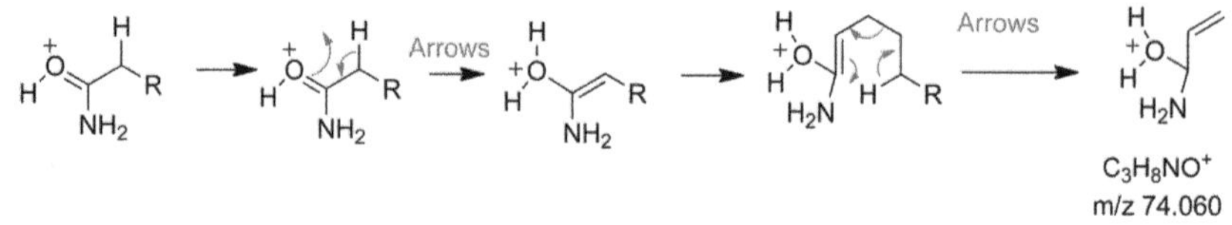

Scheme 3.1

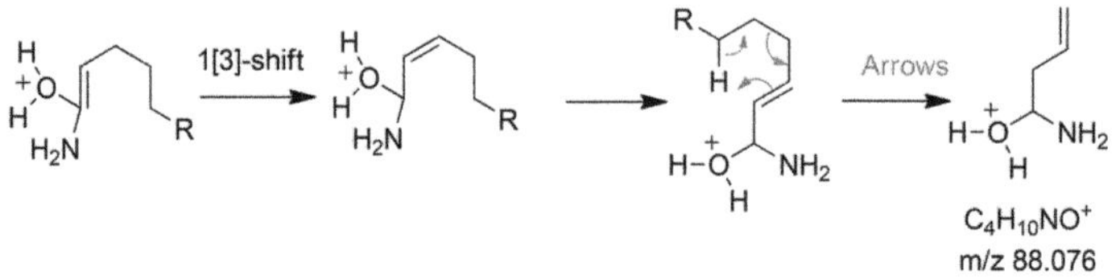

Scheme 3.2

typically analyzed as positive ions following protonation of the amide group $[M + H]^+$, likely at the carbonyl oxygen atom. These lipids are also acids and release a proton to form an anion, also with charge location on the oxygen atom after amide resonance $[M - H]^-$. In spite of this chemical nature, these primary amides do not yield abundant negative ions under normal ESI conditions.

Collisional activation of saturated amides ($[M + H]^+$) leads to a series of very low mass-to-charge ions found at m/z 74, 88, 102, and 116 as the major product ions (Figure 3.1A).[3] Unfortunately, these ions could be lost in an ion trap instrument because of the difficulty in trapping low mass product ions after collisional activation of a relatively high mass molecular ion. The formation of this low mass series has not been studied rigorously, but is likely to proceed from a 1[3]-sigmatropic shift to form a carbon–carbon double bond between C-2 and C-3. This structure can then undergo an ene-type reaction seen from other lipids described previously (Scheme 3.1).

If the initially formed carbon–carbon double bond undergoes 1 or 2 subsequent 1[3]-sigmatropic shifts, the resulting structure could undergo the same ene-reaction, which would appear at 14, 28, and 42 mass units higher to yield the observed ion series. Such an example is presented in Scheme 3.2 for a single double bond shift prior to the allylic bond cleavage to form m/z 88. The $[1,2,3,4-{}^{13}C_4]$-palmitamide (Figure 3.1B) reveals that m/z 88 retains the first four carbon atoms since it is shifted to m/z 92. The mechanism in Scheme 3.2 is consistent with this result.

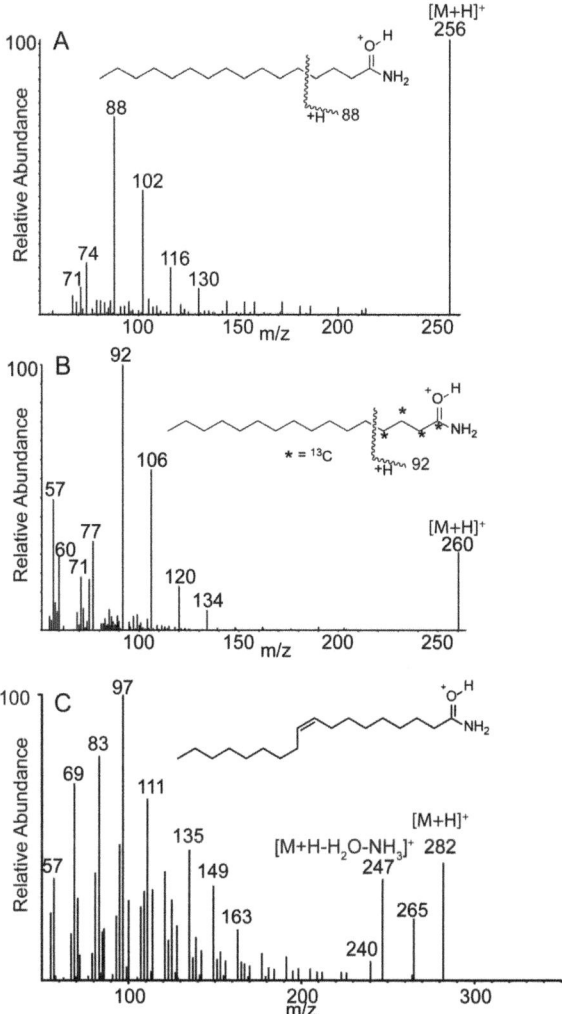

Figure 3.1 Electrospray ionization (positive ions) and tandem mass spectrometry of fatty acyl as the protonated cation. (A) Product ions following collisional activation of palmitamide [M + H]⁺ at *m/z* 356; (B) product ions obtained following collisional activation of [1,2,3,4¹³C₄]-palmitamide [M + H]⁺ at *m/z* 260; (C) product ions following collisional activation of oleamide [M + H]⁺ at *m/z* 282. These MS/MS spectra were obtained using a tandem quadrupole mass spectrometer.

A single double bond in the alkyl chain, even quite remote from the charge site, alters the observed product ions dramatically and the above abundant low series seen for the saturated amides is not even observed,[3,4] but rather a hydrocarbon series at low mass and a hydrocarbon ion corresponding to the loss of both hetero atoms of the amide group (Figure 3.1C).

The nature of these product ions suggests intimate reaction of the remote double bond which in many cases is 9-carbons removed from the amide moiety.

Considering the intermediate previously proposed for the saturated fatty acid amides (Scheme 3.1), the formation of a macrocylic ion by reaction of a nascent cationic site at C-1 with either C- 9 or C-10 in the example of oleoylamide is likely driven by initial loss of a neutral water molecule. This mechanism would result in two product ions of different ring size, both of which could subsequently lose ammonia by a 1[3]-elimination reaction (Scheme 3.3) forming two isomeric structures for the m/z 247, hydrocarbon cation. The abundance of this ion might be a result of cationic charge stabilization by the adjacent two double bonds. The rather abundant low mass ions at m/z 121 and 135 would be stable product ions expected for the 9-carbon and 10-carbon macrocyclic dieneyl cation, which would undergo olefin formation during cleavage of the alkyl chain for the macrocyclic dieneyl cation (Scheme 3.4).

The above mechanisms are consistent with high resolution measurements, the product ions observed, and limited stable isotope analogs that have been investigated.

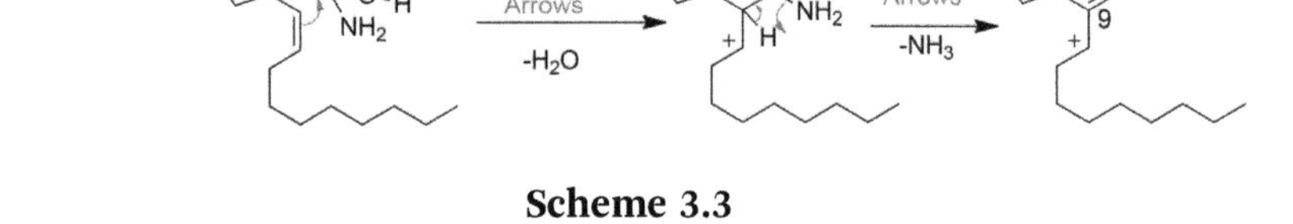

Scheme 3.3

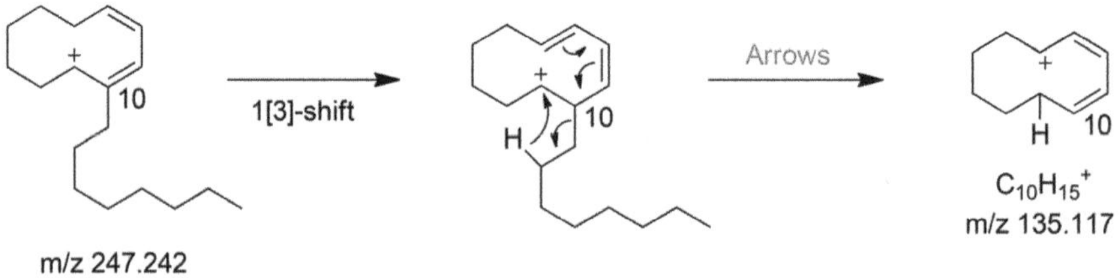

Scheme 3.4

3.1.2 Ethanolamides

A larger family of amides is the fatty acyl ethanolamides that have been structurally characterized and found to possess remarkable biological activities.[5] The first bioactive ethanolamide reported was discovered in the search for the endogenous substance in the brain that was the agonists for the cannabinoid receptor. It was revealed that the active principal was in fact arachidonoyl ethanolamide (AEA) and was given the trivial name of anandamide.[6] The biosynthesis of this amide is made by covalent attachment of arachidonate from arachidonoyl CoA ester to the free amino group of phosphatidylethanolamine catalyzed by an N-acyltransferase. This modified phospholipid is cleaved at the phosphate bond directly forming AEA by a phospholipase D or by sequential phospholipase C and lipid phosphatase to form the AEA structure.

These N-acyl ethanolamides are characterized and quantitated using LC-MS/MS techniques typically as $[M + H]^+$ ions. The collisional activation and tandem mass spectral analysis reveals that virtually all ethanolamides are characterized by an abundant product ion at m/z 62 and an ion corresponding to $[M + H-18]^+$ (Figure 3.2). There are also product ions corresponding to the fatty acyl acylium cation. These are observed at m/z 265 and 263 for oleoethanolamide (Figure 3.2A) and linoleoylethanolamide (Figure 3.2B), respectively.

The formation of m/z 62 is likely a reaction similar to that observed for the primary amides; however, the primary carbinol oxygen atom in this case could abstract the C-2 proton in a 7-membered cyclic transition leading to loss of the fatty acyl chain as a ketene with formation of the ethanolamine cation (Scheme 3.5).

While some low abundance fragments are seen for palmitoylethanolamide, oleoyl- ethanolamide (Figure 3.2A), and linoleoylethanolamide (Figure 3.2B),[7] many more product ions are observed following collisional activation of AEA (Figure 3.3A). When eight deuterium atoms are present on each double bond in the arachidonyl acyl chain as in D_8-AEA (Figure 3.3B), the ethanolamine cation becomes a doublet at m/z 62/63, revealing that the mechanism of m/z 62 formation is more complex than simply attack of the C-2 proton and the loss of ketene. A possible mechanism for this observation for D_8-AEA is that the carbinol oxygen atom on the ethanolamine can sequentially probe each of the protons along the acyl chain (before and after

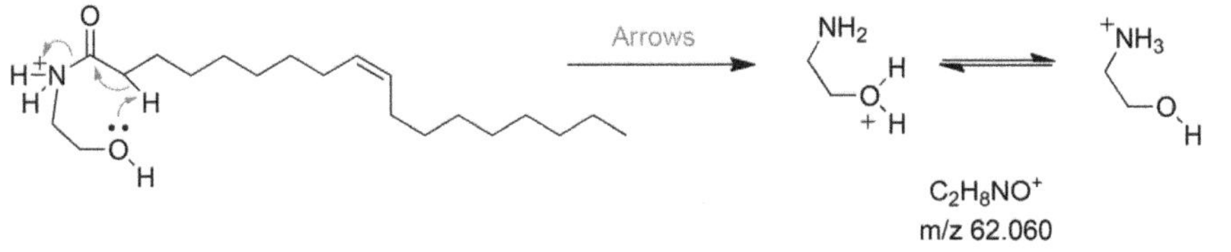

$C_2H_8NO^+$
m/z 62.060

Scheme 3.5

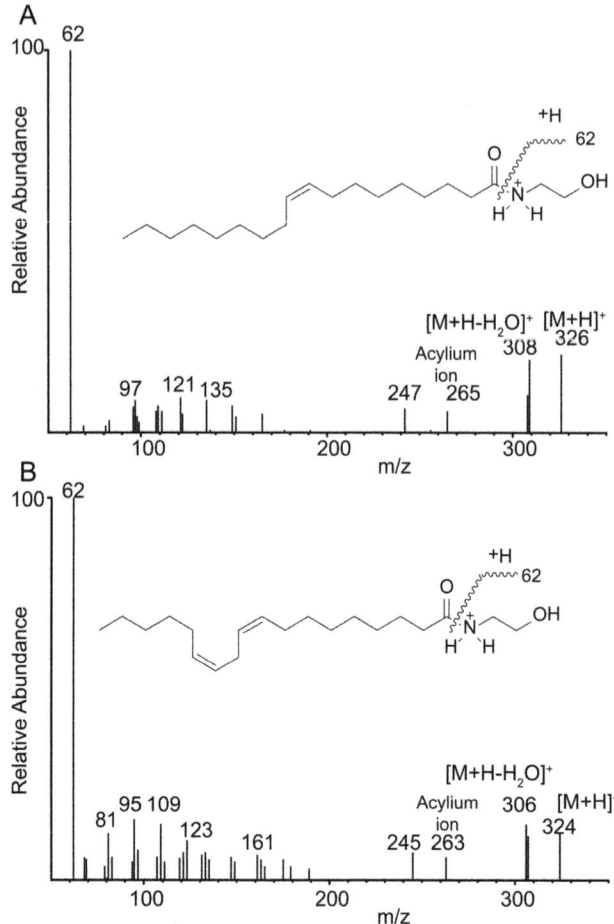

Figure 3.2 Electrospray ionization (positive ions) and tandem mass spectrometry of N-acyl ethanolamides. (A) Product ions obtained following collisional activation of N-oleoyl ethanolamide [M + H]⁺ at *m/z* 326; (B) product ions obtained following collisional activation of N-linoleoyl ethanolamide [M + H]⁺ at *m/z* 324. This figure was redrawn from data presented in ref. 7.

double bond migration) and in many cases remove a deuteron in the process. This would make the neutral product that is lost not necessary a ketene molecule, but a neutral cyclic structure. In fact, most abundant product ions in the D$_8$-AEA collisional mass spectrum are observed as multiplets (Figure 3.3B) with four to six mass unit envelopes, suggesting each of these seemingly simple fragment ions have a fairly complex mechanism of origin. The acylium ion of arachidonate can be observed at *m/z* 287 (Figure 3.3A).

Negative ions are observed after ESI (Figure 3.3C and D) and likely result from amide resonance and formation of the carbonyl oxygen-centered anion (Scheme 3.6). Collisional activation can then populate a more reactive

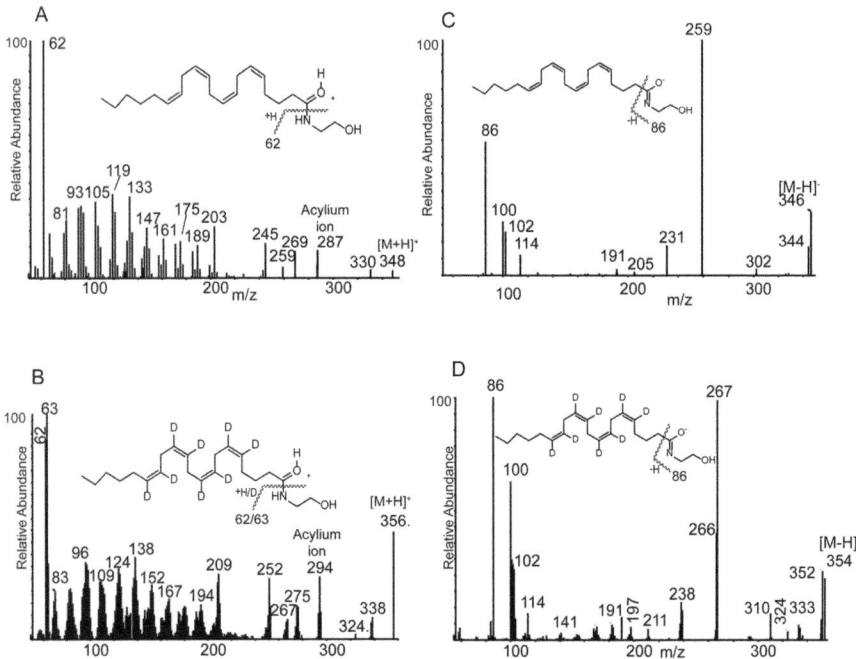

Figure 3.3 Electrospray ionization (positive and negative ions) and tandem mass spectrometry of N-arachidonoyl ethanolamide. (A) Product ions obtained following collisional activation of N-arachidonoyl ethanolamide [M + H]⁺ at *m/z* 348; (B) product ions obtained following collisional activation of N-(D₈-arachidonoyl) ethanolamide [M + H]⁺ at *m/z* 356; (C) product ions obtained following collisional activation of N-arachidonoyl ethanolamide [M − H]⁻ at *m/z* 346; (D) product ions obtained following collisional activation of N-(D₈-arachidonoyl) ethanolamide [M − H]⁻ at *m/z* 354. These MS/MS spectra were obtained using a tandem quadrupole mass spectrometer.

Scheme 3.6

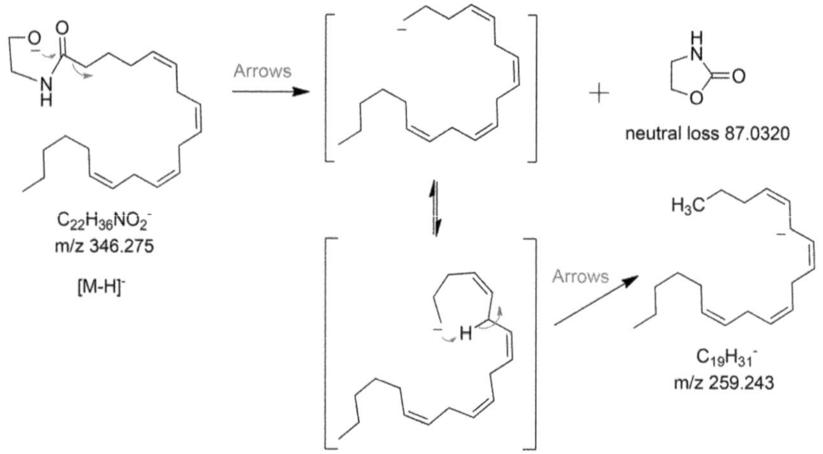

Scheme 3.7

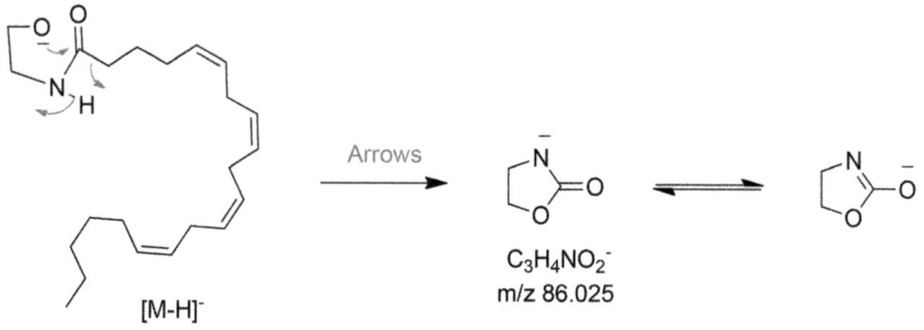

Scheme 3.8

negative ion site at the carbinol oxygen atom. This site would facilitate a favorable attack at the amide carbonyl and charge-driven cleavage of the adjacent carbon–carbon bond to form the abundant ion corresponding to neutral loss of 87 Da (Scheme 3.7). In the case of AEA, this primary carbanion is likely stabilized by removing a bis-allylic proton, leading to a charge delocalized anion observed at m/z 259. Alternatively, the nascent primary anion at carbon-2 of the fatty acyl chain could attack the proton on the amide nitrogen with formation of the abundant m/z 86 as a cyclic structure (Scheme 3.8).

3.1.3 Neurotransmitter N-Acylamides

Fatty acyl amides of various neurotransmitters such as dopamine, norepinephrine, and serotonin have been reported along with amino acid N-acyl amides (Scheme 3.9). Many of these unique lipids have been reported to have significant biological activity.[8,9] No studies of the tandem mass spectral behavior of these neutral lipids have been reported, but the serotonin fatty

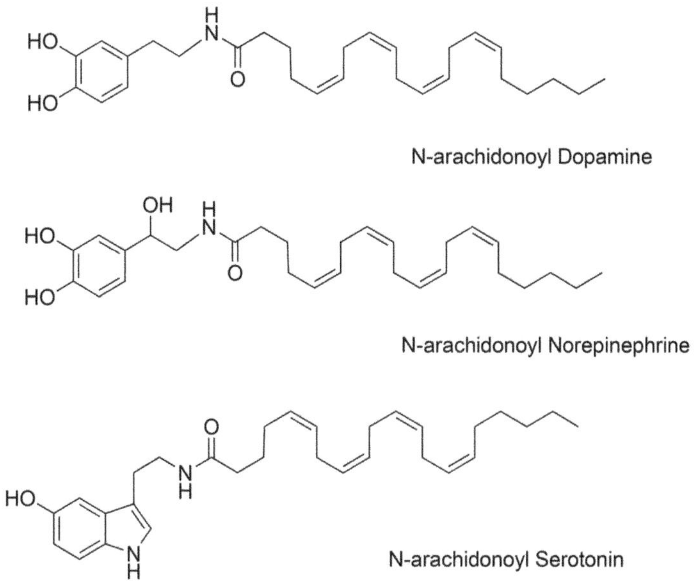

N-arachidonoyl Dopamine

N-arachidonoyl Norepinephrine

N-arachidonoyl Serotonin

Scheme 3.9

ethanolamide has an interesting property of generating a stable anion with ionization of the phenolic hydroxyl group.

The formation of positive ions is also an efficient process for the serotonin N-acylamides since the indole-nitrogen atom is fairly basic. Collisional activation of both polarities of ions yield interesting and abundant fragment ions dominated by the neurotransmitter structure, as exemplified by N-arachidonoyl and N-oleoyl serotonin (Figure 3.4A and B). The collisional activation of N-oleoyl serotonin as well as N-arachidonoyl serotonin yields very abundant ions at m/z 160 and a somewhat less abundant ion at m/z 177 when $[M + H]^+$ ions are collisionally activated. The proposed structures for these cations is consistent with the dominate nature of the basic indole nitrogen atom where m/z 160 would be a product of an ene-reaction, resulting in the loss of a neutral fatty acid amide (Scheme 3.10). The ion at m/z 177 (Scheme 3.10) would result from the loss of a neutral fatty acyl ketene structure. These ions are quite abundant in the positive ion mode and common to both N-arachidonoyl serotonin as well as N-oleoyl serotonin (Figure 3.4A and B).

The negative molecular ion species of these neurotransmitter amides are a result of the negative charge being localized on the phenolic oxygen atom. The same charge remote ene-reaction and ketene neutral loss mechanisms would form the abundant anions observed at m/z 158 and 175 from these amides (Figure 3.5A and B). There are interesting fatty acyl specific anions found at m/z 302 and 280 corresponding to charge driven ionized amides (Scheme 3.11) seen for arachidonoyl serotonin and oleoyl serotonin, respectively.

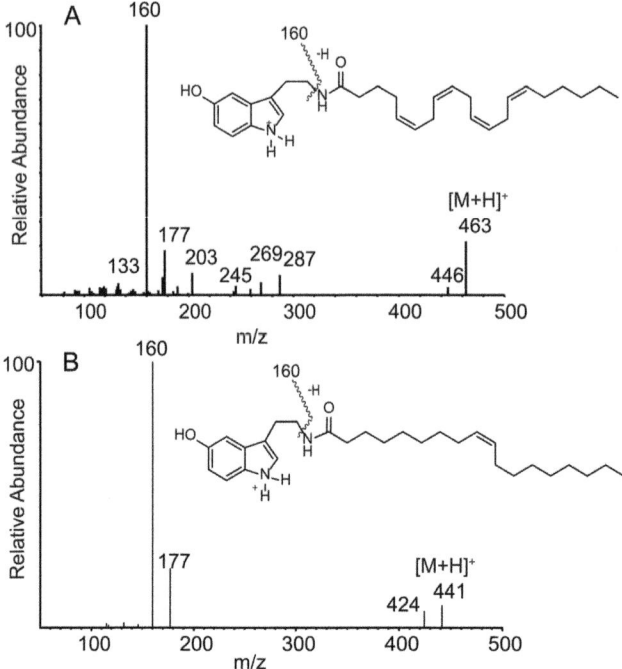

Figure 3.4 Electrospray ionization (positive ions) and tandem mass spectrometry of serotonin N-acylamides. (A) Product ions obtained following collisional activation of N-arachidonoyl serotonin [M + H]⁺ at *m/z* 463; (B) product ions obtained following collisional activation of N-oleoyl serotonin [M + H]⁺ at *m/z* 441. These MS/MS spectra were obtained using a tandem quadrupole mass spectrometer.

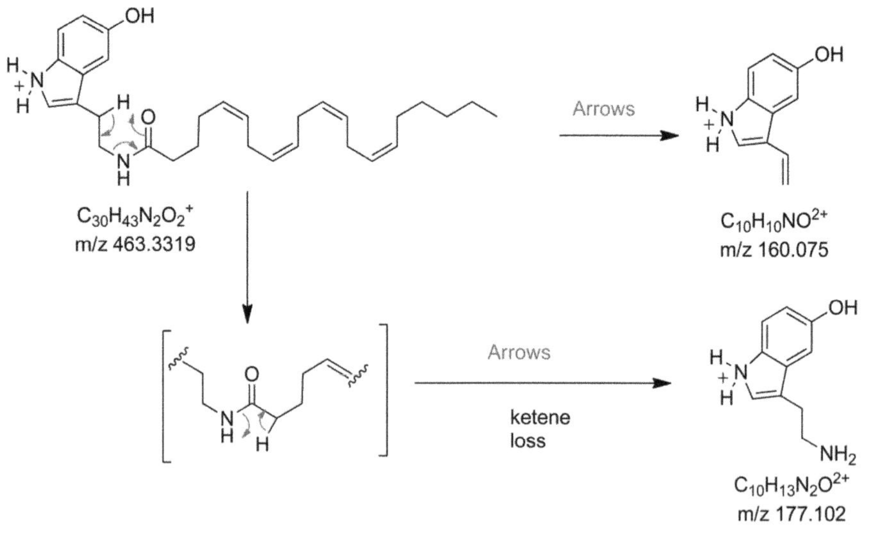

Scheme 3.10

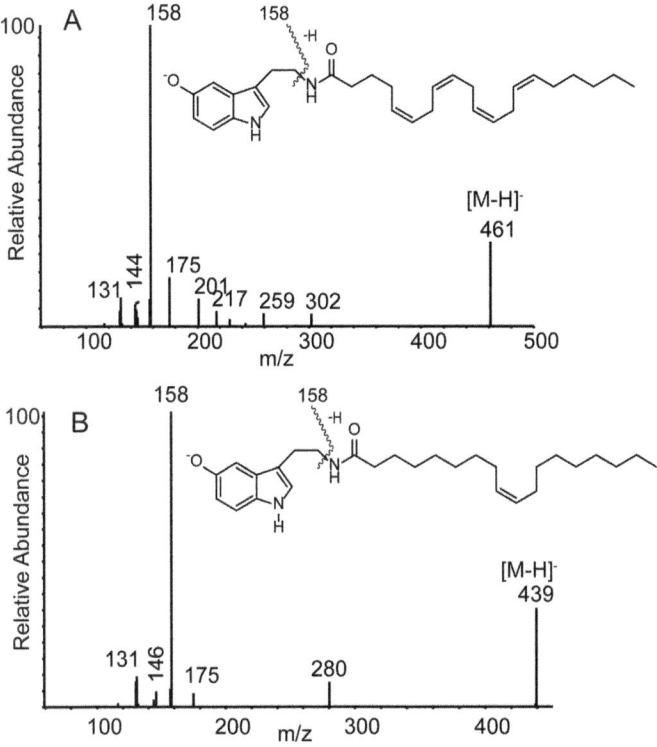

Figure 3.5 Electrospray ionization (negative ions) and tandem mass spectrometry of serotonin N-acylamides. (A) Product ions obtained following collisional activation of N-arachidonoyl serotonin [M − H]⁺ at *m/z* 461; (B) product ions obtained following collisional activation of N-oleoyl serotonin [M − H]⁻ at *m/z* 439. These MS/MS spectra were obtained using a tandem quadrupole mass spectrometer.

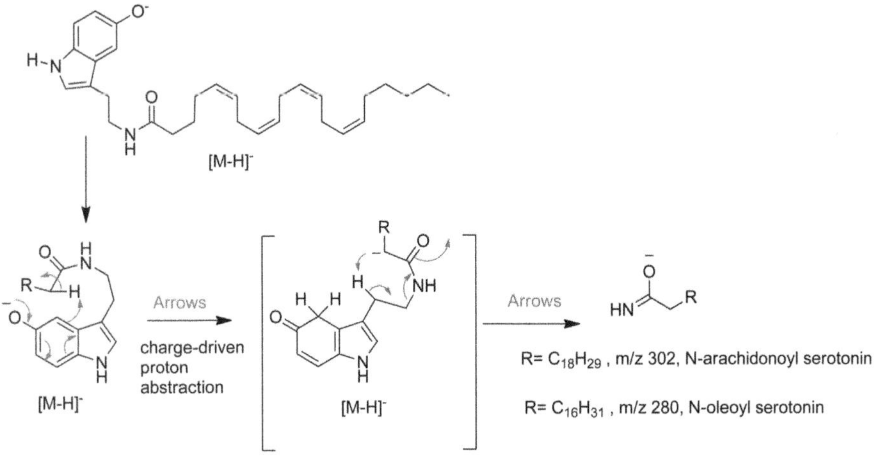

Scheme 3.11

3.2 Carnitine Esters

An important family of esters that are involved in the net transport of fatty acids into the mitochondria for β-oxidation are the carnitine esters. Carnitine is 3-hydroxy-4-trimethylaminobutyric acid and acyl carnitines arise from the esterification of the 3-hydroxyl moiety with a long-chain fatty acyl group activated as a CoA ester (Scheme 3.12). Numerous approaches using both electrospray and MALDI ionization have been reported to detect and quantitate these important esters present in cells.[10,11] In some cases measurement of the intact quaternary compounds $[M + H]^+$ is used, while in many instances the approach of using tandem mass spectrometry following collisional activation of this molecular cation has been used.[10,12]

The tandem mass spectra of long-chain fatty acyl carnitine esters are relatively simple in that all carnitine esters are dominated by the same product ion at m/z 85 (Figure 3.6). This ion is a result of two neutral losses. One loss would correspond to the neutral fatty acid after an ene-type reaction. This mechanism rationalizes the ion observed at m/z 144, which is shifted in the MS/MS of N-CD$_3$-palmitoyl carnitine (m/z 147),[12] indicating retention of the trimethyl ammonium moiety in this fragment ion (Scheme 3.13).

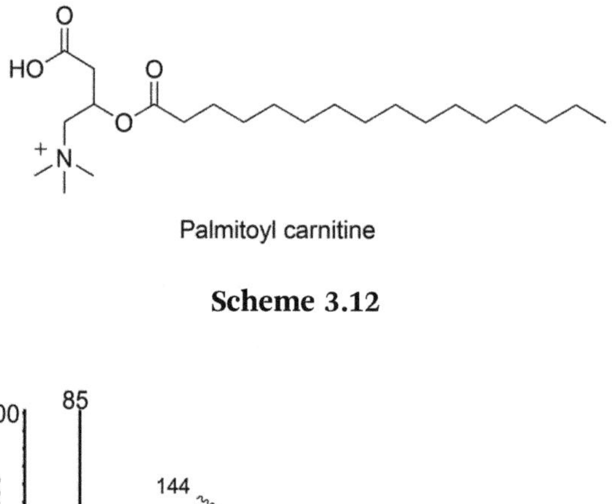

Palmitoyl carnitine

Scheme 3.12

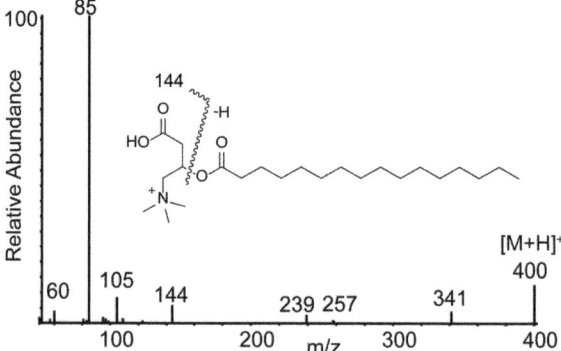

Figure 3.6   Electrospray ionization (positive ions) and tandem mass spectrometry of palmitoyl carnitine. Product ions obtained following collisional activation of palmitoyl carnitine $[M + H]^+$ at m/z 400. This MS/MS spectrum was obtained using a tandem quadrupole mass spectrometer.

The ion at *m/z* 144 then decomposes by loss of neutral trimethylamine to form the highly stable oxonium ion which is quite abundant and observed at *m/z* 85 for both the D_3-palmitoyl carnitine and carnitine (Scheme 3.14).

A product ion derived from the molecular ion $[M + H]^+$ and observed at *m/z* 341, have also been used as the quantitative signal for palmitoyl carnitine in MRM-based assays.[12] This ion likely arises from the same loss of neutral trimethylamine without prior loss of the fatty acid ester (Scheme 3.15).

3.3   Coenzyme A Thioesters (Fatty Acyl CoA)

The fatty acyl thioesters of coenzyme A (CoA) are key lipids for many biochemical processes, including fatty acid biosynthesis, elongation, desaturation as well as phospholipid, cholesterol ester, and glycerol ester biosynthesis. This lipid form is also required for mitochondrial on

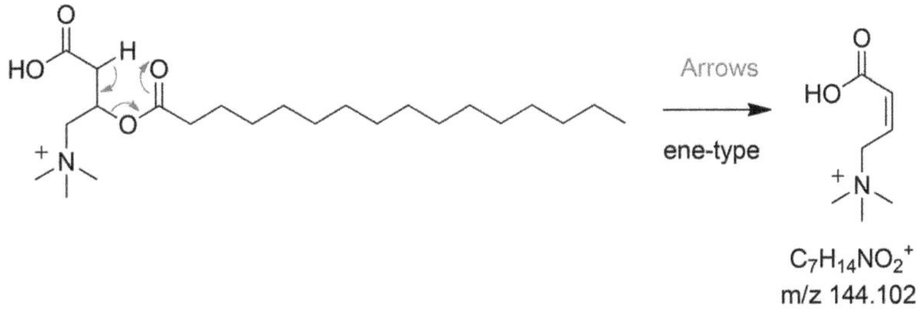

Arrows
ene-type

$C_7H_{14}NO_2^+$
m/z 144.102

Scheme 3.13

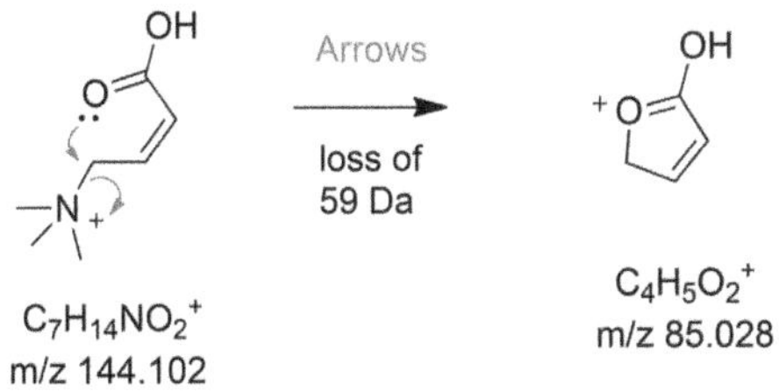

Arrows
loss of
59 Da

$C_7H_{14}NO_2^+$
m/z 144.102

$C_4H_5O_2^+$
m/z 85.028

Scheme 3.14

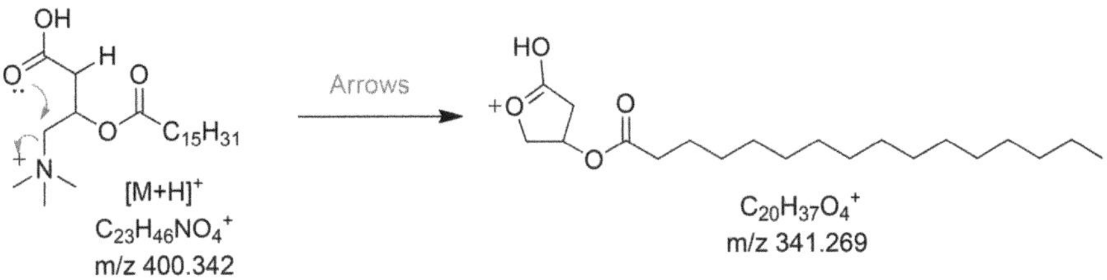

Arrows

$[M+H]^+$
$C_{23}H_{46}NO_4^+$
m/z 400.342

$C_{20}H_{37}O_4^+$
m/z 341.269

Scheme 3.15

peroxisomal β-oxidation and production of ATP during fatty acid metabolism. The first committed step of sphingolipid biosynthesis is the condensation of palmitoyl CoA ester with serine. The chemical structure of the CoA esters can be divided into three identifiable regions: (1) fatty acyl group; (2) phosphopantetheine; and (3) 3′,5′-diphosphoadenosine, which render this lipid notably water soluble due to the three acidic phosphate residues and the basic amino group on adenosine, in spite of the hydrophobic region due to the fatty acyl group. Extraction of this lipid using organic solvents can be challenging and solvent extraction protocols typically used for lipid extraction leave fatty acyl CoA esters in the upper (aqueous) layer.

This lipid is readily ionized by electrospray ionization into both positive and negative ions. Tandem mass spectrometric behavior of both polarity ions has been studied and suggestions as to the origin of fragment ions specific to the fatty acyl thioester structural region and the phosphoadenosine region have been made.[13,14] Haynes *et al.*[14] have suggested a shorthand nomenclature for the abundant and common positive and negative product ions observed after collisional activation, which are indicated in Scheme 3.16.

Several methods to optimize extraction and HPLC chromatographic separation have been reported and these approaches have made major advances in the ability to analyze fatty acyl CoA esters at very low levels. In order to maximize extraction efficiency, 2–3 sequential extractions may be required.[14] Chromatographic separation has also been carefully studied.[14–16] Interestingly, using a mobile phase at neutral or basic pH, still enables both positive and negative ionization modes to be employed. Most reports suggest that positive ionization is surprisingly more sensitive due to the higher yield of positive ions. In part this sensitivity could be due to minimizing sodium ion adduct formation by promoting ammonium ion pairing in the basic HPLC mobile phase during chromatography, as well as the fact that favorable

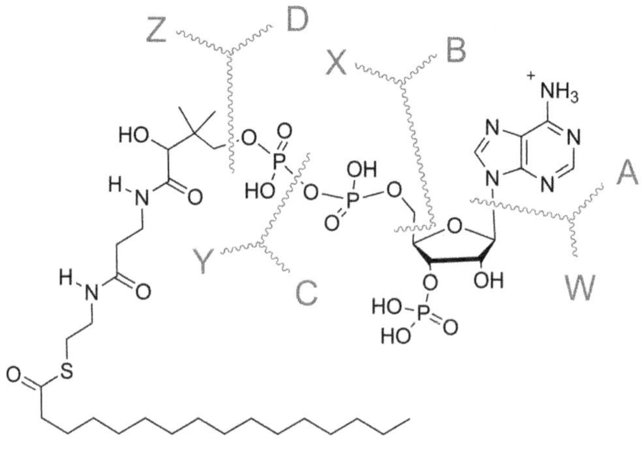

Scheme 3.16

product ions are formed following collisional activation of the $[M + H]^+$ (see below).

Electrospray of palmitoyl CoA (Figure 3.7A) in positive ion mode yields $[M + H]^+$ at *m/z* 1006.4. In the negative ion mode $[M - H]^-$ is observed at *m/z* 1004.4. Over 18 different long-chain fatty acid CoA esters and 10 short-chain CoA esters of plant origin have been studied by tandem mass spectrometry and found to behave identically to that of collisional activation of $[M + H]^+$ from palmitoyl CoA ester, as presented in Figure 3.7A, as well as in the negative ions presented in Figure 3.8B.[14,15,19] Typically, the yield of positive ions is 2–3 times higher than that observed for the corresponding negative ions.

The most abundant product ion observed following collisional activation of $[M + H]^+$ of palmitoyl CoA as well as arachidonoyl CoA ester (Figure 3.7B) is observed at *m/z* 499, which corresponds to a fragment termed a "Z-type" fragment as suggested by Haynes.[17] This ion corresponds to the neutral loss

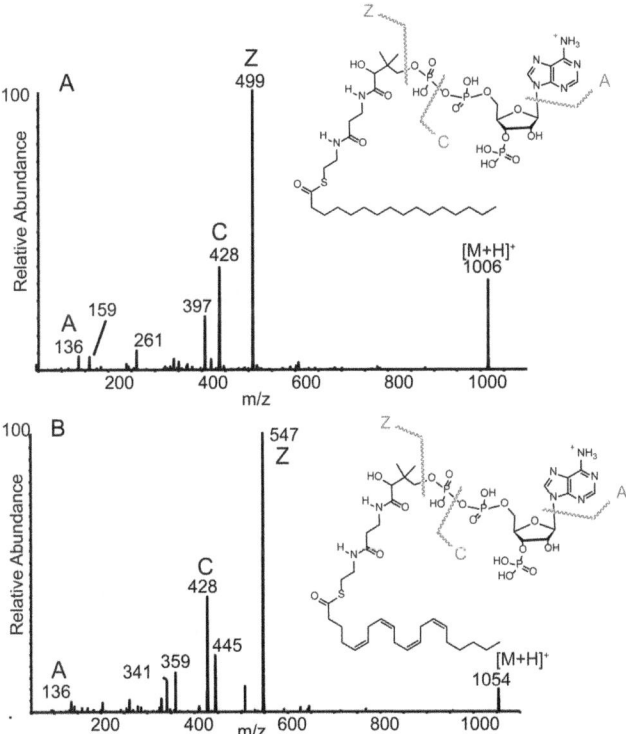

Figure 3.7 Electrospray ionization (positive ions) and tandem mass spectrometry of fatty acyl CoA thioesters. (A) Product ions obtained following collisional activation of palmitoyl CoA thioester $[M + H]^+$ at *m/z* 1006; (B) product ions obtained following collisional activation of arachidonoyl CoA thioester $[M + H]^+$ at *m/z* 1054. These MS/MS spectra were obtained using a tandem quadrupole mass spectrometer.

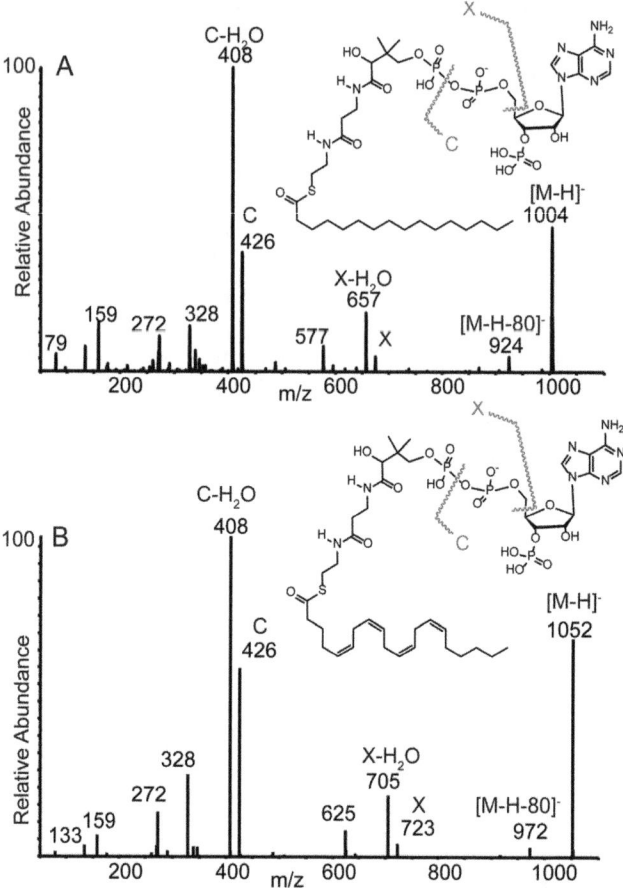

Figure 3.8 Electrospray ionization (negative ions) and tandem mass spectrometry of fatty acyl CoA thioesters. (A) Product ions obtained following collisional activation of palmitoyl CoA thioester [M − H]⁻ at *m/z* 1004; (B) product ions obtained following collisional activation of arachidonoyl CoA thioester [M − H]⁻ at *m/z* 1052. These MS/MS spectra were obtained using a tandem quadrupole mass spectrometer.

of 507 Da and this neutral loss has been used as the basis for detection of CoA esters using tandem mass spectrometry.[14] This ion was also observed as a major product ion from MALDI ionization and collisional activation of CoA esters.[15] In addition to neutral loss scanning, selected ion monitoring (SIM) (targeted MS/MS) experiment can be used to specifically detect CoA esters using the molecular anion as a precursor ion and the corresponding loss of 507 as the product ion (Table 3.1).

The origin of the Z-type positive ion is cleavage of the phosphodiester bond proximal to the thioester with charge retention on the thioester region. Since the most likely protonation site in these CoA esters would be at the adenosine

Table 3.1 Major positive and negative product ions from the collisional activation of $[M + H]^+$ and $[M − H]^−$ from common fatty acyl CoA esters.

Abbreviation[a]	$[M + H]^+$ m/z	Z-type m/z	$[M − H]^-$ m/z	X-type m/z	X-type-H_2O m/z	Y-type m/z	Y-type-H_2O m/z
12 : 0	950.290	443.294	948.276	619.223	601.213	539.256	521.246
14 : 1	976.306	469.310	974.292	645.239	627.229	565.272	547.262
14 : 0	978.321	471.325	976.307	647.254	629.244	567.287	549.277
16 : 1	1004.336	497.340	1002.322	673.269	655.259	593.302	575.292
16 : 0	1006.352	499.356	1004.338	675.285	657.275	595.318	577.308
18 : 4	1026.322	519.326	1024.308	695.255	677.245	615.288	597.278
18 : 3	1028.337	521.341	1026.323	697.270	679.260	617.303	599.293
18 : 2	1030.352	523.356	1028.338	699.285	681.275	619.318	601.308
18 : 1	1032.368	525.372	1030.354	701.301	683.291	621.334	603.324
18 : 0	1034.384	527.388	1032.370	703.317	685.307	623.350	605.340
20 : 5	1052.337	545.341	1050.323	721.270	703.260	641.303	623.293
20 : 4	1054.352	547.356	1052.338	723.285	705.275	643.318	625.308
20 : 3	1056.368	549.372	1054.354	725.301	707.291	645.334	627.324
20 : 2	1058.384	551.388	1056.370	727.317	709.307	647.350	629.340
20 : 1	1060.399	553.403	1058.385	729.332	711.322	649.365	631.355
20 : 0	1062.415	555.419	1060.401	731.348	713.338	651.381	633.371
22 : 6	1078.352	571.356	1076.338	747.285	729.275	667.318	649.308
22 : 5	1080.368	573.372	1078.354	749.301	731.291	669.334	651.324
22 : 4	1082.384	575.388	1080.370	751.317	733.307	671.350	653.340
22 : 1	1088.431	581.435	1086.417	757.364	739.354	677.397	659.387
24 : 1	1116.469	609.473	1114.455	785.402	767.392	705.435	687.425
24 : 0	1118.478	611.482	1116.464	787.411	769.401	707.444	689.434

[a]Total carbon atoms: total number of double bonds.

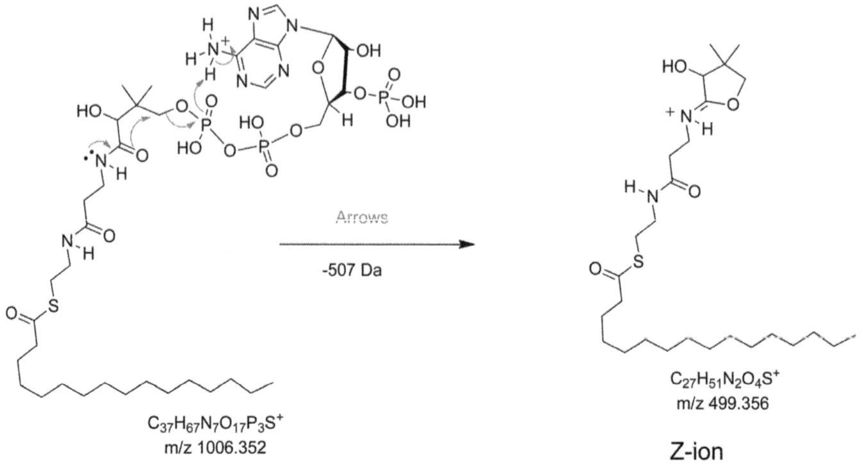

$C_{37}H_{67}N_7O_{17}P_3S^+$
m/z 1006.352

Arrows

-507 Da

$C_{27}H_{51}N_2O_4S^+$
m/z 499.356

Z-ion

Scheme 3.17

amino group, this fragment ion must arise from a charge driven process and a possible mechanism would involve the nonbonding electrons on the distal amide group to the thioester, driving attack of the amide carbonyl oxygen atom on the methylene carbon adjacent to the first phosphodiester moiety. This carbon–oxygen bond is made somewhat electron deficient by the withdrawing properties of the phosphate oxygen atoms. The phosphoryl oxygen atom could then remove the proton from the adenosine amino group, neutralizing this charge site. Molecular models support the suggestion that the ionized adenosine amino group would be in a position, due to the stereochemistry of the ribose sugar moiety, to carry out the mechanism illustrated in Scheme 3.17. An important feature of this mechanism is the stability of the 5-membered cyclic product ion and a positive charge retained on the amide nitrogen atom. Similar mechanisms of peptide fragmentation had been suggested to form *b-type and y-type* peptide cleavage ions.[17] This loss of the adenosine triphosphate residue (507 Da) is almost always the most abundant product ion following collisional activation of fatty acyl CoA thioesters $[M + H]^+$ and the resulting Z-ion quite specific in terms of the observed product mass-to-charge ratio for each CoA ester molecular species (Table 3.1).

The other product ions observed are not specific to the CoA thioester molecular species, but generic to the 3′-phospho-ADP moiety. The most abundant nonspecific ion has been termed a "C-type" ion which likely has the structure indicated in Scheme 3.18.

Many product ions are observed following collisional activation of $[M - H]^-$ of fatty acyl CoA esters (Figure 3.8). The fatty acyl specific ion termed "X-type" can be readily identified and would most likely be formed by ene-type reaction engaging a phosphoryl oxygen atom to extract a proton from the C-4′ position of ribose with concomitant cleavage of the

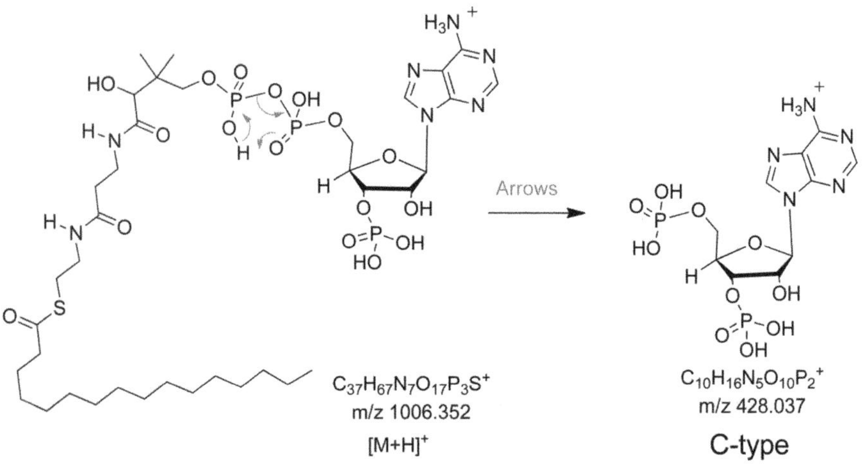

$C_{37}H_{67}N_7O_{17}P_3S^+$
m/z 1006.352
[M+H]$^+$

$C_{10}H_{16}N_5O_{10}P_2^+$
m/z 428.037

C-type

Scheme 3.18

5'-ribose carbon atom bonded to the phosphate oxygen atom (Scheme 3.19). This is the same mechanism as that for the C-type fragment (positive ions) but since this is a charge remote process, the site of charge determines which portion of the CoA complex structure is seen as a mass spectral product ion.

The more abundant fatty acid specific negative ions, which are termed "X-type" ions, correspond to X–H_2O (*m/z* 657). This X-type ion also dehydrates readily, and while the exact position of the water loss has not been studied, one suggestion is the formation of a six-membered cyclic phosphate diester (Scheme 3.19). The product ion 80 mass units lower would formally correspond to X–H_2O–HPO_3 (Scheme 3.20). However, this ion does not originate from the X–H_2O ion, but rather from initial cleavage of the ribose 5'-phosphate bond as a Y–H_2O type ion. The most abundant negative product ions correspond to charge retention on the ADP (C-type, *m/z* 426) and then formation of the cyclic ester (loss of water) C–H_2O type structure, observed at *m/z* 408 (Scheme 3.20). This is a common ion seen in the negative ion tandem mass spectra of CoA esters as in the spectrum of palmitoyl CoA ester (Figure 3.8A) and arachidonoyl CoA ester (Figure 3.8B).

The importance of the fatty acyl CoA thioesters has led to a number of precise and sensitive techniques to quantitate these molecules using LC-MS/MS approaches. Reversed phase HPLC under basic solvent conditions and detection of ions in the positive ion mode appears to be the most facile method to quantitate the CoA esters. The chromatographic separation of molecular species is important to minimize the potential overlap of low abundance unsaturated CoA esters that could be confounded by the $^{13}C_2$ isotope peaks of a CoA ester with one more double bond and of high abundance. Nonetheless, it is now quite feasible to detect CoA esters in

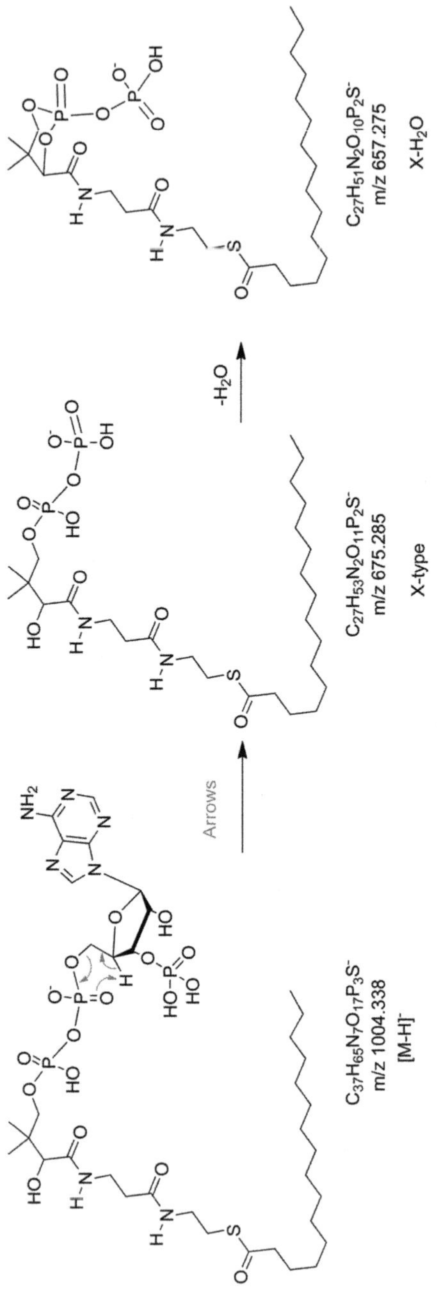

Scheme 3.19

$C_{37}H_{65}N_7O_{17}P_3S^-$
m/z 1004.338
[M-H]$^-$

$C_{27}H_{53}N_2O_{11}P_2S^-$
m/z 675.285
X-type

$C_{27}H_{51}N_2O_{10}P_2S^-$
m/z 657.275
X-H$_2$O

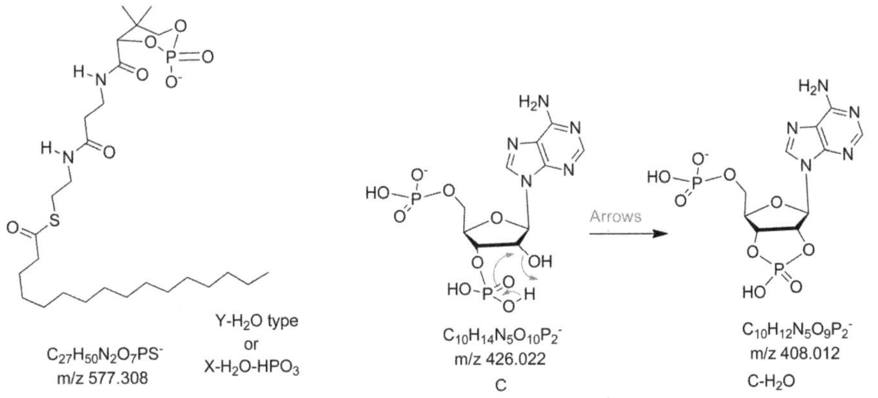

Scheme 3.20

meaningful biochemical studies using the ions described above as specific indicators of each of the CoA esters as they appear within cellular systems.

3.4 Wax Esters

The wax esters are a class of highly hydrophobic lipids consisting of a fatty alcohol esterified to a fatty acid. These rather simple neutral lipids occur widely in biology, serving very specialized roles such as preventing leaf dehydration in the cuticle of plants, secretions of sebaceous and meibomian glands, transmission of sound in whales and dolphins, protection of the skin or ocular surface from dehydration.[18] In addition to sebaceous and meibomian gland secretion, wax esters are also abundant in human hair shafts.[19] Additionally, α,ω-diesters and related lipids are part of the wax ester family (Scheme 3.21).[20]

Even though these lipids appear to be simple in structure, wax esters are found in exceedingly complex mixtures with normal, branched, and unsaturated fatty alkyl groups of both the fatty alcohol and fatty acyl chains. Electrospray ionization based analysis, including LC-MS/MS approaches, have not been developed sufficiently for full structural characterization of biologically derived wax esters found in biological mixtures due to the complexity introduced by methyl branching of alkyl chains and double bond positions/geometry. Electron ionization and capillary GC/MS of hydrolyzed fatty acids as well as intact wax esters has been previously discussed and is used to add analytical power to the characterization of such mixtures.[21,22] Complete structural characterization of wax esters still requires multiple methods and platforms of analysis, but ESI does offer an excellent method for quantitative analysis.

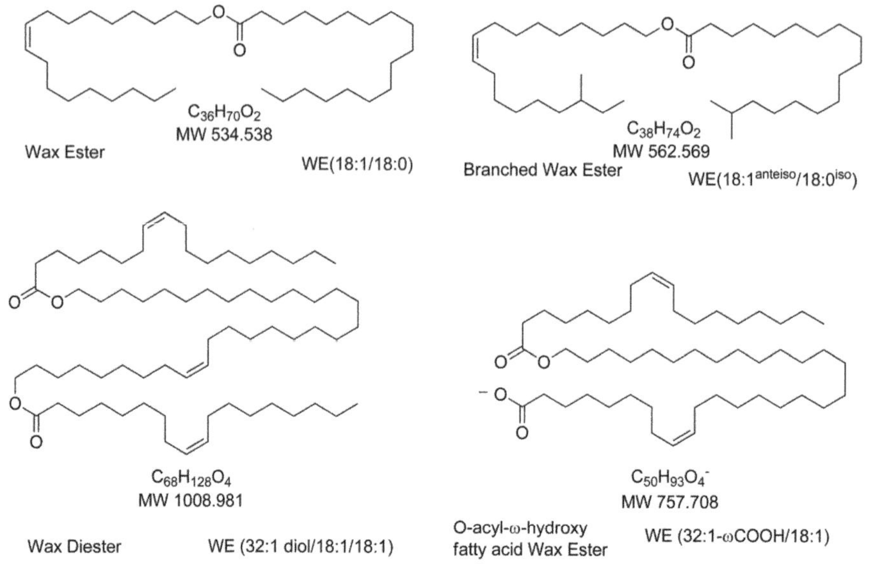

$C_{36}H_{70}O_2$
MW 534.538

Wax Ester

WE(18:1/18:0)

$C_{38}H_{74}O_2$
MW 562.569

Branched Wax Ester

WE(18:1^{anteiso}/18:0^{iso})

$C_{68}H_{128}O_4$
MW 1008.981

Wax Diester   WE (32:1 diol/18:1/18:1)

$C_{50}H_{93}O_4^-$
MW 757.708

O-acyl-ω-hydroxy   WE (32:1-ωCOOH/18:1)
fatty acid Wax Ester

Scheme 3.21

Electrospray ionization has been reported to produce positive ions as ammonium adducts $[M + NH_4]^+$ and alkali metal adducts $[M + Na]^+$ or $[M + Li]^+$, depending upon the spray solvent composition and added electrolytes. Studies had been reported discussing conditions to maximize the positive ions during electrospray ionization that are derived from wax esters isolated from biological systems such as tear fluid.[20] Those wax diesters that retain a free carboxyl group can be analyzed by negative ions.

The collisional activation and tandem mass spectrometry of $[M + NH_4]^+$ has been studied,[18–20,23] but not the collisional activation of alkali metal adduct ions. The tandem mass spectra of these positive ions are fairly straightforward but some variations and unique ions are observed when double bonds in the alcoholic alkyl group or fatty acid acyl group of the molecule occur.[19] A MS/MS study of the electrospray product ions (ammonium adducts) using several different variants of the wax ester structure supported the importance of the side chain unsaturation in driving specific fragmentation events and reactions (Figure 3.9).[23]

The predominant fragment ion from all wax esters after collisional activation of $[M + NH_4]^+$ is the expected loss of ammonia to form a $[M + H]^+$. This ion is either the precursor of all subsequent ions or in some cases product ions may directly form from the ammonium adduct ion. One tautomeric form of this ion would have the proton bonded to the bridging oxygen atom of the ester, which could facilitate an ene-type

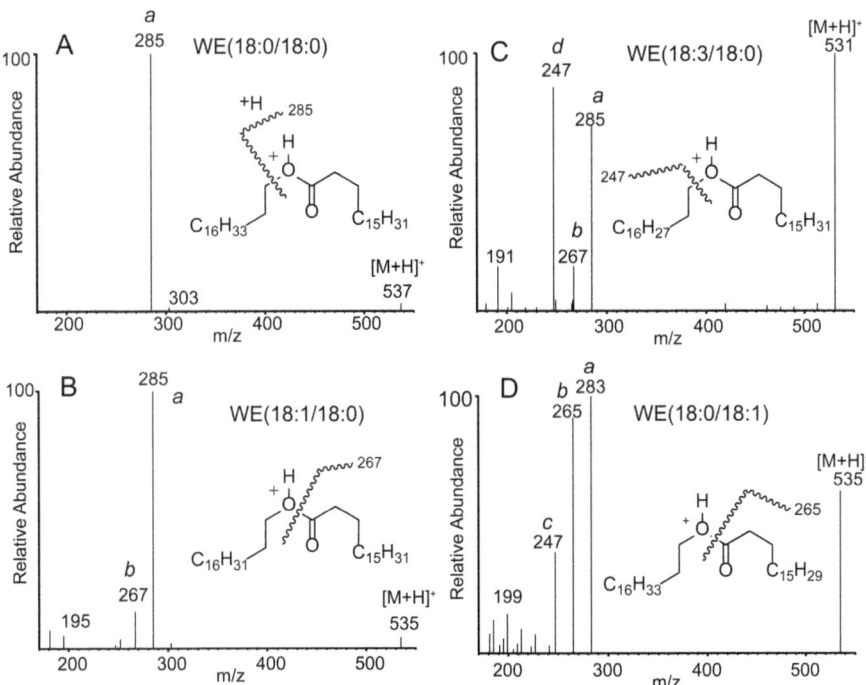

Figure 3.9 Electrospray ionization (positive ions) and tandem mass spectrometry of wax esters. (A) Product ions obtained following collisional activation of stearoyl stearate $[M + NH_4]^+$ at m/z 554; (B) product ions obtained following collisional activation of oleoyl stearate $[M + NH_4]^+$ at m/z 552; (C) product ions obtained following collisional activation of octadecatrienoyl stearate $[M + H]^+$ at m/z 548; (D) product ions obtained following collisional activation of stearoyl oleate $[M + H]^+$ at m/z 535. This figure was redrawn from the data presented in ref. 23.

reaction resulting in cleavage of the wax ester and formation of another abundant product ion $[RCO_2H_2]^+$ (Scheme 3.22). This protonated carboxylic acid product ion has been termed an "*a-type*" ion by Iven[23] and is one of the most abundant product ions seen in wax ester spectra as exemplified in Figure 3.9A–C at m/z 285 (since the wax esters in these panels were all stearate esters). This a-type ion in Figure 3.9D is observed at m/z 283 since the fatty acyl group is oleoyl.[†]

[†]The useful shorthand abbreviation used in Figure 3.9 is (x:a/Y:b), where the fatty alcohol total carbon atoms (x) is in the first part of the abbreviation followed by the total number of double bonds in the alcohol alkyl group (a). The total number of carbons in the fatty acid moiety follows the slash (Y) and the number of double bonds in the fatty acyl portion of the molecule (b).

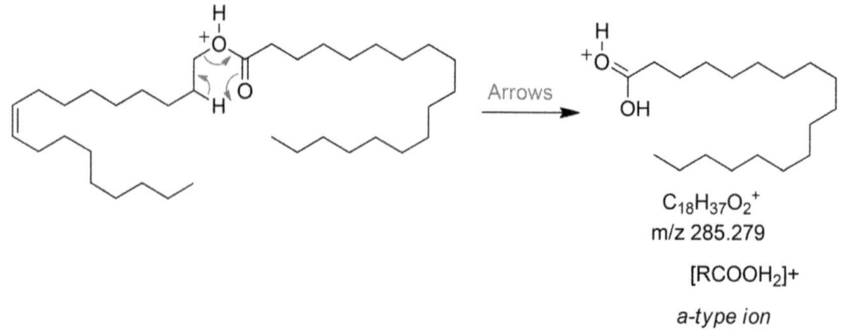
H
+O
H
O
Arrows
H
+O
OH
$C_{18}H_{37}O_2^+$
m/z 285.279
[RCOOH2]+
a-type ion

Scheme 3.22

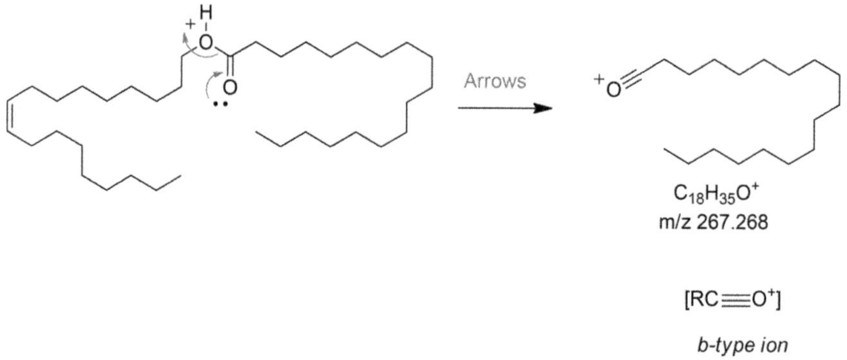
H
+O
O
Arrows
+O
$C_{18}H_{35}O^+$
m/z 267.268
[RC≡O+]
b-type ion

Scheme 3.23

The "*a-type*" fragmentation has been observed for all tandem mass spectra of wax esters and the mass difference between the collisionally activated $[M + NH_4]^+$ precursor ion and the observed *m*/*z* for the *a-type* ion corresponds to the loss of the alcohol moiety as an olefin neutral.[19] It is possible to observe all esters of a particular alcohol using a neutral loss scan for this particular mass difference observed when the wax ester extract is ionized as the ammonium adduct ion and collisionally activated.

When one or more double bonds is present on either side chain, the WE-$[M + NH_4]^+$ forms an acylium ion at $[RC{\equiv}O^+]$ as a significant product (Scheme 3.23). This was termed a "*b-type*" fragment ion by Iven.[23] Notable is that this acylium ion can become one of the most abundant product ions if double bonds are present in the fatty acyl chain (Figure 3.9D), suggesting an important involvement of these double bonds in the fatty acyl chain in the mechanism of *b-type* ion formation.

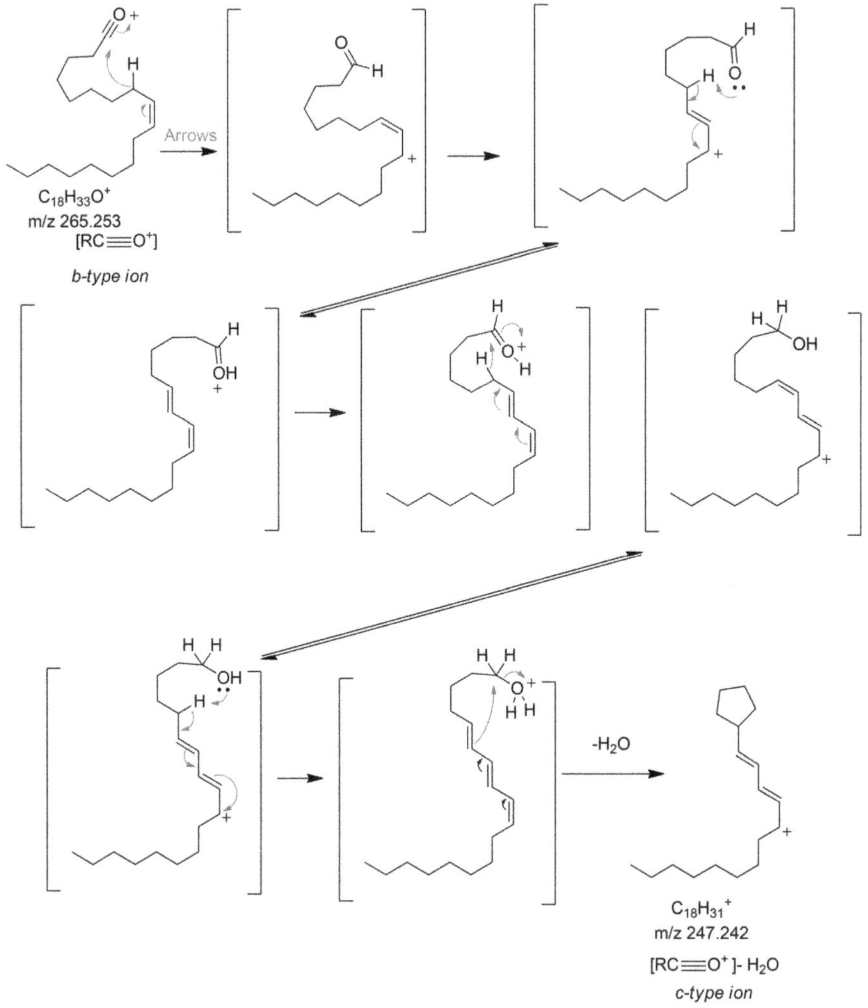

Scheme 3.24

The loss of water from this acylium ion (*c-type ion*) is also quite evident in those cases where additional double bonds are only in the fatty acyl side chain (Figure 3.9D). The loss of water from the acylium ion would require extensive proton migration, and a mechanism previously proposed by Castro *et al.*[24] can be adapted to rationalize the assistance of the fatty acyl double bonds to a dehydration mechanism (Scheme 3.24). The abundance of the dehydrated acylium ion would be consistent with a delocalization of the positive charge over 5-carbon atoms as suggested in the mechanism of proton rearrangements. With additional double bonds in the fatty acyl chain, this delocalization could be extended even further.

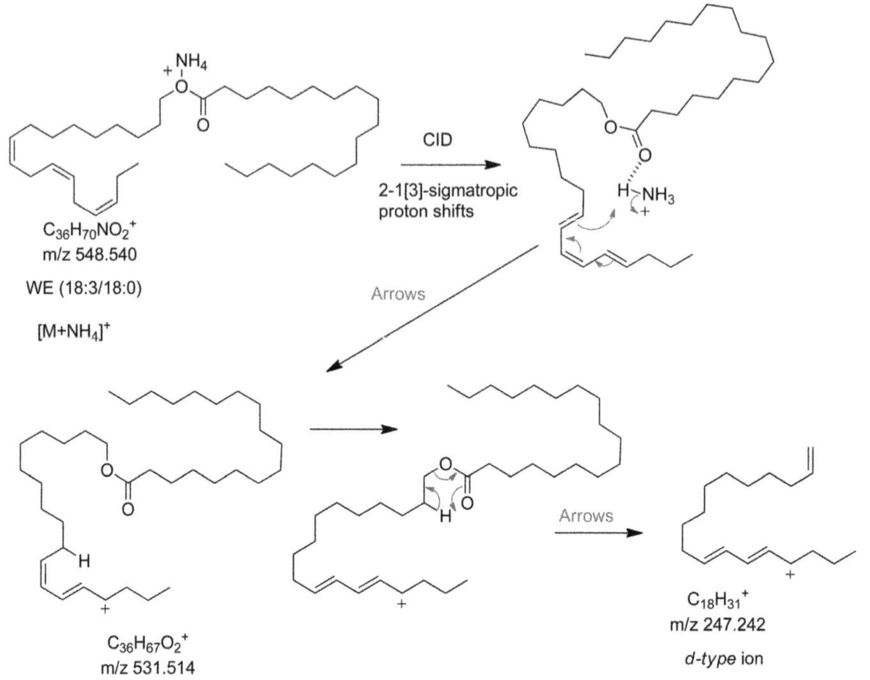

Scheme 3.25

It is interesting to note that the only instance where a CID fragment ion derived from the alcohol portion of the molecule is observed (*d-type* ion), is when multiple double bonds are present in the alcohol radyl group of the wax ester (Figure 3.9C). This suggests that multiple double bonds facilitate some population of proton attachment to this site of the wax ester rather than the ester moiety, perhaps after one or more proton 1[3]-sigmatropic shifts take place to move double bonds into conjugation (Scheme 3.25) and then attack of the conjugated π-bonds at a proton of the ammonium adduct (at the ester site) with loss of NH_3. With such a conjugated double bond system it is possible to become a charging site for proton leaving the ester neutral. Therefore, with the alcohol side chain retaining the charge site, a facile ene-type reaction could take place which would cleave the alcohol side chain as an olefin, which was shown previously in Scheme 3-21, only now the charge site is on the alcoholic alkyl carbon atoms (Scheme 3.25), resulting in the *d-type* ion.

The tandem mass spectrometry of the vary large wax diesters (Scheme 3.26) that are formed by acylation of a long chain α,ω-diol by fatty acyl CoA esters are quite similar in behavior to the $[M + NH_4]^+$ observed following collisional activation and tandem mass spectrometry.[25] The most abundant ions are the "*a-type*" and the "*b-type*" acylium ions described in Scheme 3.22 and 3.23,

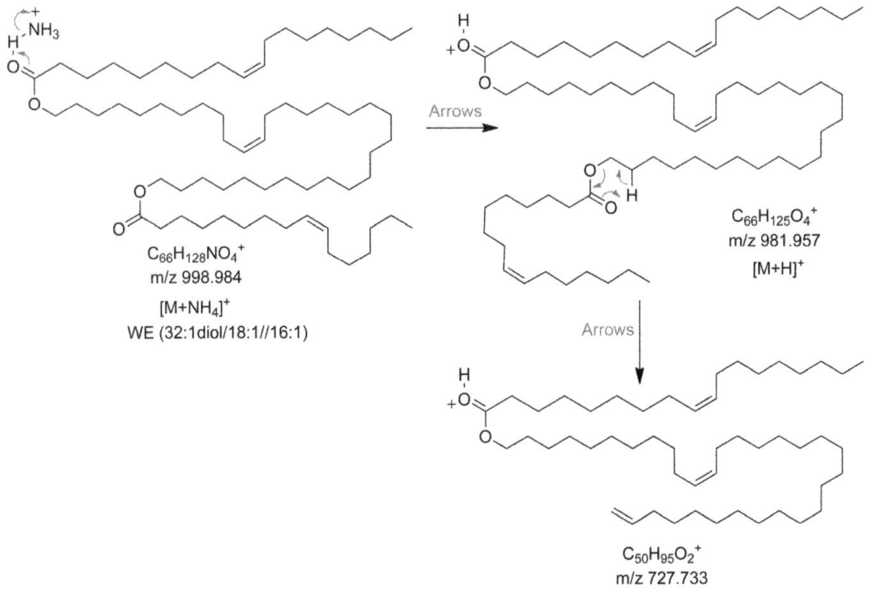
NH3
H
O
O
O
O
$C_{66}H_{128}NO_4^+$
m/z 998.984
$[M+NH_4]^+$
WE (32:1diol/18:1//16:1)
Arrows
H
+O
O
O
H
O
$C_{66}H_{125}O_4^+$
m/z 981.957
$[M+H]^+$
Arrows
H
+O
O
$C_{50}H_{95}O_2^+$
m/z 727.733

Scheme 3.26

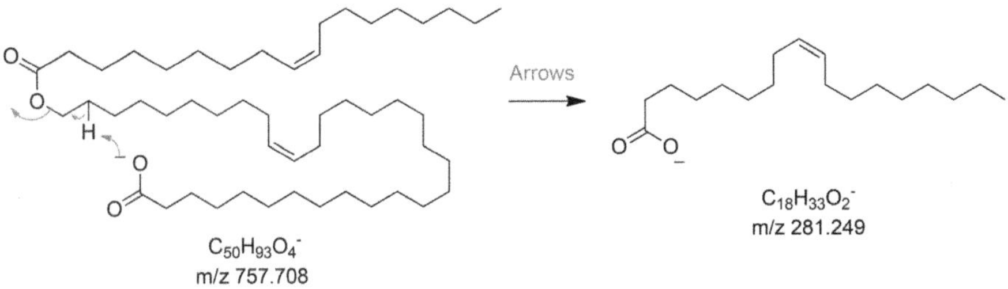
O
O
H
O
O
$C_{50}H_{93}O_4^-$
m/z 757.708
Arrows
O
O
$C_{18}H_{33}O_2^-$
m/z 281.249

Scheme 3.27

respectively. In addition, abundant ions corresponding to a loss of long chain ketene plus ammonia and loss of long chain fatty acid neutrals plus ammonia are abundant (Scheme 3.26).

Another member of the wax ester family involves acylation of long chain ω-hydroxy fatty acids by a long chain acid. This wax ester is not neutral, but readily forms negative ions during electrospray ionization due to the free carboxylic acid moiety (Scheme 3.27). Collisional activation of this negative ion leads to the production of ions corresponding to the loss of the long chain fatty acid as a ketene and cleavage of the ester bond with formation of the carboxylate anion from the fatty acid ester.[26] The mechanism of this latter reaction is most likely a charge driven attack of the carboxylate anion on a remote proton, leaving a fairly reactive carbanion that could attack the carbon adjacent to the ester, making a stable carboxylate anion as a good leaving group as the observed product and a long chain cyclic neutral fatty acid, or direct attack of a proton on the C'-2 resulting in loss of a neutral α,β-unsaturated long chain fatty acid (Scheme 3.27).

References

1. E. Fahy, S. Subramaniam, H. A. Brown, C. K. Glass, A. H. Merrill Jr, R. C. Murphy, C. R. H. Raetz, D. W. Russell, Y. Seyama, W. Shaw, T. Shimizu, F. Spener, G. van Meer, M. S. Van Nieuwenhze, S. H. White, J. L. Witztum and E. A. Dennis, A comprehensive classification system for lipids, *J. Lipid Res.,* 2005, **46**, 839–862.
2. E. B. Divito and M. Cascio, Metabolism, physiology, and analyses of primary fatty acid amides, *Chem. Rev.,* 2013, **113**, 7343–7353.
3. E. B. Divito, A. P. Davic, M. E. Johnson and M. Cascio, Electrospray ionization and collision induced dissociation mass spectrometry of primary fatty acid amides, *Anal. Chem.,* 2012, **84**, 2388–2394.
4. K. K. Nichols, B. M. Ham, J. J. Nichols, C. Ziegler and K. B. Green-Church, Identification of fatty acids and fatty acid amides in human meibomian gland secretions, *Invest. Ophthalmol. Visual Sci.,* 2007, **48**, 34–39.
5. N. Ueda, K. Tsuboi and T. Uyama, Enzymological studies on the biosynthesis of N-acylethanolamines, *Biochim. Biophys. Acta,* 2010, **1801**, 1274–1285.
6. W. A. Devane, L. Hanus, A. Breuer, R. G. Pertwee, L. A. Stevenson, G. Griffin, D. Gibson, A. Mandelbaum, A. Etinger and R. Mechoulam, Isolation and structure of a brain constituent that binds to the cannabinoid receptor, *Science,* 1992, **258**, 1946–1949.
7. J. Palandra, J. Prusakiewicz, J. S. Ozer, Y. Zhang and T. G. Heath, Endogenous ethanolamide analysis in human plasma using HPLC tandem MS with electrospray ionization, *J. Chromatogr. B: Anal. Technol. Biomed. Life Sci.,* 2009, **877**, 2052–2060.
8. H. S. Hansen, B. Moesgaard, G. Petersen and H. H. Hansen, Putative neuroprotective actions of N-acyl-ethanolamines, *Pharmacol. Ther.,* 2002, **95**, 119–126.
9. K. C. Verhoeckx, T. Voortman, M. G. Balvers, H. F. Hendriks, M. Wortelboer and R. F. Witkamp, Presence, formation and putative biological activities of N-acyl serotonins, a novel class of fatty-acid derived mediators, in the intestinal tract, *Biochim. Biophys. Acta,* 2011, **1811**, 578–586.
10. D. Sun, M. G. Cree, X. J. Zhang, E. Boersheim and R. R. Wolfe, Measurement of stable isotopic enrichment and concentration of long-chain fatty acyl-carnitines in tissue by HPLC-MS, *J. Lipid Res.,* 2006, **47**, 431–439.

11. K. M. Ostermann, R. Dieplinger, N. M. Lutsch, K. Strupat, T. F. Metz, T. P. Mechtler and D. C. Kasper, Matrix-assisted laser desorption/ionization for simultaneous quantitation of (acyl-)carnitines and organic acids in dried blood spots, *Rapid Commun. Mass Spectrom.*, 2013, **27**, 1497–1504.

12. Z. Liu, A. E. Mutlib, J. Wang and R. E. Talaat, Liquid chromatography/mass spectrometry determination of endogenous plasma acetyl and palmitoyl carnitines as potential biomarkers of β-oxidation in mice, *Rapid Commun. Mass Spectrom.*, 2008, **22**, 3434–3442.

13. J. A. Hankin and R. C. Murphy, MALDI-TOF and electrospray tandem mass spectrometric analysis of fatty acyl-CoA esters, *Int. J. Mass Spectrom. Ion Processes*, 1997, **165/166**, 467–474.

14. C. A. Haynes, J. C. Allegood, K. Sims, E. W. Wang, M. C. Sullards and A. H. Merrill Jr, Quantitation of fatty acyl-coenzyme as in mammalian cells by liquid chromatography-electrospray ionization tandem mass spectrometry, *J. Lipid Res.*, 2008, **49**, 1113–1125.

15. T. Mauriala, K. H. Herzig, M. Heinonen, J. Idziak and S. Auriola, Determination of long-chain fatty acid acyl-coenzyme A compounds using liquid chromatography-electrospray ionization tandem mass spectrometry, *J. Chromatogr. B: Anal. Technol. Biomed. Life Sci.*, 2004, **808**, 263–268.

16. M. A. Perera, S. Y. Choi, E. S. Wurtele and B. J. Nikolau, Quantitative analysis of short-chain acyl-coenzymeAs in plant tissues by LC-MS-MS electrospray ionization method, *J. Chromatogr. B: Anal. Technol. Biomed. Life Sci.*, 2009, **877**, 482–488.

17. V. H. Wysocki, G. Tsaprailis, L. L. Smith and L. A. Breci, Mobile and localized protons: a framework for understanding peptide dissociation, *J. Mass Spectrom.*, 2000, **35**, 1399–1406.

18. I. A. Butovich, E. Uchiyama and J. P. McCulley, Lipids of human meibum: mass-spectrometric analysis and structural elucidation, *J. Lipid Res.*, 2007, **48**, 2220–2235.

19. M. Fitzgerald and R. C. Murphy, Electrospray mass spectrometry of human hair wax esters, *J. Lipid Res.*, 2007, **48**, 1231–1246.

20. J. Chen, K. B. Green-Church and K. K. Nichols, Shotgun lipidomic analysis of human meibomian gland secretions with electrospray ionization tandem mass spectrometry, *Invest. Ophthalmol. Visual Sci.*, 2010, **51**, 6220–6231.

21. R. C. Murphy, *Mass Spectrometry of Lipids*, Plenum Press, New York, 1993.

22. K. Urbanova, V. Vrkoslav, I. Valterova, M. Hakova and J. Cvacka, Structural characterization of wax esters by electron ionization mass spectrometry, *J. Lipid Res.*, 2012, **53**, 204–213.

23. T. Iven, C. Herrfurth, E. Hornung, M. Heilmann, P. Hofvander, S. Stymne, L. H. Zhu and I. Feussner, Wax ester profiling of seed oil by nano-electrospray ionization tandem mass spectrometry, *Plant Methods*, 2013, **9**, 24.

24. J. Castro-Perez, T. P. Roddy, N. M. Nibbering, V. Shah, D. G. McLaren, S. Previs, A. B. Attygalle, K. Herath, Z. Chen, S. P. Wang, L. Mitnaul, B. K. Hubbard, R. J. Vreeken, D. G. Johns and T. Hankemeier, Localization of fatty acyl and double bond positions in phosphatidylcholines using a dual stage CID fragmentation coupled with ion mobility mass spectrometry, *J. Am. Soc. Mass Spectrom.*, 2011, **22**, 1552–1567.
25. J. Chen, K. B. Green and K. K. Nichols, Quantitative profiling of major neutral lipid classes in human meibum by direct infusion electrospray ionization mass spectrometry, *Invest. Ophthalmol. Visual Sci.*, 2013, **54**, 5730–5753.
26. I. A. Butovich, J. C. Wojtowicz and M. Molai, Human tear film and meibum. Very long chain wax esters and (O-acyl)-omega-hydroxy fatty acids of meibum, *J. Lipid Res.*, 2009, **50**, 2471–2485.

Glyceryl Esters

Six major families make up the glycerol ester lipids: triacylglycerols (TAGs), diacylglycerols (DAGs), monoacylglycerols (MAGs), and plant derived glycosyl mono-, diacylglycerols (MGDG and DGDG), and sulfoquinovosyl diacylglycerols (SQDG). Each of these lipid groups has a very important role in lipid biochemistry that has warranted a considerable number of studies by electrospray and tandem mass spectrometry (Scheme 4.1). In addition, the triacylglycerols are plant and animal "oil products" that have considerable economic value. The characterization of these natural lipids by mass spectrometry is made somewhat more challenging as ionization of these neutral lipids by ESI must occur prior to analysis, except for sulfonic acid ester species. In most instances, the formation of ions involves attachment of a charging species such as H^+, NH_4^+ or alkali metal (Na^+ or Li^+). Each of these adducted forms of glycerol lipids has unique behavior that provides information relevant to the structural features of the lipid. Analysis of each lipid charged adduct also engages specific strengths and weaknesses which need to be considered when choosing the type of adduct that will be used for the analytical investigation. None behave perfectly and all adduct forms have great utility.

4.1 Triacylglycerols (TAGs)

The number of TAG molecular species found in both prokaryotic and eucaryotic cells are quite large, since these molecules serve important roles in delivery of fatty acids to cellular biochemistry processes of energy metabolism and ATP production. These neutral lipids are derived from acylation of diacylglycerols which are either metabolites of other lipids (*e.g.* phospholipids by the action of phospholipases) or from *de novo* biosynthesis.[1] Acylation is a function of available acyl-CoA esters and estimates have been made

New Developments in Mass Spectrometry No. 4
Tandem Mass Spectrometry of Lipids: Molecular Analysis of Complex Lipids
By Robert C Murphy
© Robert C Murphy 2015
Published by the Royal Society of Chemistry, www.rsc.org

that 15–20 fatty acyl-CoA esters could readily be available, which would lead to more than 3000–8000 possible permutations of TAG molecules. While the number of TAG molecular species that have been identified in biological samples is much lower, nevertheless, over 500 individual molecular species were recently identified in butter fat using advanced tandem mass spectrometric and computational methods, and over 200 of these TAGs were found at concentrations of 1 mM or above.[2]

Electrospray analysis of TAGs has been carried out as the NH_4^+, Na^+ and Li^+-adducts and abundant ions corresponding to the $[M + adduct]^+$ are readily observed. This ion reveals the molecular weight of the TAG molecular species and from this information it is possible to determine total number of fatty acyl carbon atoms and double bonds, especially when high resolution mass spectrometric analysis (>5 ppm mass accuracy) is employed so that unambiguous assignment of total carbon, hydrogen, and oxygen atoms as well as the charging adduct atom(s) is made. Still, structural characterization assumes the presence of a glycerol backbone ($C_2H_4O_3$) in the ion species with the structure (R_1)O–CH_2–CHO(R_2)–CH_2O(R_3), where R_1, R_2, and R_3 are fatty acyl groups esterified at the corresponding glycerol *sn* (stereospecific nomenclature) sites. With high resolution measurements the identity of such ether-linked radyl groups can also be made, rather than being confounded by the presence of odd chain fatty acyl groups.

4.1.1 Ammonium Adducts $[M + NH_4]^+$

Collisional activation of ammonium ion adducts yields a very specific product ion spectrum composed of $[M + H]^+$ and diglyceride-like (DAG-like) product ions for each of the unique fatty acyl group present. The structure of

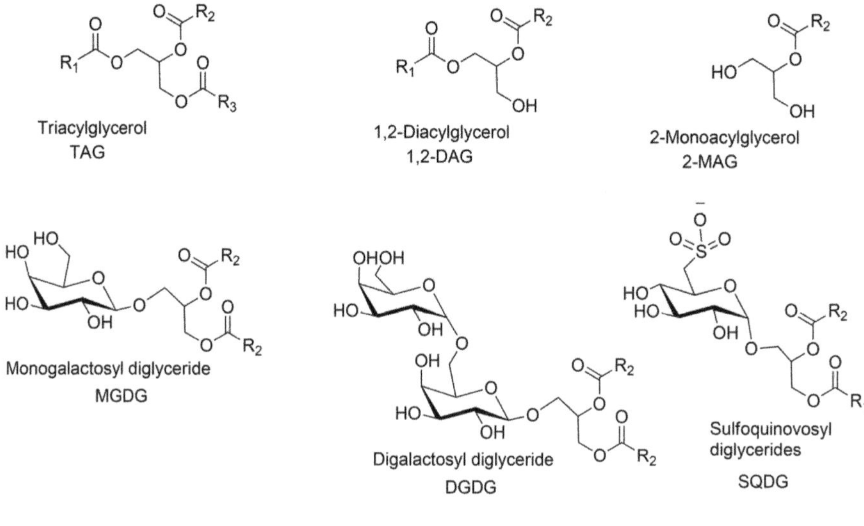

Scheme 4.1

the $[M + H]^+$ is typically indicated as a proton accepting one pair of nonbonding electrons from one of the three carboxyl oxygen atoms of the ester moieties in TAG molecular species. However, the abundance of the $[M + H]^+$ as a product ion is remarkably related to the total number of double bonds in the fatty acyl groups and totally saturated TAGs have little or no $[M + H]^+$ abundance.[3,4] One possibility for this finding is that the proton attachment site could be with π-bond orbitals, perhaps being stabilized by initial hyperconjugation and then formation of a methylene group and carbon-centered cation delocalized over a conjugated system (Scheme 4.2).

For a TAG with two different fatty acyl chains, two DAG-like ions appear and for three different fatty acyl groups, three DAG-like ions appear (Figure 4.1). These DAG-like ions are a result of each acyl group being lost as a neutral carboxylic acid and neutral ammonia (NH_3). Thus, the charge site on the DAG-like ion now appears at an oxygen atom rather than the nitrogen atom of NH_4^+. Several mechanisms have been proposed and resulting ion structures for the $[M + NH_4 - NH_3-R_xCOOH]^+$ (Scheme 4.3), but a general feature previously considered was the role of the loss of the neutral ammonia in driving the mechanism of this ion formation. The mechanism shown is supported by various stable isotope labeled analogs and ND_4^+ adducts.[5]

An early observation, that has been verified many times, is that the loss of the fatty acyl group at the second carbon of glycerol (*sn-2* R_2COOH) is somewhat lower in abundance than either loss of the R_1COOH or the R_3COOH (+NH_3), which has led to the suggestion that the relative abundance of the resulting diglyceride ions can indicate which fatty acyl group is at *sn-2*.[6] This appears to be valid over a large number of different molecular species,[7] even those containing polyunsaturated fatty acyl groups.[2,4] However, this assumption of acyl group position assignment by the relatively lower abundance of the ion corresponding to the loss of R_2COOH can be an over

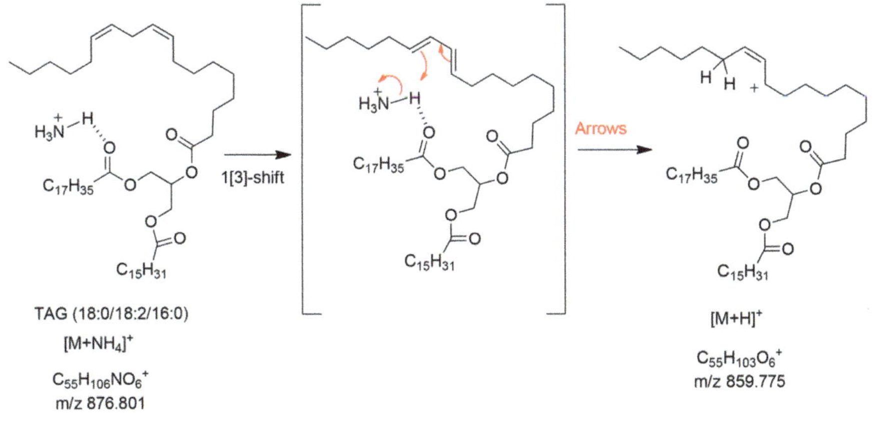

TAG (18:0/18:2/16:0)
$[M+NH_4]^+$
$C_{55}H_{106}NO_6^+$
m/z 876.801

1[3]-shift

Arrows

$[M+H]^+$
$C_{55}H_{103}O_6^+$
m/z 859.775

Scheme 4.2

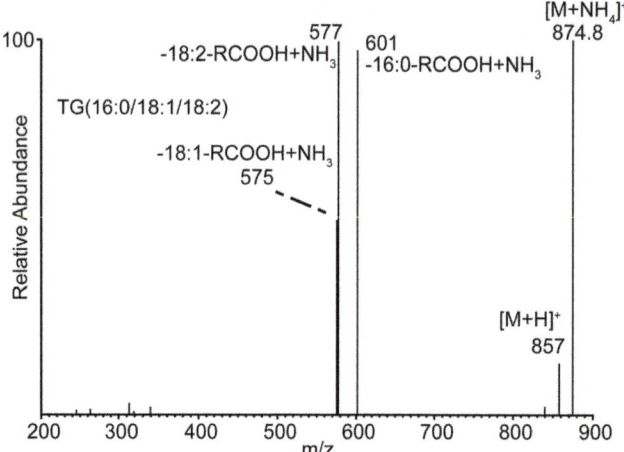

Figure 4.1 Electrospray ionization (positive ions) and tandem mass spectrometry of the triglyceride TG(16 : 0/18 : 1/18 : 2). Product ions following collisional activation of the ammonium ion adduct [M + NH₄]⁺ at *m/z* 874.8. This MS/MS spectrum was obtained using a tandem quadrupole mass spectrometer.

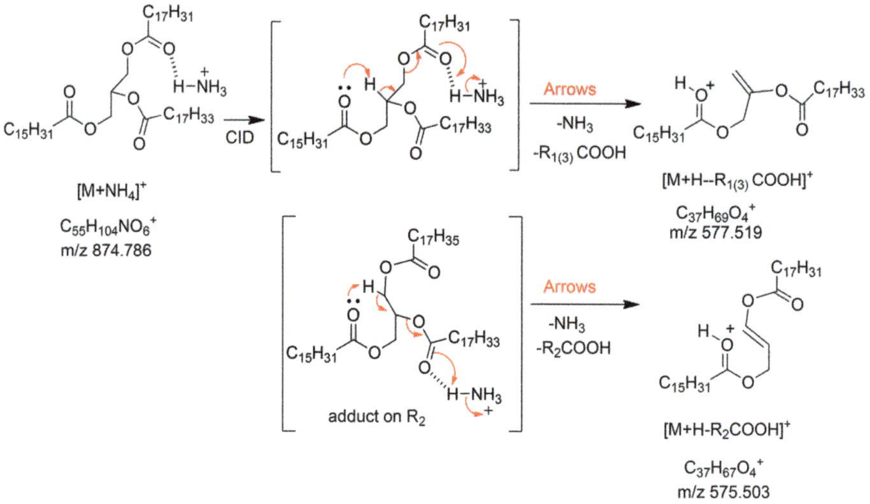

Scheme 4.3

simplification due to the effects of double bond positions and collisional energy employed.[4,5,7]

Observation of more than the loss of three fatty acids plus ammonia (DAG-like ions) is typically observed when TAGs are isolated from biological sources. This is because any specific mass-to-charge ratio corresponding to a [M+ adduct]⁺, is the result of multiple TAG molecular species being present that are completely isobaric (same elemental composition), but differ by the specific fatty acyl groups that add up to the same total number of fatty acyl carbons and double bonds. For example, collision activation of the m/z corresponding to the TAG (48 : 2) isolated from a mammalian cell line, yielded neutral fatty acid losses for 14 : 0, 15 : 0, 15 : 1, 16 : 0, 16 : 1, 17 : 0, 17 : 1, 18 : 1, and 18 : 2 that indicated 11 different TAG molecular species present in this cell extract.[5] This exemplifies one of the problems with TAG analyses that led many investigators to use various chromatographic separation strategies to simplify these complex mixtures of isobaric molecules.

Subsequent collisional activation of the DAG-like ions obtained following MS/MS of $[M + NH_4]^+$ (MS³) yields sufficient information to deduce the two remaining fatty acyl groups of the DAG-like MS² ion (Figure 4.2). Two different product ions are typically generated. The first is formation of an acylium ion derived from either of the remaining fatty acyl groups in a proton rearrangement at the charge site (Scheme 4.4).

The second MS³ product ion corresponds to the acylium ion +74 Da. A possible mechanism of formation is a charge remote ene-type proton shift that involves loss of a neutral ketene containing one of the two remaining fatty acyl chains of the DAG-like ions (Scheme 4.5). Both DAG-like ion structures proposed (Scheme 4.3) can undergo such a loss.[5] The acylium ions are seen at m/z 239 (16 : 0) and 265 (18 : 1) and acylium ion +74 at m/z 313 (16 : 0) and 339 (18 : 1) in the MS³ of these two DAG-like ions following collisional activation of m/z 874 from the TAG 52 : 3 (Figure 4.2). The abundance of the acylium ion +74 that lost an additional H_2O (m/z 319, 18 : 2) is much greater than the acylium ion +74 for the 18 : 2 fatty acyl group.

These ion-types are readily distinguishable in the MS³ spectra as the most abundant product ions. The combination of the MS² and the MS³ product ion spectra provides sufficient information to define each TAG molecular species as to fatty acid (acyl carbon atoms and number of double bonds). However, there is no information concerning fatty acyl chain branching or position/geometry of double bonds. The loss of the initial fatty acyl group as a neutral carboxylic acid (plus NH_3) specifically identifies the first acyl group. The MS³ product ion spectrum (acylium or acylium +74 ion) identifies the second acyl group (Table 4.1). Thirdly, the mass difference between the DAG-like ion that was collisionally activated and the resulting acylium ion enables

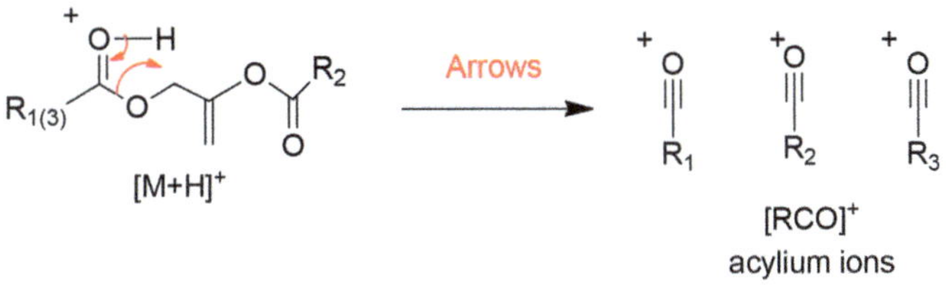

Scheme 4.4

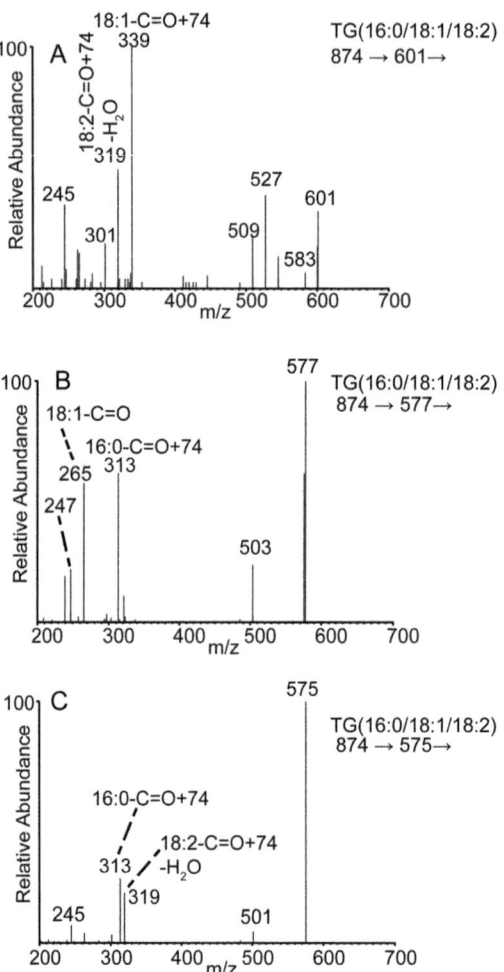

Figure 4.2 MS[3] of the major product ions formed following the collisional activation of the triacylglycerol TG(16 : 0/18 : 1/18 : 2). The losses of each fatty acyl ester as a free carboxylic acid plus ammonia were further selected and collisionally activated in this ion trap experiment. (A) Product ions obtained following collisional activation of m/z 601 derived from the loss of palmitic acid and ammonia at m/z 601; (B) product ions obtained following collisional activation of m/z 577 derived from the loss of linoleic acid plus ammonia; (C) product ions obtained following collisional activation of m/z 575 corresponding to the loss of oleic acid plus ammonia. These MS[3] spectra were obtained using a triple quadrupole mass spectrometer with the last sector as a linear ion trap.

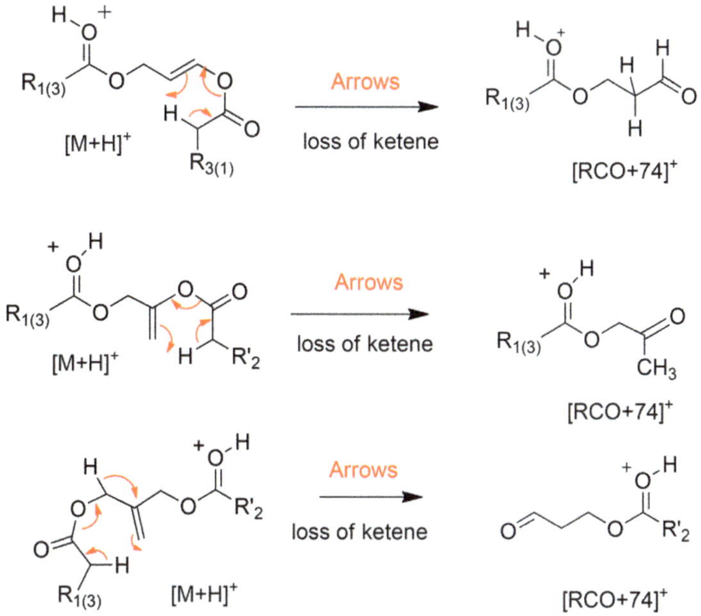

Scheme 4.5

determination of the final fatty acyl group in the TAG molecular species (Table 4.1). The exact mechanisms that control whether the acylium ion or the acylium ion +74 is the most abundant species observed have not been carefully studied, but it has been noted that the acylium ion +74 is often observed to be quite abundant for the fatty acyl group at the *sn-2* position.[6] Nonetheless, some care should be exercised when using abundance information to assign fatty acyl esterification site on the glycerol backbone. The presence of multiple isobaric species present in biological samples (of different abundances) can yield low and high abundance ions to confound a straightforward interpretation.

4.1.2 Alkali Metal Adduct Ions-[M + Na]⁺ and [M + Li]⁺

Alkali metal ions such as Na^+, K^+, and Li^+ readily form charged adducts with TAGs and collisional activation of these [M + metal]⁺ readily undergo loss of each fatty acyl group as a neutral carboxylic acid. Of much lesser abundance is the loss of the metalated carboxylic acid (RCOONa or RCOOLi) as a neutral species (Figure 4.3). This behavior is remarkably different from that of the [M + NH₄], but has been used in lipidomic studies to uniquely detect Li-adducts of TAGs.[8]

The losses of the neutral carboxylic acid indicate those fatty acyl groups present in the TAG molecular species being analyzed. The same rank order of abundance of the metalated DAG-like product ions are observed to that of the [M + NH₄]⁺ derived DAG-like ions in that the loss of the fatty acid from the

Table 4.1 Common fragment ions from the collisional activation of $[M + NH_4]^+$ from triacylglycerols.

Abbreviation[a]	RCOOH + NH₃ neutral loss[b]	Acylium ion[c] m/z	R' + 74[d] m/z
12 : 0	217	183	257
14 : 1	243	209	283
14 : 0	245	211	285
16 : 1	271	237	311
16 : 0	273	239	313
18 : 4	293	259	333
18 : 3	295	261	335
18 : 2	297	263	337
18 : 1	299	265	339
18 : 0	301	267	341
20 : 5	319	285	359
20 : 4	321	287	361
20 : 3	323	289	363
20 : 2	325	291	365
20 : 1	327	293	367
20 : 0	329	295	369
22 : 6	345	311	385
22 : 5	347	313	387
22 : 4	349	315	389
22 : 1	355	321	395
24 : 1	383	349	423
24 : 0	385	351	425

[a]Total carbon atoms: total number of double bonds.
[b]Neutral loss mass from $[M + NH_4]^+$ to form a "DAG-like" ion.
[c]$RC{\equiv}O +$ ion type. [d]See Scheme 4.5 for structure of ions.

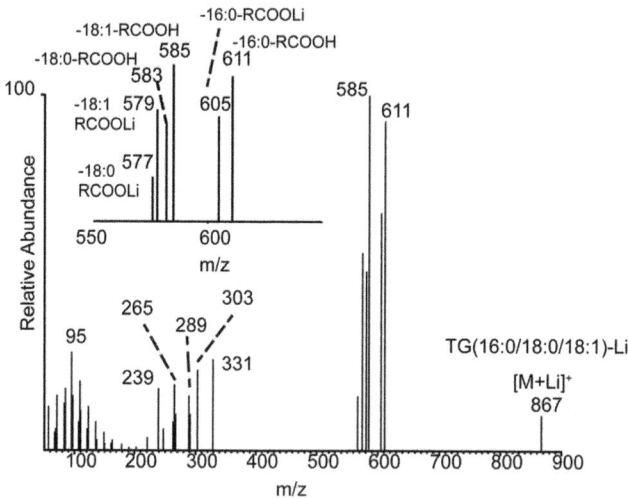

Figure 4.3 Electrospray ionization (positive ions) and tandem mass spectrometry of the lithiated adduct of the triacylglycerol molecular species TG(16 : 0/18 : 0/18 : 1). Product ions obtained following collisional activation of $[M + Li]^+$ at m/z 867. This figure was redrawn from data presented in ref. 9.

sn-2 position is less preferred.[9] This behavior has been suggested to be of value in assignment of fatty acyl positions,[10] as was the case for the ammoniated adducts, but again some care must be exercised not to over-interpret relative abundance information. Collisional activation of the sodiated diglyceride ion does not yield abundant ions that reveal the other two fatty acyl groups, which is unlike MS^3 of the DAG-like ions from $[M + NH_4]^+$.[11] However, the MS^3 of the lithiated DAG-like ions form product ions quite indicative of the remaining fatty acyl groups and even the possible location of double bonds in the fatty acyl chain (Figure 4.4). The possible explanation provided for this behavior of the Li^+-adducts centers around the stronger Li^+-carbonyl bond relative to the strength of the Na^+, especially when compared to the NH_4^+ adduct bond. This renders charge remote fragmentation mechanisms more favorable for the Li^+-adducts since charge driven mechanisms do not dominate the observed product ions. Another way to look at this is that the inability of the collisionally excited ion to release energy by loss of the small neutral species (like NH_3), enables additional energy to be available to the lithium adduct so that charge remote reactions take place and yield the product ions by bond cleavage reactions. The proposed structure of the product ion remaining after loss of one fatty acyl group from the Li^+-adduct is remarkably different from the DAG-like ions proposed for the CID of $[M + NH_4]^+$ (Scheme 4.6).

Charge remote reactions have been suggested to account for rather unique mass losses observed when these MS^2 product ions are activated in a MS^3 experiment.[11] For example, the experiment m/z 867 → 611 (Figure 4.4), the ion observed at m/z 331 would result from a loss of the *sn-2* fatty acyl groups as an unsaturated alcohol ester (Scheme 4.7) by a charge remote fragmentation (CRF) yielding ion structure *a* and *b*.

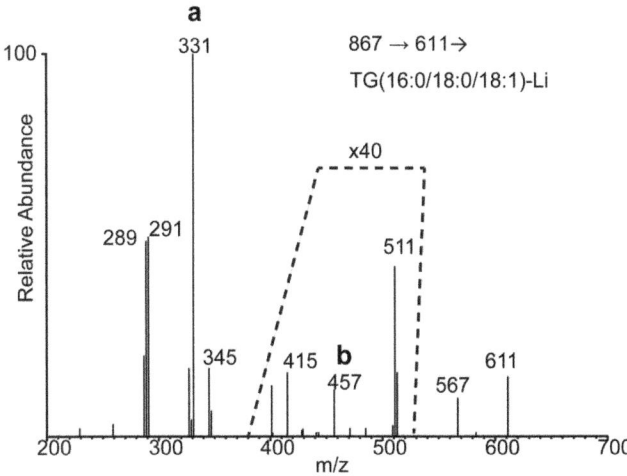

Figure 4.4 MS^3 tandem mass spectrometry of the lithiated adduct of TG(16 : 0/18 : 0/18 : 1). Product ions obtained after trapping and collisional activation of m/z 611. This figure was redrawn from data presented in ref. 11.

Very interesting ions are also observed in the MS^3 product ion spectrum, suggesting the position of the double bond in the single unsaturated fatty acyl group of this TG (16 : 0/18 : 1/18 : 0) molecular species. Isomerization of the 9-10 double bond in the R$_2$ fatty acyl chain and a proton rearrangement (ene-type reaction), would form the ion observed at *m/z* 457 (Scheme 4.7). Other product ions observed suggest that the double bond resides at carbon 9-10 portion of the R$_2$ fatty acyl chain. There is considerable evidence to support these double bond driven charge remote fragmentation reactions, but there have not been many TAG-molecular species where double bond isomers or deuterium labeled analogs corresponding to the proton transferred in the mechanisms proposed (Scheme 4.7) are specifically labeled. However, there is evidence to support the suggestion that MS^3 analysis of the

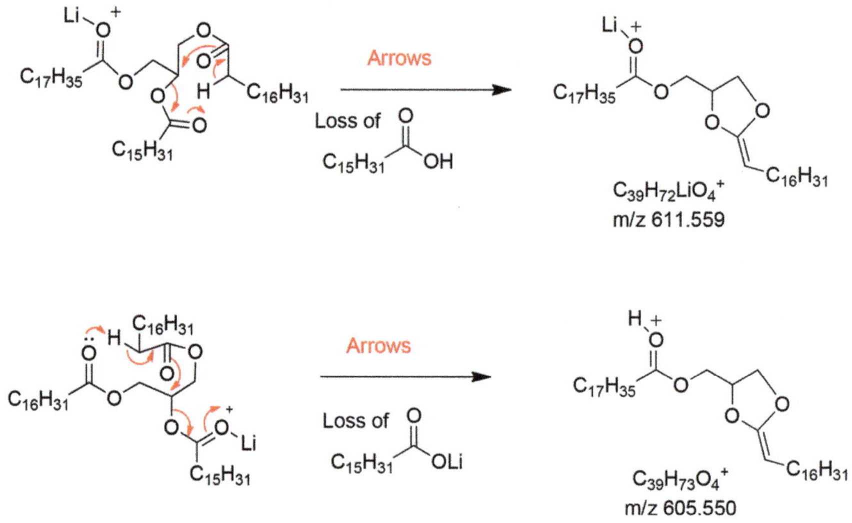

Scheme 4.6

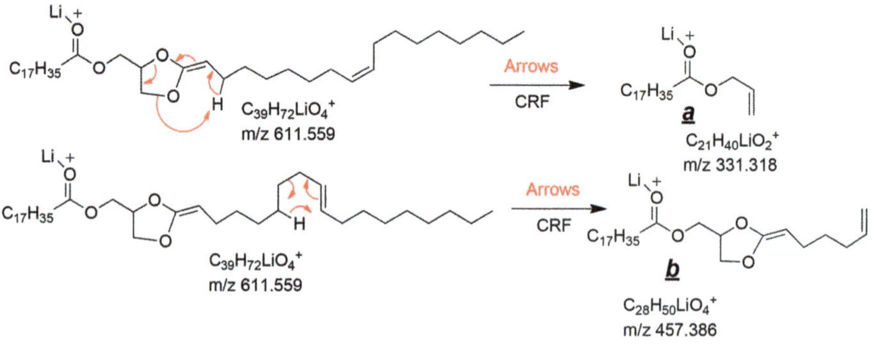

Scheme 4.7

Li$^+$-adducts can reveal double bond location of even polyunsaturated fatty acids due to the unique characteristic of the Li$^+$-adduct that facilitates charge remote fragmentation.[12]

4.2 Diacylglycerols

Diacylglycerols (DAGs) are quite important lipids, aside from being precursors or products of TAG synthesis and metabolism. These species are intimately involved in cell-cell signaling events, being the product of phospholipase C hydrolysis at the cellular membrane of phosphatidylinositol and polyphosphatidylinositols (PIP$_2$, PIP$_3$, and PIP$_4$). Specific signaling events linked to G-proteins result in DAG production that can activate protein kinase C as well as binding to very important G-protein factors.[13,14] Accurate and sensitive quantitation of these neutral lipids has been an important analytical task. Diglycerides exist as three regioisomers, *i.e.* 1,2-DAGs, 1,3-DAGs, and 2,3-DAGs. Considering chirality at the *sn*-2 carbon atom (Scheme 4.8) the 1,2-DAGs are stereoisomers of the 2,3-DAGs.

Electrospray ionization requires adduct formation and tandem mass spectrometry of these adducts has been studied for $[M + NH_4]^+$, $[M + Na]^+$, and $[M + Li]^+$ forms. The general behavior of these adduct ions is precisely that observed for the TAG adducts with very few exceptions related to the presence of the free carbinol moiety and only two fatty acyl substituents.

4.2.1 Ammonium Ion Adducts $[M + NH_4]^+$

The collisional activation of $[M + NH_4]^+$ results in the loss of either fatty acyl group as a free carboxylic acid driven the concerted loss of neutral ammonia as was described in the mechanisms outlined in Scheme 4.3. Thus two product ions are observed for unique 1,3- or 1,2-DAGs by this mechanism. The only other product ion observed corresponds to the loss of 35 Da ($H_2O + NH_3$) and this loss does not involve any of the glycerol backbone protons.[15] A likely mechanism that would involve a concerted loss of both small neutral molecules involves direct interaction of the carbinol hydroxyl with the ammonium ion driven by the loss of these two neutral species (Scheme 4.9).

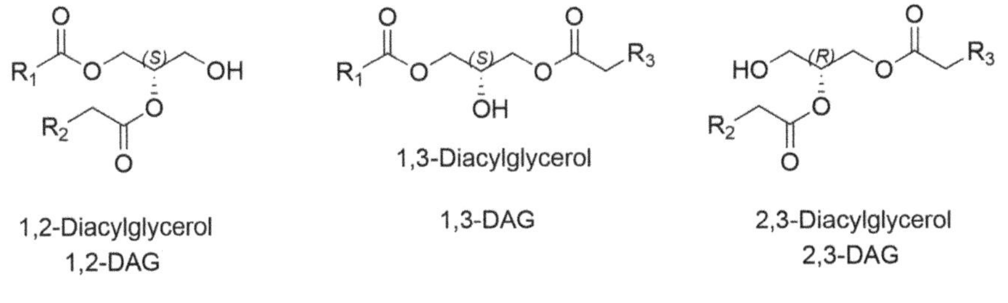

1,2-Diacylglycerol
1,2-DAG

1,3-Diacylglycerol
1,3-DAG

2,3-Diacylglycerol
2,3-DAG

Scheme 4.8

4.2.2   Alkali Metal Adducts [M + Li]⁺ and [M + Na]⁺

The sodiated adduct of DAGs also behaves similarly to that described for TAG species after collisional activation with major product ions corresponding to the loss of each fatty acid as a neutral. The loss of the sodiated carboxylic acid can be almost equally abundant under some conditions of collisional activation and the structure of the resulting product ions has been suggested.[16] An example is provided for the 1,3-DAG corresponding to DG(32 : 2). Also, the loss of either fatty acyl group as a neutral ketene is observed as a reasonably abundant ion. The loss of water is not an ion typically observed following collisional activation of the sodiated adduct.

The lithiated DAGs also behave as expected due to the very large binding energy for the Li⁺ and the carbonyl oxygen atom on either ester moiety. The loss of both fatty acid esters as neutral carboxylic acids or neutral lithiated carboxylic acid salt can be observed (Scheme 4.10) along with each fatty acyl group loss as a ketene neutral. The latter ions can be quite significant products after collisional activation of the lithiated adduct.[17] Neither the sodiated or lithiated adducts MS/MS mass spectra were found to yield ions

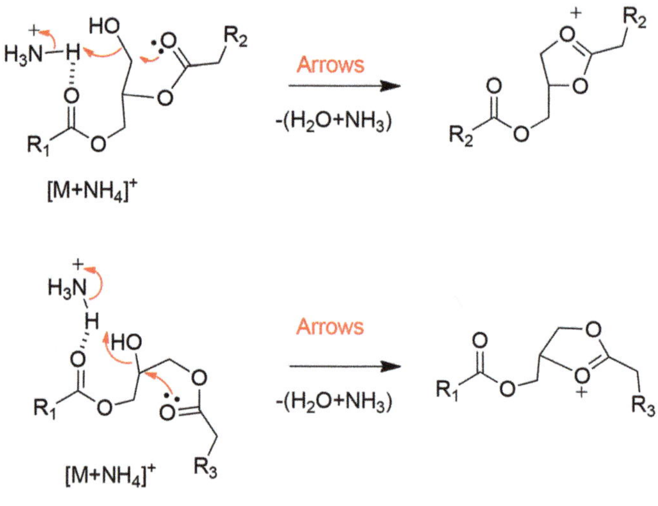

Scheme 4.9

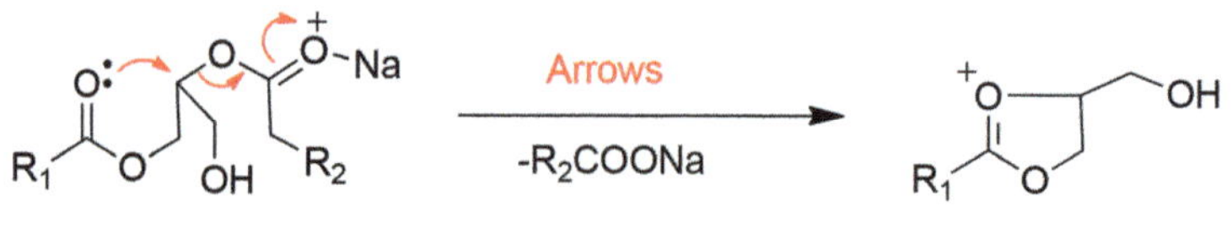

Scheme 4.10

corresponding to the loss of neutral water, suggesting that this very important ion process observed for the ammonium adduct was a unique feature driven by the weak adduct bond strength of the ammonium ion and the ability to drive such a reaction by the loss of neutral ammonia.

4.2.3 Derivatization of Diacylglycerols

Derivatization of DAGs has been reported as a means to increase detection of these important neutral lipids. One quite successful strategy involved conversion of the diglyceride alcohol moiety into an ester of betaine, which has a quaternary nitrogen atom and a fixed positive charge (Scheme 4.11). This improved sensitivity of detection of the DAG molecular species by positive ions isolated from biological samples by two orders of magnitude. However, tandem mass spectrometry was not particularly valuable since it did not yield any structurally significant ions.[18]

Another useful derivative involved making a urethane ester of the DAG using 2,4-difluorophenylisocyanate to generate the difluorophenylurethane (DFPU). This facile reaction does not lead to an ionized derivative, but it does impart favorable chromatographic and tandem mass spectrometric properties (Figure 4.5). The 1,3-derivatized DAG can be readily separated from the 1,2 DAG derivative by normal phase HPLC, facilitating a method to detect these regioisomers isolated from a biological system (Figure 4.5A).[19] The collisional activation of the $[M + NH_4]^+$ adduct ions attained following electrospray ionization from both 1,2- and 1,3-DAG DFPU yields identical product ions, corresponding to each fatty acyl group lost as a free carboxylic acid plus ammonia (as outlined in Scheme 4.3).

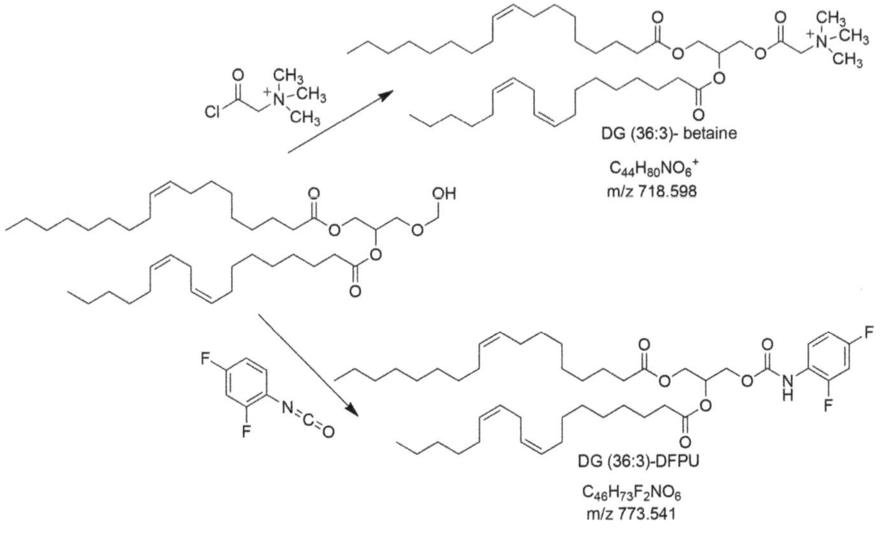

DG (36:3)- betaine
$C_{44}H_{80}NO_6^+$
m/z 718.598

DG (36:3)-DFPU
$C_{46}H_{73}F_2NO_6$
m/z 773.541

Scheme 4.11

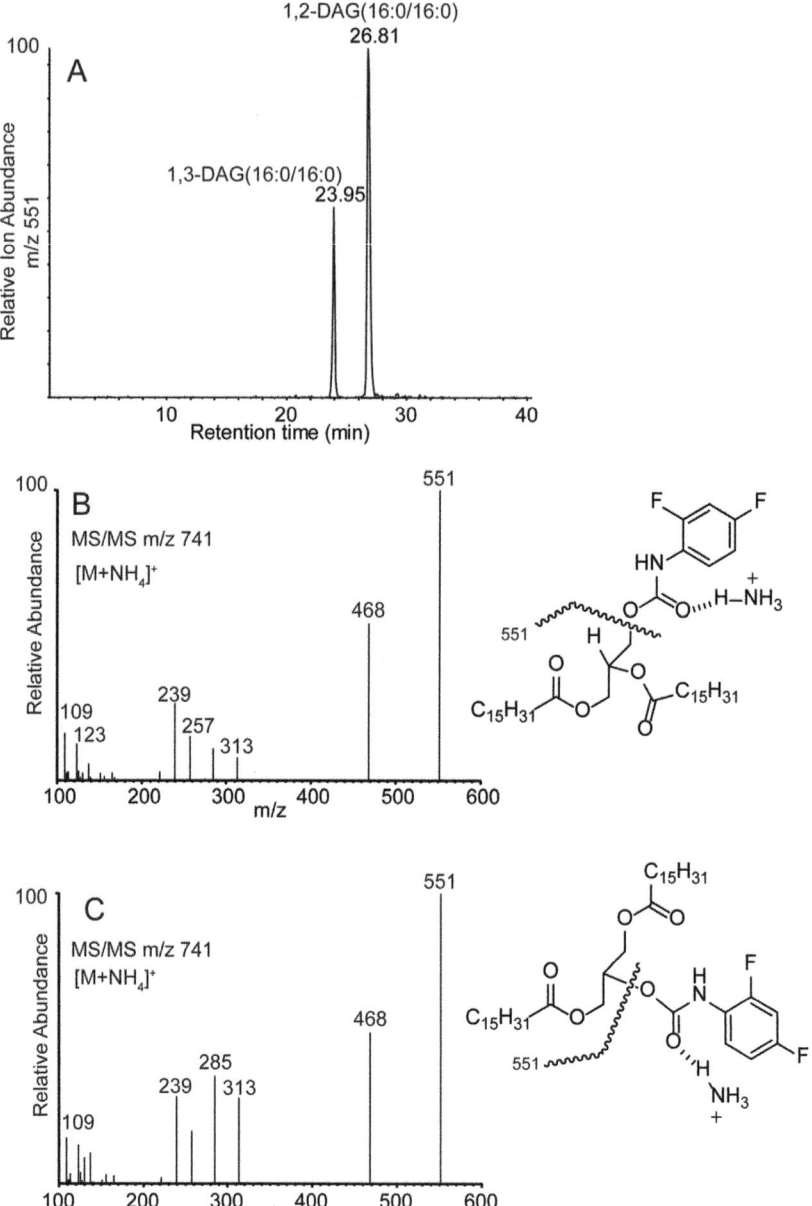

Figure 4.5 Electrospray ionization (positive ions), reversed phase HPLC separation and tandem mass spectrometry of the 2,4-diphenylfluorourethane (DPFU) derivative of the 1,2-DAG (16 : 0/16 : 0) and 1,3-DAG (16 : 0/16 : 0). (A) Reversed phase HPLC separation of the DPFU derivative of 1,3- and 1,2-DAG molecular species. Separation of these regioisomers was readily achieved as the derivatization prevented acyl group migration; (B) product ions obtained following collisional activation of the DPFU derivative of the 1,2-DAG [M + NH$_4$]$^+$ at m/z 551; (C) Product ions obtained following collisional activation of the DPFU derivative of 1,3-DAG [M + NH$_4$]$^+$ at m/z 551. This MS/MS spectrum was obtained using a tandem quadrupole mass spectrometer.

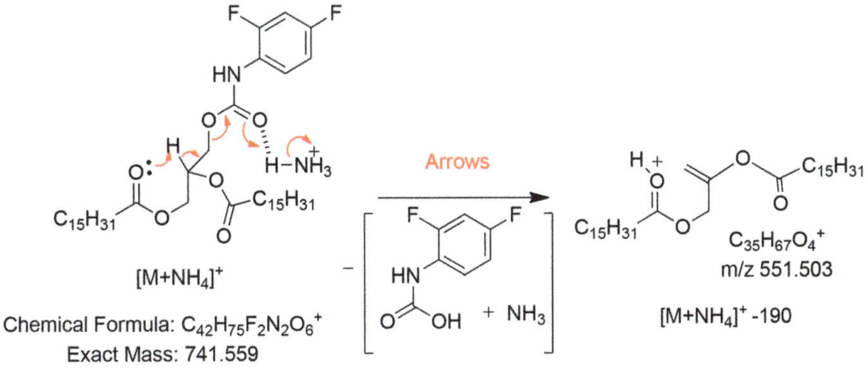

Scheme 4.12

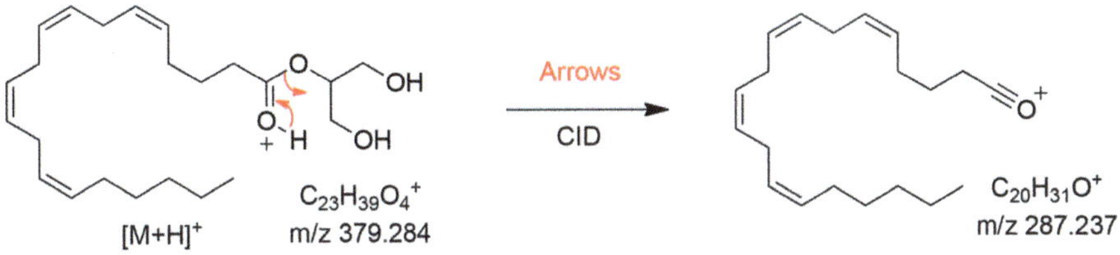

Scheme 4.13

The urethane is lost as the difluorophenylcarbamic acid plus ammonia (Scheme 4.12). A convenient strategy to detect and quantitate these derivatives that are separated by HPLC is to use a constant neutral loss of 190 Da, which is the unique mass loss for difluorophenylurethane derivatives. This strategy also works well for monoacylglycerides that are obtained after derivatization. The MAG derivatives are easily separated from diglycerides using normal phase HPLC.[19]

4.3 Monoacylglycerols

The only monoacylglycerol tandem mass spectrometry studied in detail has been 2-arachidonoyl glycerol (2-AG), largely because of its biological activity as an endocannabinoid.[20] Collisional activation of the $[M + H]^+$ leads to a large product ion at *m/z* 287 corresponding to the acylium ion of arachidonate (Scheme 4.13).

4.4 Glycodiacyldiglycerides

A major lipid species in plants and algae are diacyldiglycerides that have a carbohydrate, typically galactose at the *sn-3* position.[21] These lipids have diversity in the fatty acyl groups as well as the carbohydrate moiety. Common neutral lipids are monogalactosyl diacylglycerols (MGDG), digalactosyl diacylglycerols (DGDG), and acidic diacylglycerides such as sulfoquinovosoyl diacyldiglycerides (SQDGs) (Scheme 4.1). While the neutral glyco-diacyldiglycerides have not been the topics of fundamental studies by tandem mass spectrometry of electrospray generated ions, many of the mechanisms described in this chapter for the $[M + NH_4]^+$ and $[M + Na]^+$ adducts of triglycerides appear to be operating after collisional activation of these adduct ions of the MGDG and DGDG class.[22,23]

4.4.1   Monogalactosyl Diglycerides and Digalactosyl Diglycerides

Positive ions derived from monogalactosyl diglycerides can be analyzed as either the ammonium adduct or alkali metal adduct. Collisional activation of the $[M + NH_4]^+$ results in product ions that correspond to $[M + H\text{-}162]^+$, which are the loss of neutral ammonia (NH_3), and galactosyl moiety. This product ion is identical to the diglyceride-like ions described for the TAG MS/MS spectra (Scheme 4.3), and is observed at m/z 585 from MGDG(18 : 3/16 : 3) in Figure 4.6A. A second abundant product ion corresponds to $[M + H\text{-}180]^+$ as the loss of NH_3 and cleavage of the sugar hemiacetal oxygen bond with proton transfer to render a diacylglycerol structure seen at m/z 567 (Figure 4.6A). There are also product ions corresponding to each fatty acyl group as an acylium ion plus 74 (RCO + 74) seen at m/z 307 and 335.

For the ammonium ion adduct of digalactosyl diglycerides (Figure 4.7A), these same losses of the carbohydrate portion are observed as the loss of 341 and 359 Da.[22] Collisional activation of these DAG-like ions result in formation of acylium product ions [RCO] and [RCO + 74] that can be used to identify the fatty acyl substituents.[23]

The major product ions following the collisional activation of $[M + Na]^+$ from MGDG (Figure 4.6B) and DGDG (Figure 4.7B) correspond to the loss of each fatty acyl group as a free carboxylic acid neutral forming the ion $[M + Na\text{-}R_xCOOH]^+$ by the mechanism previously described (Scheme 4.6), however, no studies have been made to prove that a proton from the 2′-fatty acyl position participates in this rearrangement reaction.[24,25] Another mechanism could involve a proton on a glycerol carbon atom in a favorable charge remote ene-like reaction illustrated for a DGDG species (Scheme 4.14).

One of the interesting product ions from CID of the $[M + Na]^+$ is observed at m/z 243 from MGDG and m/z 405 from DGDG[25] (Scheme 4.15). These ions correspond to the elemental composition of the ion structure indicated $(C_9H_{16}O_6Na)$ as measured by high resolution mass spectrometry.[26] This ion at m/z 243 has been used as a specific diagnostic ion for both MGDG and DGDG since higher collision energies have been found to decompose m/z 405 into the ion m/z 243 as a result of the loss of the terminal galactosyl group (Scheme 4.15).[26]

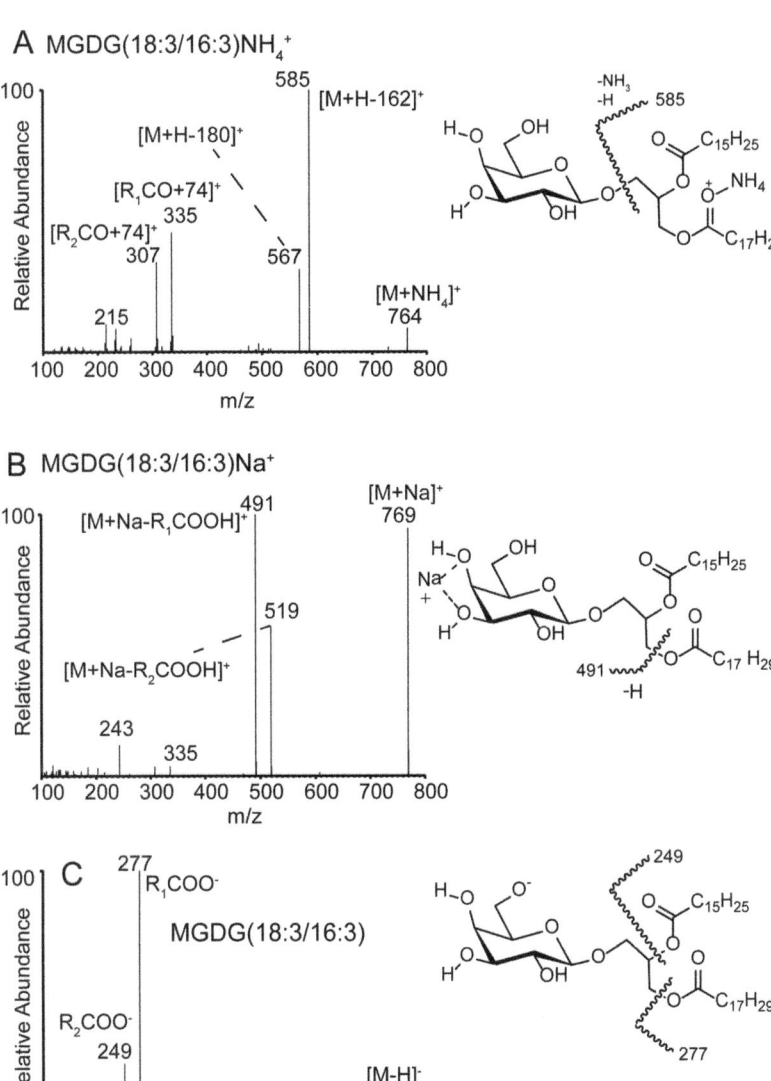

Figure 4.6 Electrospray ionization (positive ions and negative ions) and tandem mass spectrometry from the monogalactosyl diglyceride (MGDG) molecular species 1-octadecatrienoyl-2-hexadecatrienoyl-3-O-galactosylglycerol. (A) Product ions obtained following collisional activation of the MGDG(18 : 3/16 : 3) $[M + NH_4]^+$ at m/z 764; (B) product ions obtained following collisional activation of MGDG(18 : 3/16 : 3) $[M + Na]^+$ at m/z 759; (C) product ions obtained following collisional activation of MGDG(18 : 3/16 : 3) $[M - H]^-$ at m/z 745. These MS/MS spectra were obtained using a tandem quadrupole mass spectrometer.

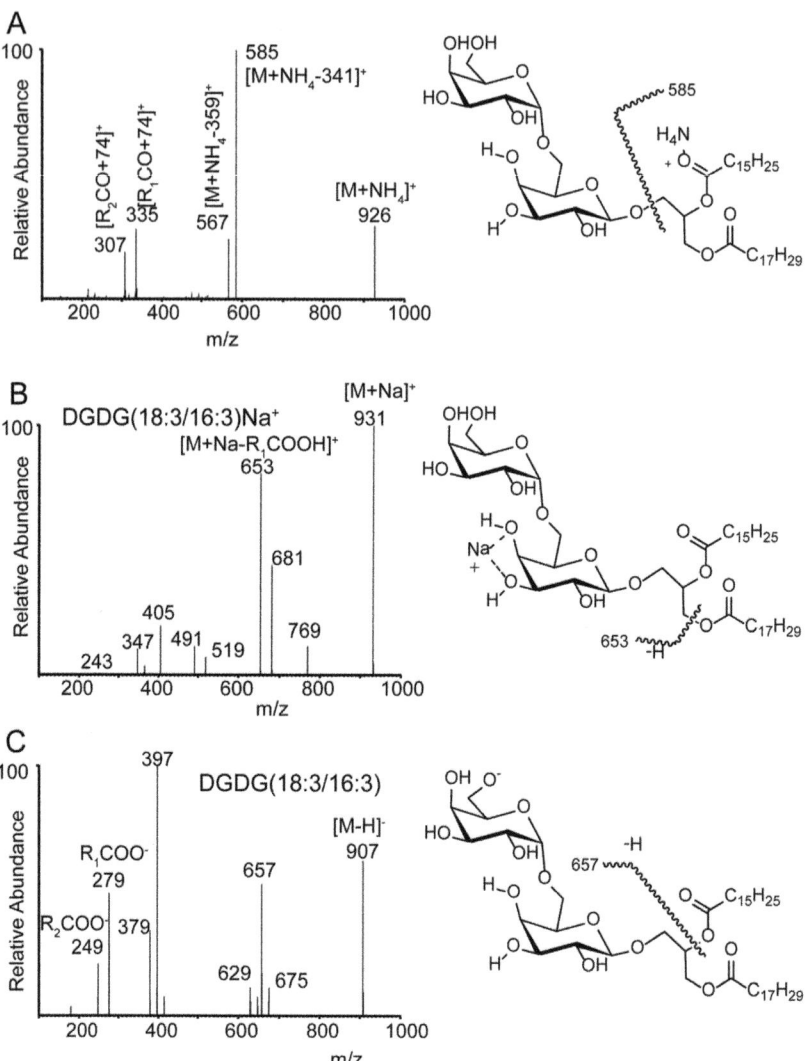

Figure 4.7 Electrospray ionization (positive ions and negative ions) and tandem mass spectrometry from the digalactosyl diglyceride (DGDG) molecular species 1-octadecatrienoyl-2-hexadecatrienoyl-3-O-(α-galactosyl-6)-β-galactosyl-glycerol. (A) Product ions obtained following collisional activation of the DGDG(18 : 3/16 : 3) [M + NH$_4$]$^+$ at *m/z* 926; (B) product ions obtained following collisional activation of DGDG(18 : 3/16 : 3) [M + Na]$^+$ at *m/z* 931; (C) product ions obtained following collisional activation of DGDG(18 : 3/16 : 3) [M − H]$^−$ at *m/z* 907. These MS/MS spectra were obtained using a tandem quadrupole mass spectrometer.

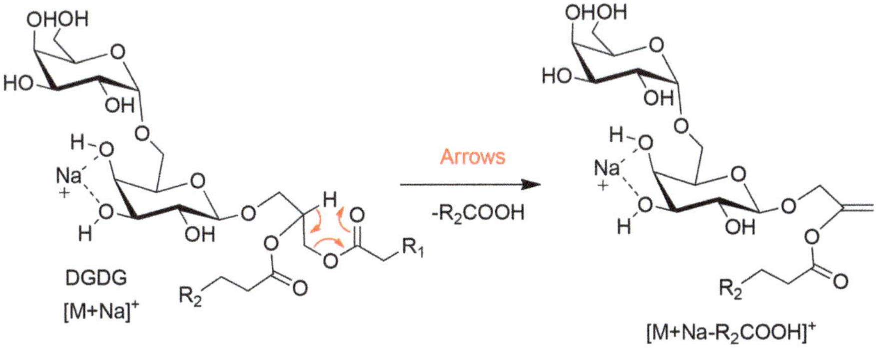

Scheme 4.14

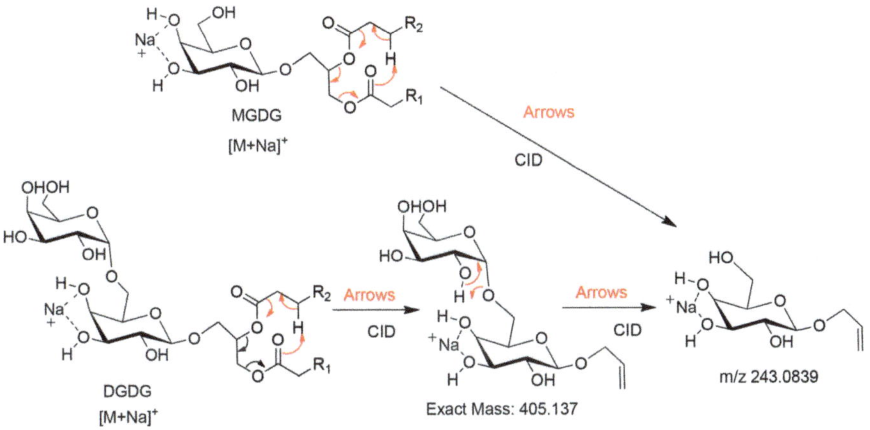

Scheme 4.15

The ratio of the abundance of these ions from the *sn-1* position relative to the *sn-2* position suggests a preference for the loss of the *sn-1* as a carboxylic acid. This observation was confirmed using lipase XI from the fungus *Rhizopus arrhizus,* which is known to hydrolyze *sn-1* esters preferentially. The ratio of these product ion abundances (*sn-1/sn-2* loss) has been suggested to assign esterification positions of the fatty acyl groups of MGDG and DGDG from collisional activation of $[M + Na]^+$.[23]

Analysis of digalactosyl and monogalactosyl diglycerides by negative electrospray ionization is also possible due to the ability of negative ions to form with alkoxide anions on carbohydrate moiety, leading to $[M - H]^-$.[27] This is most likely a summation of the tendency of each seven (or four hydroxyl) groups of digalactosyl (or monogalactosyl) moieties to become an oxygen-centered anion. Examples of negative ionization of MGDG and

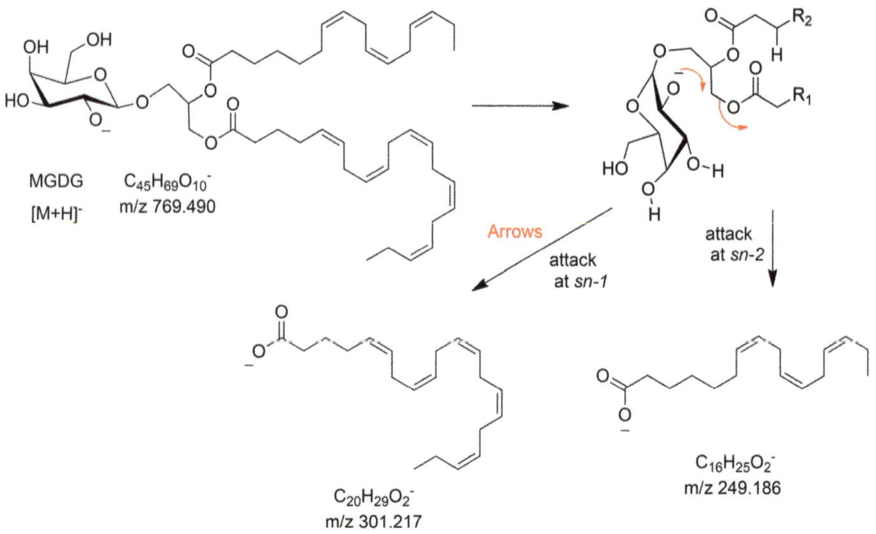

Scheme 4.16

DGDG lipids followed by tandem mass spectrometry are presented in Figures 4.6C and 4.7C. Collisional activation of these $[M - H]^-$ species yielded abundant carboxylate anions from the diglyceride portion of the molecule (Scheme 4.16).

The loss of the fatty acyl groups as a neutral ketene is a common gas phase reaction for these neutral lipids. The loss of the carboxylate anions is most likely a charge-driven process (Scheme 4.16), while the loss of ketene would be a result of charge remote process seen even with positive ions in those situations where the charge site is strongly localized and this is certainly the case with the negative ions (Scheme 4.17) illustrated for the molecular species DGDG(18 : 3/16 : 3) (Figure 4.7C). In either case the signal corresponding to the *sn-1* carboxylate anion was more abundant than the *sn-2* carboxylate-derived anion for a number of MGDG and DGDG molecular species which had the fatty acyl positions confirmed by enzymatic hydrolysis by *Rhizopus arrhizus*.[23] This is also consistent with MGDG and DGDG molecular species presented in Figures 4.6C and 4.7C.

4.4.2 Sulfoquinovosyl Diglycerides

Major acidic glycodiacyldiglycerides are the sulfoquinovosyl diglycerides (SQDG) found in the chloroplast of plants and algae. As expected for such sulfonic acid containing compounds during electrospray ionization, these lipids yield quite abundant negative ions due to the strongly acidic character of the sulfonic acid. The tandem mass spectrometry of SQDGs has been carefully studied using isotope labeling and high resolution analysis

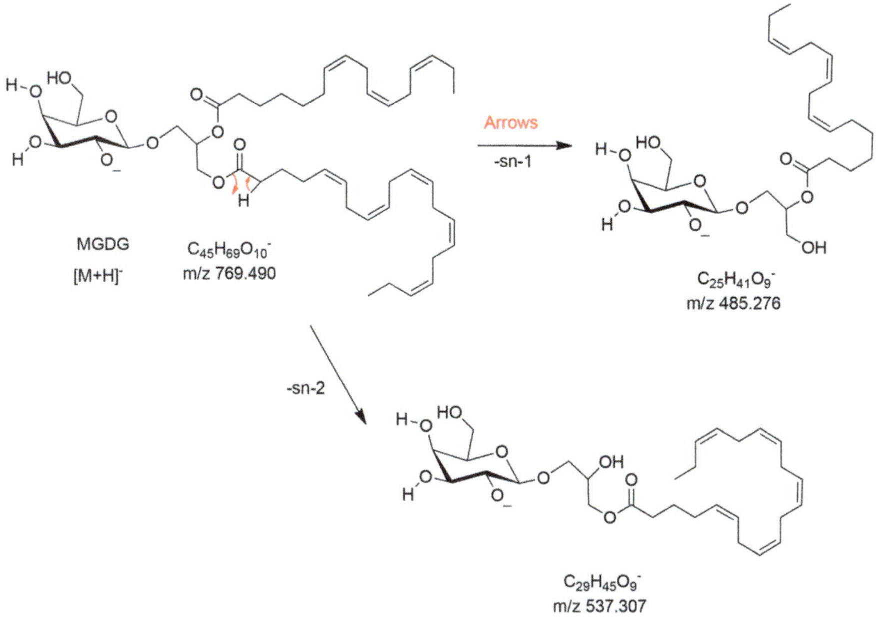

Scheme 4.17

(Figure 4.8).[27] Unlike the negative ions from DGDG and MGDG, the formation of product carboxylate anions is not a major process following CID of [M − H]⁻ and it has been suggested that carboxylate anions are only formed either after secondary collisions following a neutral loss or require higher collision energies.[27] With the negative charge localized at carbon-6 of the galactose residue, there may be hindered access to the *sn-1* and *sn-2* carbon atoms which would be required for such a charge-driven reaction.

The neutral loss of each fatty acyl group as a ketene or free carboxylic acid (-R$_x$COOH) are typically observed as reasonably abundant ions, yet there have not been sufficient synthetic SQDGs of known regioisomeric structure and specific fatty acyl groups to determine if the ratio of these neutral loss ions reveal *sn-1/sn-2* fatty acyl positions. The mechanism of formation of these neutral losses is likely identical to that found by the corresponding losses from the alkali metal adducted TAGs. However, regiochemical assignment has been suggested to be possible.[28]

The predominant and distinguishing negative ion derived from SQDG is observed at *m/z* 225. The abundance of this ion relative to a very low mass ion observed at *m/z* 81 (SO$_3$⁻) is likely related to CID energy and instrument-related conditions. Nonetheless, *m/z* 225 is typically the most abundant product ion in a tandem quadrupole mass spectrometer and has been used as a diagnostic ion to detect SQDQ in complex biological samples.[26] The

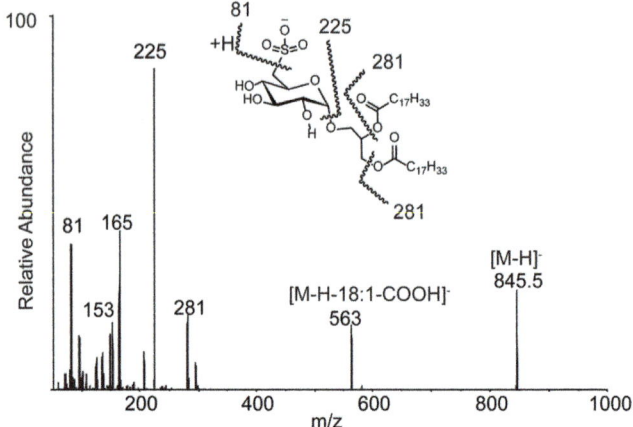

Figure 4.8 Electrospray ionization (negative ions) and tandem mass spectrometry of a sulfoquinovosyl diacylglycerol as the molecular anion. Product ions obtained following collisional activation of $[M - H]^-$ at m/z 845. This MS/MS spectrum was obtained using a tandem quadrupole mass spectrometer.

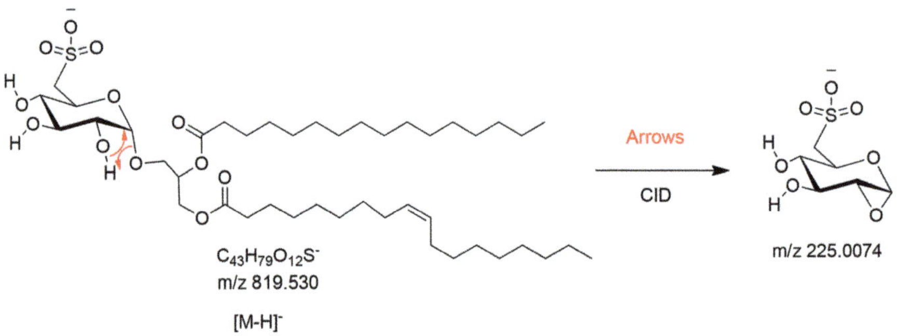

Scheme 4.18

formation of this ion has been suggested to be a charge remote loss of the lipid portion of the molecule with concomitant formation of a sulfogalactosyl epoxide (Scheme 4.18). This ion structure has been supported by high resolution and deuterium isotope labeling studies and likely is formed directly from $[M - H]^-$ or could be formed as a secondary product ion after neutral loss of either fatty acyl or carboxylic acid as suggested by MS^3 studies.[27,29]

References

1. V. A. Zammit, Hepatic triacylglycerol synthesis and secretion: DGAT2 as the link between glycaemia and triglyceridaemia, *Biochem. J.*, 2013, **451**, 1–12.
2. K. Nagy, L. Sandoz, F. Destaillats and O. Schafer, Mapping the regioisomeric distribution of fatty acids in triacylglycerols by hybrid mass spectrometry, *J. Lipid Res.*, 2013, **54**, 290–305.
3. X. Li, E. J. Collins and J. J. Evans, Examining the collision-induced decomposition spectra of ammoniated triglycerides as a function of fatty acid chain length and degree of unsaturation. II. The PXP/YPY series, *Rapid Commun. Mass Spectrom.*, 2006, **20**, 171–177.
4. R. Gakwaya, X. Li, Y. L. Wong, S. Chivukula, E. J. Collins and J. J. Evans, Examining the collision-induced decomposition spectra of ammoniated triglycerides. III. The linoleate and arachidonate series, *Rapid Commun. Mass Spectrom.*, 2007, **21**, 3262–3268.
5. A. M. McAnoy, C. C. Wu and R. C. Murphy, Direct qualitative analysis of triacylglycerols by electrospray mass spectrometry using a linear ion trap, *J. Am. Soc. Mass Spectrom.*, 2005, **16**, 1498–1509.
6. L. A. Marzilli, L. B. Fay, F. Dionisi and P. Vouros, Structural characterization of triacylglycerols using electrospray ionization-MS^n ion-trap MS, *J. Am. Oil Chem. Soc.*, 2003, **80**, 195–202.
7. E. Hvattum, Analysis of triacylglycerols with non-aqueous reversed-phase liquid chromatography and positive ion electrospray tandem mass spectrometry, *Rapid Commun. Mass Spectrom.*, 2001, **15**, 187–190.
8. X. Han and R. W. Gross, Quantitative analysis and molecular species fingerprinting of triacylglyceride molecular species directly from lipid extracts of biological samples by electrospray ionization tandem mass spectrometry, *Anal. Biochem.*, 2001, **295**, 88–100.
9. F. F. Hsu and J. Turk, Structural characterization of triacylglycerols as lithiated adducts by electrospray ionization mass spectrometry using low-energy collisionally activated dissociation on a triple stage quadrupole instrument, *J. Am. Soc. Mass Spectrom.*, 1999, **10**, 587–599.
10. L. C. Herrera, M. A. Potvin and J. E. Melanson, Quantitative analysis of positional isomers of triacylglycerols via electrospray ionization tandem mass spectrometry of sodiated adducts, *Rapid Commun. Mass Spectrom.*, 2010, **24**, 2745–2752.
11. F. F. Hsu and J. Turk, Electrospray ionization multiple-stage linear ion-trap mass spectrometry for structural elucidation of triacylglycerols: assignment of fatty acyl groups on the glycerol backbone and location of double bonds, *J. Am. Soc. Mass Spectrom.*, 2010, **21**, 657–669.
12. J. A. Bowden, C. J. Albert, O. S. Barnaby and D. A. Ford, Analysis of cholesteryl esters and diacylglycerols using lithiated adducts and electrospray ionization-tandem mass spectrometry, *Anal. Biochem.*, 2011, **417**, 202–210.

13. W. R. Bishop and R. M. Bell, Functions of diacylglycerol in glycerolipid metabolism, signal transduction and cellular transformation, *Oncog. Res.,* 1988, **2**, 205–218.

14. M. J. Caloca, M. L. Garcia-Bermejo, P. M. Blumberg, N. E. Lewin, E. Kremmer, H. Mischak, S. Wang, K. Nacro, B. Bienfait, V. E. Marquez and M. G. Kazanietz, Beta2-chimaerin is a novel target for diacylglycerol: binding properties and changes in subcellular localization mediated by ligand binding to its C1 domain, *Proc. Natl. Acad. Sci. U. S. A.,* 1999, **96**, 11854–11859.

15. R. C. Murphy, P. F. James, A. M. McAnoy, J. Krank, E. Duchoslav and R. M. Barkley, Detection of the abundance of diacylglycerol and triacylglycerol molecular species in cells using neutral loss mass spectrometry, *Anal. Biochem.,* 2007, **366**, 59–70.

16. H. L. Callender, J. S. Forrester, P. Ivanova, A. Preininger, S. Milne and H. A. Brown, Quantification of diacylglycerol species from cellular extracts by electrospray ionization mass spectrometry using a linear regression algorithm, *Anal. Chem.,* 2007, **79**, 263–272.

17. B. M. Ham and R. B. Cole, Determination of apparent decomposition threshold energies of lithium adducts of acylglycerols using tandem mass spectrometry and a novel derived effective reaction path length approach, *J. Mass Spectrom.,* 2008, **43**, 1482–1493.

18. Y. L. Li, X. Su, P. D. Stahl and M. L. Gross, Quantification of diacylglycerol molecular species in biological samples by electrospray ionization mass spectrometry after one-step derivatization, *Anal. Chem.,* 2007, **79**, 1569–1574.

19. T. J. Leiker, R. M. Barkley and R. C. Murphy, Analysis of Diacylglycerol Molecular Species in Cellular Lipid Extracts by Normal-Phase LC-Electrospray Mass Spectrometry, *Int. J. Mass Spectrom.,* 2011, **305**, 103–109.

20. M. Y. Zhang, Y. Gao, J. Btesh, N. Kagan, E. Kerns, T. A. Samad and P. K. Chanda, Simultaneous determination of 2-arachidonoylglycerol, 1-arachidonoylglycerol and arachidonic acid in mouse brain tissue using liquid chromatography/tandem mass spectrometry, *J. Mass Spectrom.,* 2010, **45**, 167–177.

21. N. Mizusawa and H. Wada, The role of lipids in photosystem II, *Biochim. Biophys. Acta,* 2012, **1817**, 194–208.

22. R. A. Moreau, D. C. Doehlert, R. Welti, G. Isaac, M. Roth, P. Tamura and A. Nunez, The identification of mono-, di-, tri-, and tetragalactosyl-diacylglycerols and their natural estolides in oat kernels, *Lipids,* 2008, **43**, 533–548.

23. G. Guella, R. Frassanito and I. Mancini, A new solution for an old problem: the regiochemical distribution of the acyl chains in galactolipids can be established by electrospray ionization tandem mass spectrometry, *Rapid Commun. Mass Spectrom.,* 2003, **17**, 1982–1994.

24. A. Napolitano, V. Carbone, P. Saggese, K. Takagaki and C. Pizza, Novel galactolipids from the leaves of *Ipomoea batatas L.*: Characterization by liquid chromatography coupled with electrospray ionization-quadrupole time-of-flight tandem mass spectrometry, *J. Agric. Food Chem.*, 2007, **55**, 10289–10297.

25. Y. H. Kim, J. S. Choi, J. S. Yoo, Y. M. Park and M. S. Kim, Structural identification of glycerolipid molecular species isolated from cyanobacterium Synechocystis sp. PCC 6803 using fast atom bombardment tandem mass spectrometry, *Anal. Biochem.*, 1999, **267**, 260–270.

26. R. Welti, X. Wang and T. D. Williams, Electrospray ionization tandem mass spectrometry scan modes for plant chloroplast lipids, *Anal. Biochem.*, 2003, **314**, 149–152.

27. X. Zhang, C. J. Fhaner, S. M. Ferguson-Miller and G. E. Reid, Evaluation of ion activation strategies and mechanisms for the gas-phase fragmentation of sulfoquinovosyldiacylglycerol lipids from Rhodobacter sphaeroides, *Int. J. Mass Spectrom.*, 2012, **316–318**, 100–107.

28. R. Zianni, G. Bianco, F. Lelario, I. Losito, F. Palmisano and T. R. Cataldi, Fatty acid neutral losses observed in tandem mass spectrometry with collision-induced dissociation allows regiochemical assignment of sulfoquinovosyl-diacylglycerols, *J. Mass Spectrom.*, 2013, **48**, 205–215.

29. M. Keusgen, J. M. Curtis, P. Thibault, J. A. Walter, A. Windust and S. W. Ayer, Sulfoquinovosyl diacylglycerols from the alga Heterosigma carterae, *Lipids,* 1997, **32**, 1101–1112.

CHAPTER 5

Glycerophospholipids

Glycerophospholipids are phosphodiesters of glycerol and an alcohol referred to as the polar headgroup. The two glycerol carbinol groups are typically esterified with long chain fatty acyl groups, which impart hydrophobic character to these otherwise polar molecules. The simplest glycerophospholipid is phosphatidic acid, which is a monophosphoester of diacylglycerol. The common polar headgroups are choline, ethanolamine, serine, inositol, and glycerol, which divide the phospholipids into different classes. Cardiolipins are the largest glycerophospholipids and correspond to condensation of two phosphatidylglycerols. Since glycerol esters can be chiral molecules, a designation of each of the specific carbon atoms of this 3-carbon sugar approved by IUPAC, is employed largely due to specific enzymatic reactions that take place at each carbon of glycerol during biosynthesis and metabolism. This IUPAC system uses the stereospecific numbering system (sn).[1] The glycerol carbon atoms containing the fatty acyl groups are typically referred to as sn-1 and sn-2, where the sn-2 position is chiral. By this convention the sn-3 position is where the phosphate residue is esterified (Scheme 5.1).

Another convenient nomenclature is to refer to acyl groups at the sn-1 carbon of glycerol as R_1 and the sn-2 carbon acyl group as R_2. Often the carboxylate (ester) is implied with this nomenclature. This book is rather explicit and uses R_1 and R_2 to refer only to alkyl groups. These structural features of the glycerophospholipids make these molecules amphipathic substances. This unique property is undoubtedly the reason for the importance of these lipids in cellular biochemistry and the fact that they are self-assembling molecules that make up the outer and inner lipid bilayers of living cells and bacteria.

New Developments in Mass Spectrometry No. 4
Tandem Mass Spectrometry of Lipids: Molecular Analysis of Complex Lipids
By Robert C Murphy
© Robert C Murphy 2015
Published by the Royal Society of Chemistry, www.rsc.org

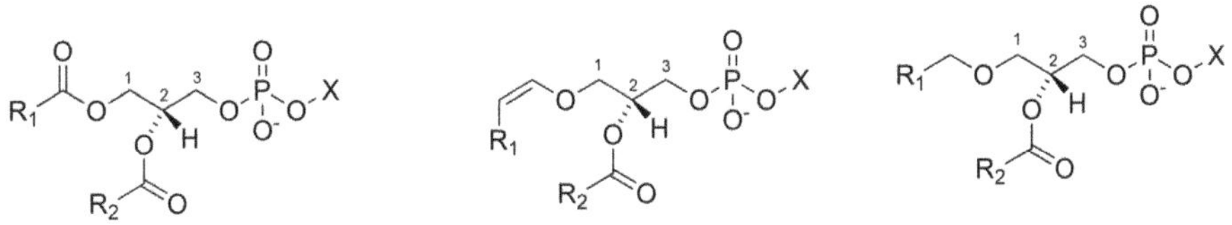

| | 1,2-diacyl | plasmenyl (plasmalogen) Abbreviation-p | plasmanyl (ether) Abbreviation-e |

X= class
choline $CH_2CH_2N(CH_3)_3$
ethanolamine $CH_2CH_2NH_2$
inositol $C_6H_{11}O_5$
serine $CHCH(NH_2)COOH$
glycerol $CH_2CHOHCH_2OH$
phosphatidic acid   H

Scheme 5.1

Another structural complexity occurs in some biological systems where the substituent at the *sn*-1 carbon is not an ester, but an alkyl ether or a vinyl ether. These unique phospholipids are termed plasmanyl (ether) and plasmenyl (plasmalogen) phospholipids (Scheme 5.1), and are found most abundantly in glycerophosphocholine and glycerophosphoethanolamine lipid classes. The glycerophospholipids from the archaea kingdom are ethers at both *sn*-1 and *sn*-2.[2] This complexity in phospholipid structure introduces another variant in nomenclature to describe such ether lipids and introduction of the term "radyl" which refers to hydrophobic group attached to either *sn*-1 or *sn*-2 glycerol positions, but does not designate if they are either esters or ether bonded moieties. The commonly used shorthand term "phosphatidyl" has been reserved specifically (IUPAC rules) for the 1,2-diacylphospholipids as a convenience to shorten an otherwise very long term "diacylglycerol phospholipid".[1] However, phosphatidyl should not be used for referencing plasmalogen or ether glycerophospholipids.

Electrospray and MALDI ionization readily form both positive and negative molecular ion species $[M + H]^+$ and $[M - H]^-$ for most glycerophospholipids, but the abundance of these ions with each polarity is determined by the characteristics of the polar headgroup. Phosphatidylcholine (PC) has a quaternary trimethylammonium cation moiety and positive $[M + H]^+$ ions predominate in electrospray ionization for PC, where the addition of the proton neutralizes the negative phosphate anion. The phospholipid classes generally form abundant $[M - H]^-$ when a proton is not present on the phosphate hydroxyl group, yielding the phosphate anion (the case for PC will be discussed below). In addition to these molecular ion species, alkali metals such as K^+, Na^+, or Li^+ can form adducts that yield $[M + K]^+$, $[M + Na]^+$, or $[M + Li]^+$ positive ion species. The first two alkali metal adducts are observed when phospholipid extracts from biological systems are analyzed and no effort is made to control alkali metal content.

HPLC separation can remove alkali metals from crude lipid extracts by mass action when the chromatographic solvent systems include cationic modifiers such as ammonium salts. Normal phase LC can separate phospholipid classes by polarity and the observed molecular ion species used to determine total fatty acyl carbon atoms and total number of double bonds (or cyclic structures) when the elements that make up glycerophosphate and the polar headgroup are assumed.[3] Accurate mass measurement by advanced high resolution mass spectrometry removes many assumptions as to the total fatty acyl groups and polar headgroup. This is a very helpful feature when dealing with modified phospholipids such as oxidized phospholipids. However, branch chained radyl groups, double bond geometry, and double bond position are not easily gleaned from mass spectrometric data.

Structural details of phospholipids can be revealed by collisional activation of either the $[M + H]^+$, $[M + Li]^+$, $[M + Na]^+$ or $[M - H]^-$ ions. The decomposition of the positive ion species reveals information typically about the polar headgroup since each of the phospholipid classes specifically decompose to unique product ions. Analysis of product ions generated by collisional activation of $[M - H]^-$ yields information about the fatty acyl and other radyl groups. The detailed mechanisms involved in the product ion formation from phospholipids were first investigated when ions were generated by fast atom bombardment ionization.[4] The previous book, *Mass Spectrometry of Lipids*,[5] described much of that ion chemistry but substantial additional information has been added with electrospray generation of phospholipid ions for collisional activation. The general behavior of the major phospholipid classes will be discussed here using electrospray ionization formation of $[M + H]^+$, $[M + Li]^+$ or $[M - H]^-$ at low collisional activation to yield product ions found in the triple quadrupole or trap-based instruments.

There are several common pathways of glycerophospholipid molecular ion decomposition following collisional activation of either positive or negative molecular ion species. These pathways are related to the universal structural features of glycerophosphate esters and the pathways often lead to the most abundant product ions in the tandem mass spectrum. The mechanisms of formation of product ion formation will be presented in generic form and then referenced when each class of glycerophospholipid is examined in detail.

5.1   Charge Remote Diglyceride Fragmentation

The cleavage of the *sn-3* carbon–oxygen bond (phosphoester bond) simultaneous with transfer of a proton to the polar head group is a major ion process seen following collisional activation of $[M + H]^+$ at lower energies and, in some cases, seen as a product of negative ion decomposition. This is the origin of the dominant ion *m/z* 184 from PC lipids and the "diglyceride-like" ion from acidic phospholipids such as PG, PS, PA, PI, and even the aminophospholipid PE.

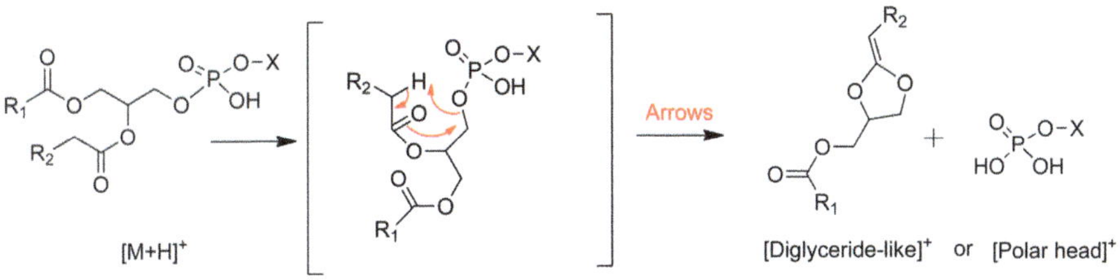

Scheme 5.2

While the mechanism of formation of this ion-type was initially thought to involve transfer of a glycerol-backbone hydrogen atom, the observation that D_3-PAF (where the deuterium atoms were on the *sn-2* acetate moiety) formed only *m/z* 185 suggested involvement of the protons from the *sn-2* acyl group.[6] Subsequent studies with phospholipids with fully deuterated fatty acyl groups at *sn-2* revealed the involvement of the 2'-hydrogen atoms in this transfer. The mechanism for this remote cleavage mechanism is presented in Scheme 5.2. There is evidence that a similar mechanism involving the *sn-1* 2'-protons also operates to a lesser extent in the formation of this ion.[6] The site at which the ionization resides in the intact phospholipid determines which fragment that is the observed ion and which is not detected as the neutral product. For [M + H]$^+$ from PC and [M – H]$^-$ in PI(3)P, the charge is strongly localized to the polar head group and the polar head group ions are observed. For phospholipids that are protonated or sodiated at a fatty acyl ester, the ion is observed as the diglyceride-like ion.

5.2 Charge Driven Carboxylate Anion Formation (Negative Ions Only)

Prominent ions formed following collisional activation of [M – H]$^-$ from phospholipids (or [M – 15]$^-$ from PC) are derived from the ester groups as carboxylate anions. These are most likely product ions of direct attack of the phosphate anion, in a 5- or 6-membered transition state, to displace carboxylate anions (Scheme 5.3). In many cases, but not all, the most abundant anions observed are derived from the *sn-2* position; however, strict interpretation of this rule is compromised by experimental conditions.[7] The important exception is PA and phospholipids that form PA-like ions.[8]

5.3 Charge Driven Loss of Neutral Ketene $(R_xCH=C=O)$

In a process similar to that seen for the positive ions, the phosphate anionic site can attack a 2'-fatty acyl proton on either the *sn-1* or *sn-2* ester moiety, to drive loss of a corresponding neutral ketene. Likely the charge site is initially

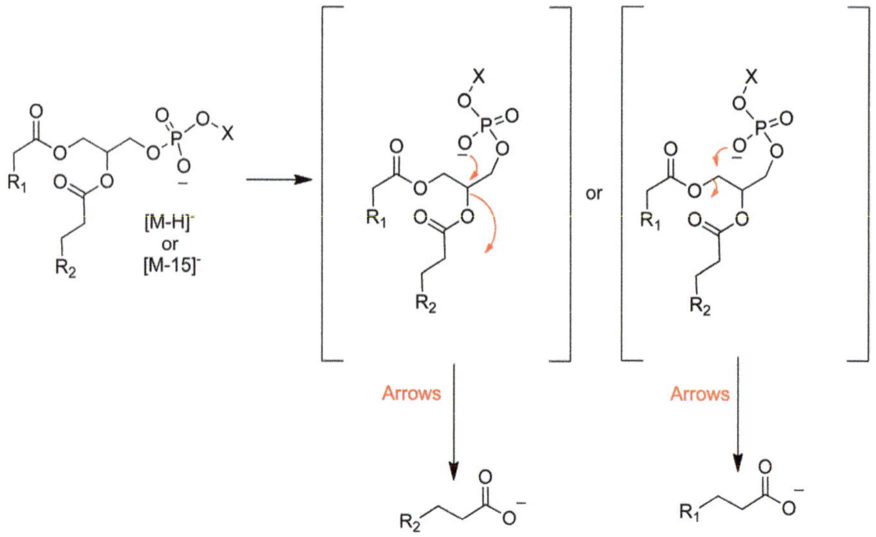

Scheme 5.3

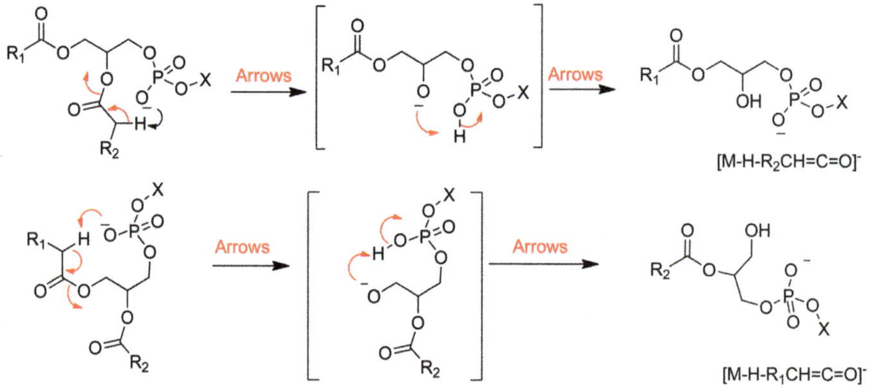

Scheme 5.4

transferred to a glycerol alkoxide ion that is sufficiently energetic that it abstracts a proton from the neutral phosphate to form a more stable anion as shown in Scheme 5.4.

5.4   Charge Remote Loss of R_xCOOH (Negative Ions)

Both positive and negative ions can lose R_xCOOH, however, this is not a particularly abundant product ion from positive ion decomposition. For negative ions, this process has been shown, using deuterium labeled

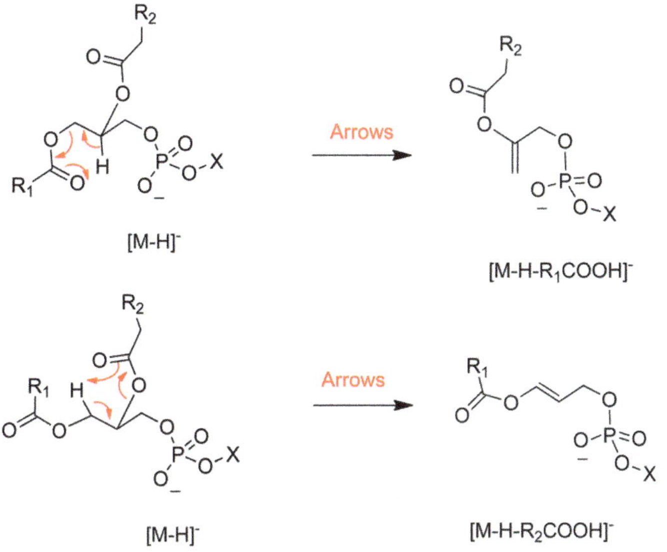

Scheme 5.5

phospholipids, to proceed by way of two distinct routes.[9] One involves removing a glycerol backbone proton (Scheme 5.5). However, as will be illustrated for certain phospholipid classes and positive ions, a loss of R_xCOOH does not involve the glycerol protons at all, but rather a proton from the phosphate hydroxyl group of PA or from an exchangeable proton on the polar head group of other phospholipids.

5.5 Glycerophosphocholine Lipids (PC)

5.5.1 Positive Ion $[M + H]^+$ PC Adducts

Phosphatidylcholine is a unique phospholipid in that the polar headgroup has a permanently charged quaternary ammonium group that substantially enhances electrospray ionization to form abundant $[M + H]^+$. The tandem mass spectrometry of this $[M + H]^+$ (Figure 5.1A) were first studied decades ago when these ions were observed by fast atom bombardment ionization and that an abundant m/z 184 was derived from all PC species.[4] The mechanism involved in formation of this ion was thought to engage the glycerol backbone protons, but studies with stable isotope labeled platelet activating factor (1-hexadecyl-2-trideuteroacetyl-PC) clearly revealed involvement of protons at the *sn*-2 side chain. When these protons were labeled with deuterium atom, the deuterated PC yielded an abundant product ion at m/z 185.[6] This finding has been confirmed by electrospray based tandem mass spectrometry studies using dipalmitoyl-PC when all hydrogen atoms on the *sn*-1 and *sn*-2 fatty acyl carbon atoms were deuterated, which was found to yield only m/z 185.[10]

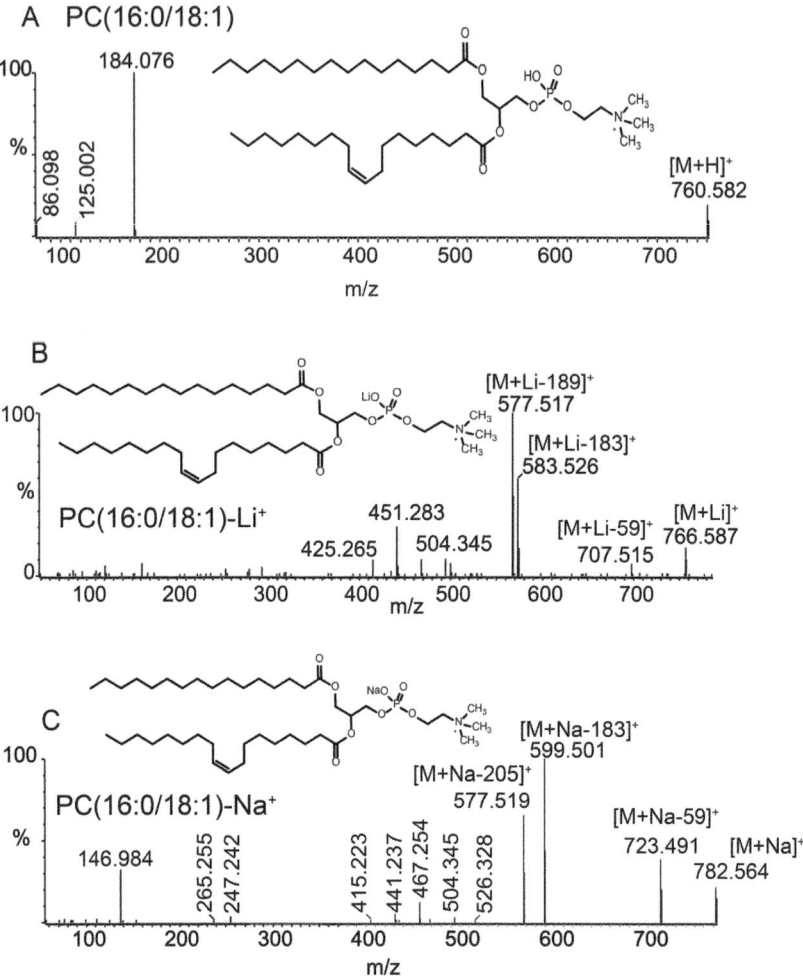

Figure 5.1   Electrospray ionization (positive ions) and tandem mass spectrometry of 1-palmitoyl-2-oleoyl glycerophosphocholine. (A) Product ions following collisional activation of PC(16 : 0/18 : 1) [M + H]⁺ at *m/z* 760; (B) product ions obtained following collisional activation of PC(16 : 0/18 : 1) [M + Li]⁺ at *m/z* 766; (C) product ions following collisional activation of PC(16 : 0/18 : 1) [M + Na]⁺ at *m/z* 782. These high resolution MS/MS spectra were obtained using a quadrupole time-of-flight mass spectrometer.

The mechanism of formation of *m/z* 184 (seen as a general mechanism in Scheme 5.2) clearly must take advantage of the somewhat acidic 2′ protons found on both the *sn*-1 and *sn*-2 fatty acyl groups. When the *sn*-1 fatty acyl group was perdeuterated, but the *sn*-2 fatty acid contained only protons; both *m/z* 184 and 185 were observed with an abundance ratio of 2/3.[10] This ratio was reversed to 3/2 when the deuterium labeled acyl group was only at the

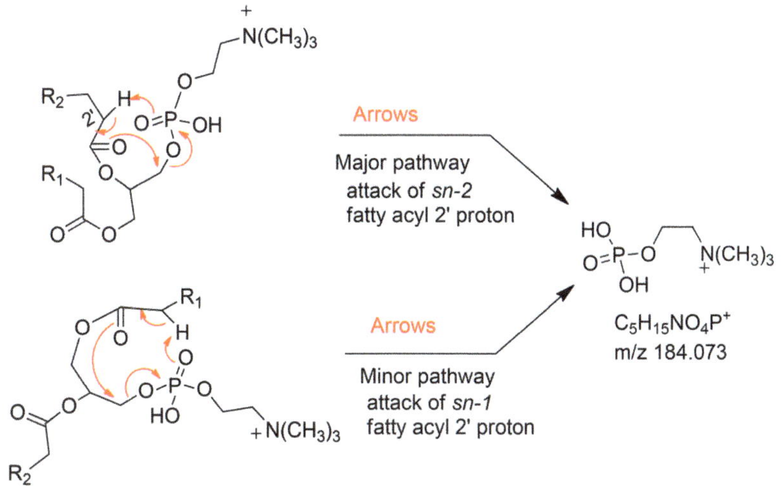

Scheme 5.6

sn-2 position, indicating some preference for this 2′ protons from this acyl group for this rearrangement. The formation of this dominant product ion following activation of the $[M + H]^+$ is likely a charge remote event involving participation of the phosphate oxygen atoms to attack the 2′ proton and a concerted loss of an olefin with cleavage of the *sn*-3 phosphoester bond (Scheme 5.6).[10]

This product ion *m/z* 184 has been used widely as a generic signature of phosphatidylcholine molecular species. Since sphingomyelin (a sphingolipid), also contains phosphocholine, it is not surprising that it also generates an abundant *m/z* 184 product ion after collisional activation of its $[M + H]^+$.

Other product ions of low abundance are observed following collisional induced dissociation of the $[M + H]^+$ from PC molecular species. These product ions also involve charge remote proton abstraction (general mechanism in Scheme 5.3) with breaking and simultaneous making a phosphorus oxygen bond. The loss of either *sn*-1 or *sn*-2 fatty acyl groups as a neutral ketene (R_1-CH=C=O and R_2-CH=C=O) have been observed with a rather more favorable loss of R_2-CH=C=O ketene for most PC molecular species studied, including the most abundant PC species found in common biological systems.[10,11] More advanced mass spectrometers containing multiple collision cells and ion mobility have focused on these ion losses from PC to determine fatty acyl esterification position at *sn*-1 or *sn*-2.[12] The mechanism for the loss of either ketenes has been proposed to involve attack of the 2′ fatty acyl proton by the phosphate carbonyl oxygen atom with cleavage of the ester bond and reformation of a phosphate carbonyl moiety by abstraction of a proton on the phosphate residue by the alkoxide ion intermediate shown for the loss of R_2-CH=C=O in Scheme 5.7.

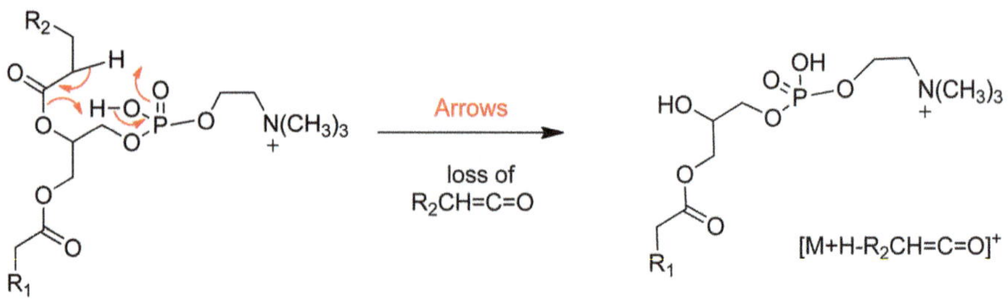

Scheme 5.7

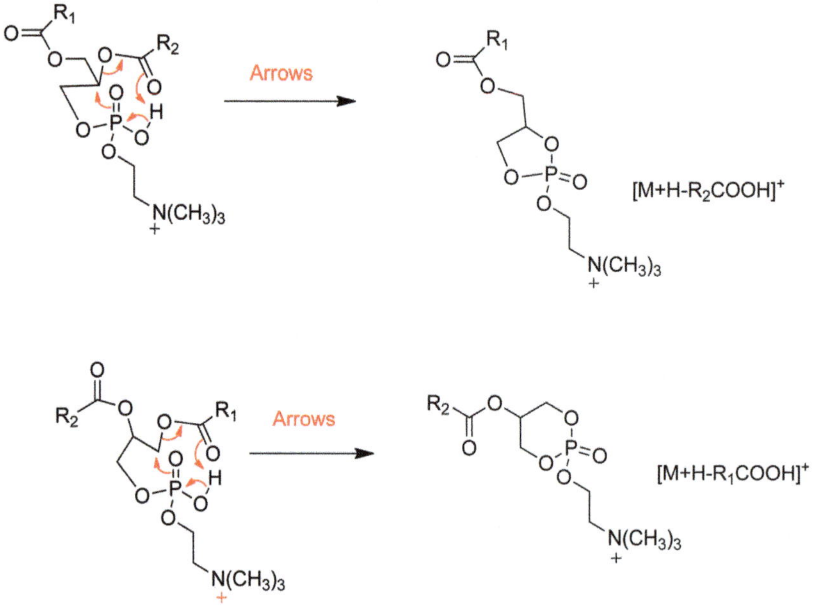

Scheme 5.8

The ion corresponding to loss of either fatty acyl groups as a neutral carboxylic acid is typically even lower in abundance and there does not appear to be a preference for either fatty acyl loss.[10] The mechanism proposed for this loss has been suggested to involve a charge remote formation of a cyclic phosphoester in either a 5- or 6-membered ring, driven by attack of the phosphate carbonyl oxygen atom at either the *sn*-1 or *sn*-2 carbon atoms to form a cyclic phosphate ester, followed by proton abstraction by the ester carbonyl (Scheme 5.8). This mechanism has been supported by isotope labeling studies.[10]

The tandem mass spectrometry of [M + H]$^+$ ions from plasmenyl PCs (plasmalogen PCs) are quite similar to that seen for the 1,2-diacylphosphocholines with dominance of *m/z* 184 and a very low abundance of some fragments ions from the *sn*-1 and *sn*-2 position.[13] The alkyl ether PC $[M + H]^+$ behave identically.[14]

However, the lyso-plasmenyl, plasmanyl, and acyl subclass PCs behave surprisingly differently because of the influence of the *sn*-1 radyl group (Figure 5.2). The 1-fatty acyl-2-lysoPC generates an abundant $[M + H]^+$ and

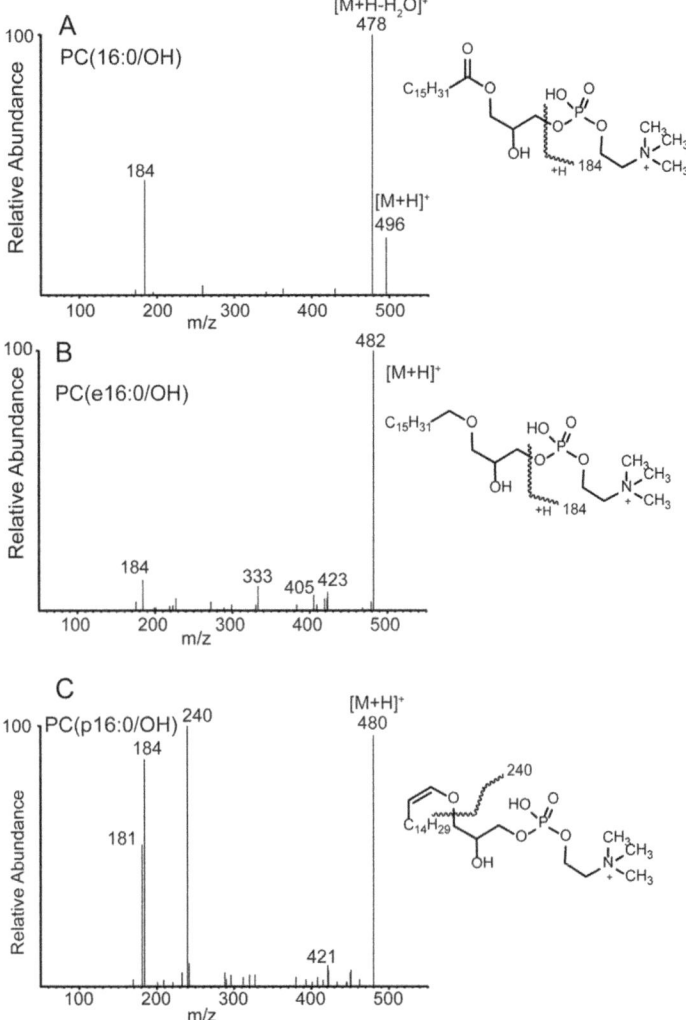

Figure 5.2 Electrospray ionization (positive ions) and tandem mass spectrometry of acyl, ether, and plasmalogen lysophosphatidylcholines. (A) Product ions obtained following collisional activation of PC(16 : 0/OH) $[M + H]^+$ at *m/z* 496; (B) product ions obtained following collisional activation of PC(e16 : 0/OH) $[M + H]^+$ at *m/z* 482; (C) product ions obtained following collisional activation of PC(p16 : 0/OH) $[M + H]^+$ at *m/z* 480. This figure was redrawn from data presented in ref. 15.

product ions at the expected m/z 184 (Figure 5.2A). But the most abundant product ion, corresponds to $[M + H - H_2O]^+$ likely due to protonation of the ester moiety as a charge site (sn-1).[15] The charge-driven, concerted mechanism for loss of H_2O is proposed as Scheme 5.9. With the alkyl ether subclass (Figure 5.2B), the secondary charge site is not present and m/z 184 and the loss of trimethylamine (-59 Da) is observed (Scheme 5.10). The plasmalogen vinyl ether subclass has an even different major product ion, corresponding to the loss of the vinyl ether radyl group at sn-1 as an aldehyde (Figure 5.2C). Most likely this ion results from proton attachment at the sn-1 oxygen vinyl ether moiety and attack of the phosphate anion to form m/z 240 (Scheme 5.11). This

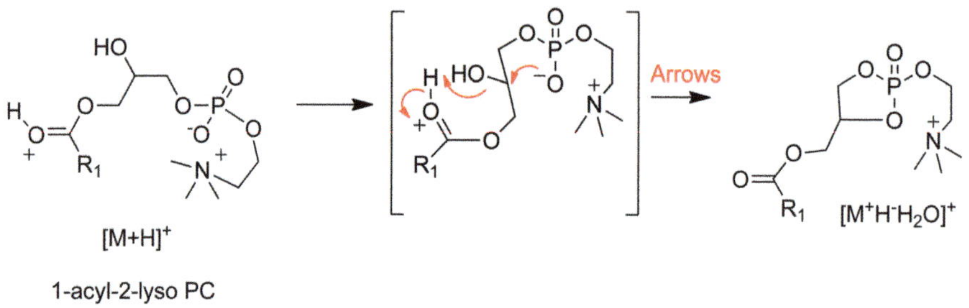

Scheme 5.9

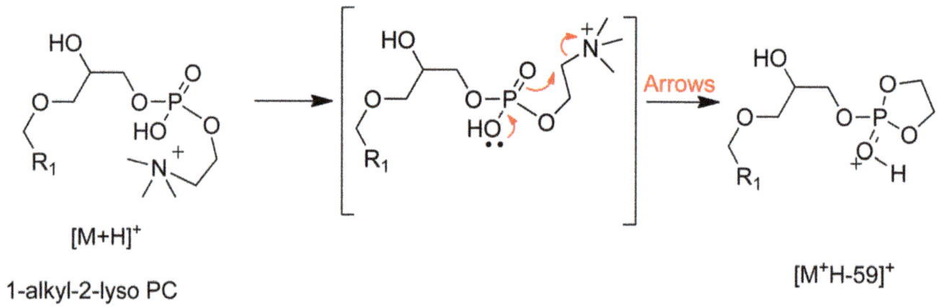

Scheme 5.10

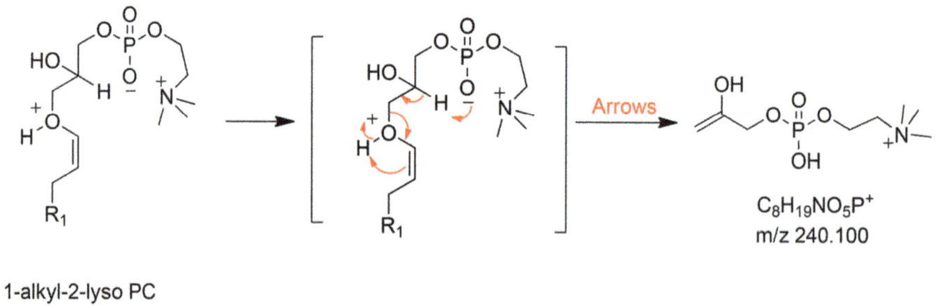

Scheme 5.11

ion can further lose 59 Da to form *m/z* 181. These latter two ions are rather generic product ions for all plasmalogen PCs in the positive ion mode.

5.5.2 Positive Ion Lithiated PC

The formation of an alkali metal ion adduct using salts of Li^+ or Na^+ readily takes place and leads to the formation of $[M + Li]^+$ or $[M + Na]^+$, respectively, by electrospray ionization (Figure 5.1B and C). The position at which the adduct forms for PC undoubtedly determines the pathway from which product ions are formed. One position of alkali metal adduct would be to neutralize the phosphate anion, resulting in a single positive charge remaining on the choline residue. Alternatively, the metal cation could coordinate with the partially negative character of the carbonyl moiety on either of the two ester groups or even the phosphoryl carbonyl. In these cases, the charge site is at the site of metal ion attachment, leaving the phosphate anion (negatively charged) to neutralize the quaternary trimethylammonium cation (Scheme 5.12). It is also possible to add two metal adducts $[M - H + 2Me]^+$ if the phosphate group is not protonated. The mass spectrometry of these adducts have been studied,[10] but the singly metalated adducts are more widely employed for analysis, such as in lipidomic studies.

The product ions formed by low energy collisional activation of $[M + Li]^+$ are remarkably different (Figure 5.1B) from those formed following decomposition of PC derived $[M + H]^+$. The most abundant ions correspond to neutral loss of 59, 183, and 189 Da. No common product ion (at an observed *m/z*) is characteristic of the lithium ion adduct species in contrast to that of decomposition of $[M + H]^+$. Understanding the events taking place after collision induced dissociation has been greatly aided by the work of Hsu and Turk, with various molecular species of phosphocholine studied, including deuterium labeled fatty acyl phosphatidylcholines.[10,16]

The abundant high mass ion common for both Na^+ and Li^+ after collisional activation, corresponds to loss of 59 Da which is loss of trimethylamine. The most likely mechanism involves a loss of this neutral amine driven by the phosphate anion when the metal adduct occurs at a remote ester site (Scheme 5.13). If the metal ion adduct site is a salt form of the phosphate anion, then the nonbonded electrons or a phosphate bound oxygen atom could form the metalated O,O'-dimethylenephosphate cation with loss of $N(CH_3)_3$ (Scheme 5.14).

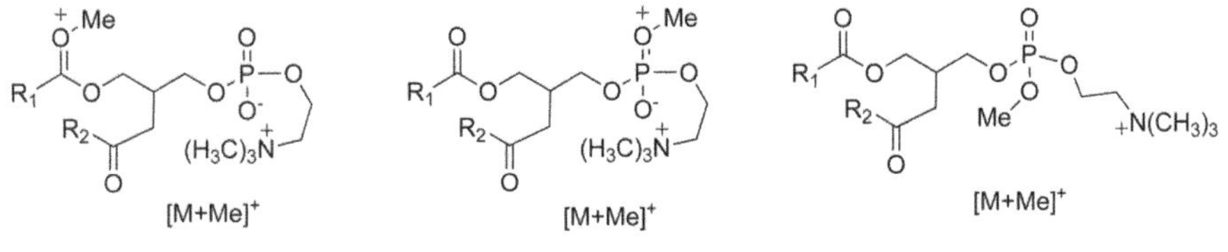

Scheme 5.12

The most abundant product ions following collisional activation of [M + Li]$^+$ adducts of PC are the neutral loss of 183 and 189 Da, which correspond to loss of either the phosphocholine headgroup, or lithium plus the phosphocholine headgroup. These ions have been described as being formed from the cyclic O,O'-dimethylenephosphate [M + Li − 59]$^+$.[10] Based upon synthetic PC species with 2' deuterium atoms in either the *sn*-1 or *sn*-2 fatty acyl groups, the loss of 183 involves participation of this fatty acyl proton which is lost in the deuterated species with the corresponding loss of 184 Da. The 2' proton preferentially lost is from the *sn*-2 fatty acyl group as illustrated in Scheme 5.15 with formation of the neutral cyclic O,O'-dimethylenephosphoric acid.

The loss of 189 Da as a lithiated O,O'-dimethylenephosphoric acid is not shifted in mass when the *sn*-1 or *sn*-2 fatty acyl chains are deuterated, but no doubt involves participation of the 2'-protons remote from the attachment site of lithium on the phosphate residue (Scheme 5.16). Perhaps the electron density of the nonbonded electrons from the *sn*-1 ester group would attack the 2' proton; but now with the lithium adducted to the cyclic phosphoric acid, the cyclic phosphoric acid is lost as a lithium adduct. While for many diacyl PC molecular species, the [M + Li − 189]$^+$ abundance is higher than the abundance for the [M + Li − 183]$^+$, suggesting that the location of the lithium adduct site maybe preferentially as a phosphate salt. An important exception is that for the polyunsaturated PC species where the ion at [M + Li − 183]$^+$ is much more abundant,[16] suggesting participation of the double bonds in the polyunsaturated fatty acyl group in adducting the Li$^+$ charging species and promoting charge remote rearrangements (Scheme 5.15).

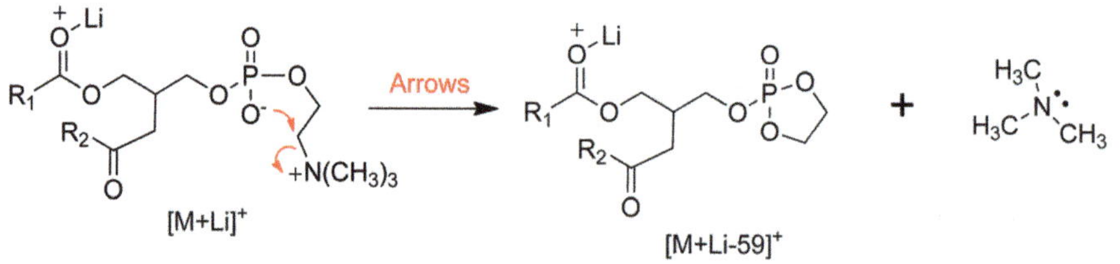

Scheme 5.13

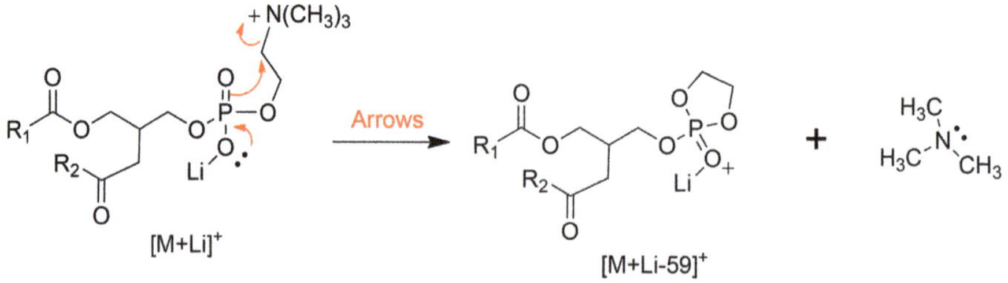

Scheme 5.14

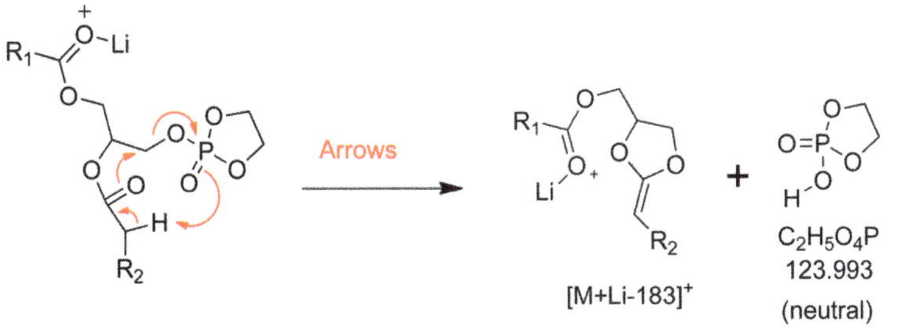

Scheme 5.15

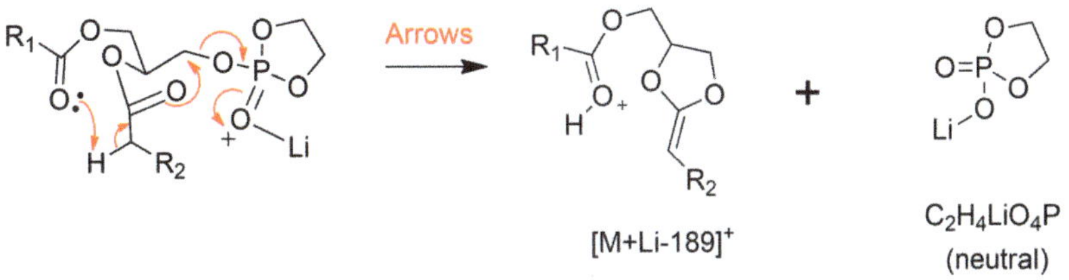

Scheme 5.16

Additional ions are observed in the diacyl $[M + Li]^+$ product ion spectra and these correspond to the loss of either R_1COOH or R_2COOH from the $[M + Li]^+$ species as well as R_1COOLi or R_2COOLi. These ions are much more abundant for the lithium adducts than that observed for the protonated PC and likely operate by the same charge remote processes described in Schemes 5.3 and 5.4. In addition, ions corresponding to the loss of *sn*-1 or *sn*-2 fatty acids from the $[M + Li - 59]^+$ species are formed. While some suggestion has been made that the ratios of these neutral losses can be informative as to the position of the fatty acyl groups at either *sn*-1 or *sn*-2, these ions are not particularly abundant.[10] If phosphatidylcholines are being examined from extracts of naturally occurring biological sources, there could be multiple isobaric species at each specific *m/z* being collisionally activated to confound this region on the mass-to-charge scale, since the abundance of these ion-types would be split by the mole fraction of each isobaric PC present and the ion abundance for the loss of $R_{1,2}COOH$ from each species would be reduced. Furthermore, not a large number of synthetic PC species have been examined where the *sn*-1 and *sn*-2 fatty acyl groups are known without question to confirm and extend these suggestions as to the value of the ratio of neutral losses.

The lithiated plasmalogen PC species yield a somewhat different product ion spectrum as the diacyl species following collisional activation (**Figure 5.3**). Quite abundant $[M + Li - 59]^+$ and $[M + Li - 189]^+$ are observed, but there is no 2′ acyl proton at the *sn*-1 group to drive the loss of the *sn*-2 as a protonated fatty acid from the $[M + Li - 189]^+$ as suggested by Scheme 5.12.

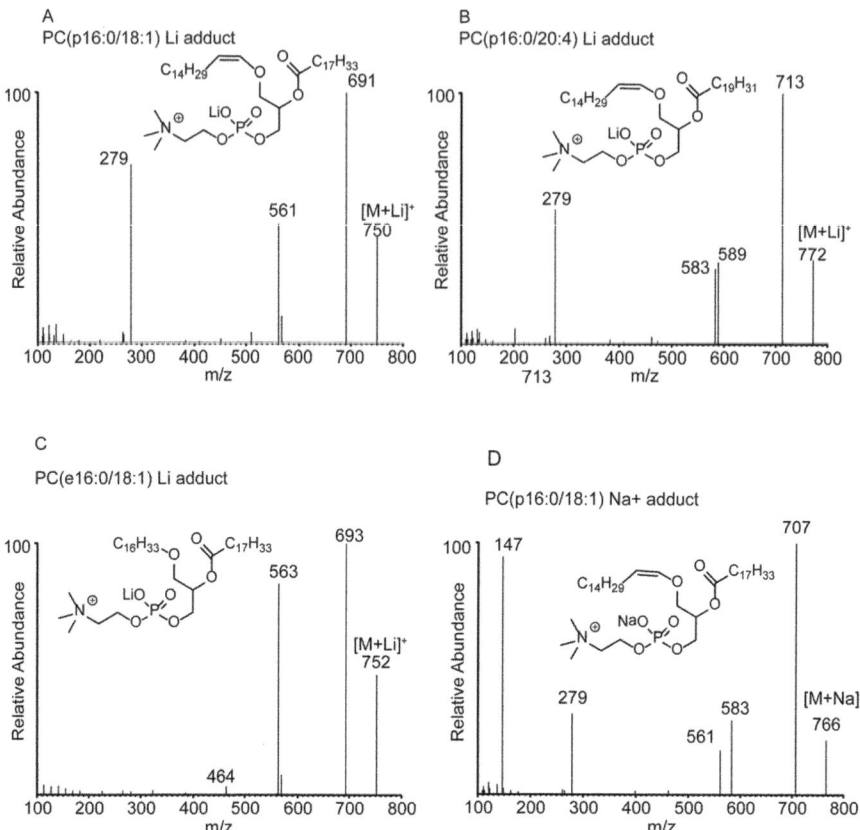

Figure 5.3 Electrospray ionization (positive ions) and tandem mass spectrometry of diacyl, ether, and plasmalogen glycerophosphocholine lipids as lithium and sodium adducts; (A) Product ions obtained following collisional activation of PC(16 : 0/18 : 1) [M + Li]⁺ at *m/z* 750; (B) product ions obtained following collisional activation of PC(p16 : 0/20 : 4) [M + Li]⁺ at *m/z* 772; (C) product ions obtained following collisional activation of PC(e16 : 0/18 : 1) [M + Li]⁺ at *m/z* 752; (D) product ions obtained following collisional activation of PC(p16 : 0/18 : 1) [M + Na]⁺ at *m/z* 766. This figure was redrawn from data presented in ref. 16.

Rather, the vinyl ether at *sn*-1 significantly alters the pathway of decomposition and the ion corresponding to loss of 189 Da and loss of the *sn*-2 group as a carboxylic acid, are quite abundant (Figure 5.3A and B). This uniquely abundant loss of a carboxylic acid from the plasmalogen ion undoubtedly involves the vinyl ether group participating in this reaction as suggested in Scheme 5.17.

The corresponding plasmanyl, alkyl ether lithium salt of PC (e18 : 0/18 : 1) does not produce this ion upon collisional activation (Figure 5.3C), but rather unique ions likely derived from the alkyl ether residue that are observed as an alkene series at low mass.[10]

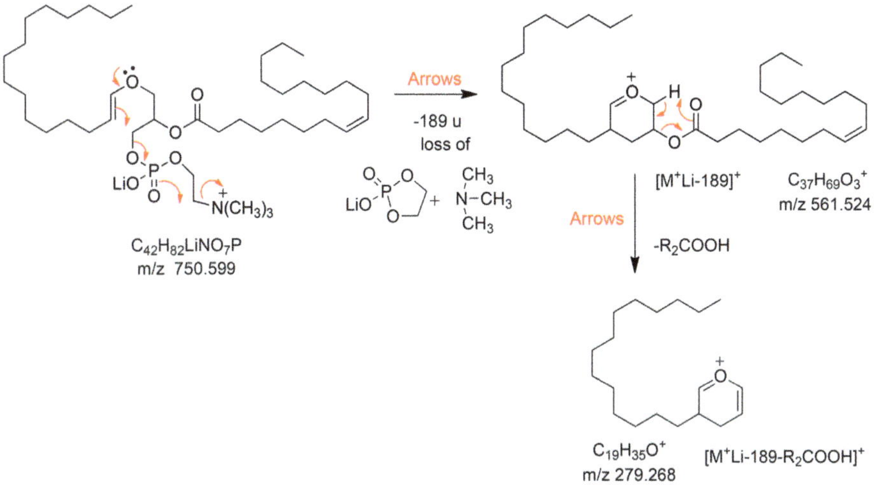

Scheme 5.17

5.5.3 Positive Ion Sodiated PC Adducts

The collisional activation of sodiated 1,2-diacyl PC yields product ions that are very similar to the lithiated species discussed above, except for those ions containing the alkali metal imbedded in the ion structure. The product ions obtained by collisional activation of the sodium ion adducts of plasmalogen PC (Figure 5.3D) also have some of the same characteristics of the lithium ion adducts, for example, a major product ion from PC (p16 : 0/18 : 1) Na adduct $[M + Na]^+$ at m/z 766 has a major neutral loss of trimethylamine (59 Da) found at m/z 707 as well as the loss of 183 Da (m/z 583) and the loss of 183 + Na from the $[M + Na]^+$ at m/z 561. These undoubtedly involve the same mechanisms illustrated in Schemes 5.11 and 5.12, substituting Na^+ for Li^+. The rather abundant ions due to the loss of R_2COOH from m/z 561 (observed at m/z 279) would arise from the same mechanism for the Li^+ adduct involving the *sn*-1 vinyl ether double bond (Figure 5.3D). For both 1,2-diacyl and plasmalogen PC, an abundant ion at m/z 147 is quite characteristic of the tandem mass spectrometry of sodiated PC adducts in the positive ion mode and likely arises as a stable sodium ion salt of O,O'-dimethylenephosphoric acid (Scheme 5.18).

5.5.4 Negative PC Ions

Negative ions can be formed by electrospray ionization of PC despite the permanently charged quaternary ammonium moiety present in the choline group, which cannot be altered by simple loss or gain of a proton (Figure 5.4). These ions were first demonstrated as demethylated PC in fast atom bombardment ionization and are known to originate from an adduct ion such

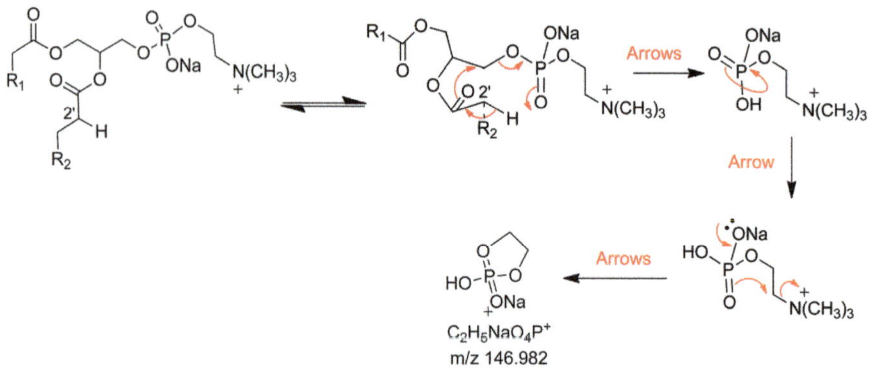

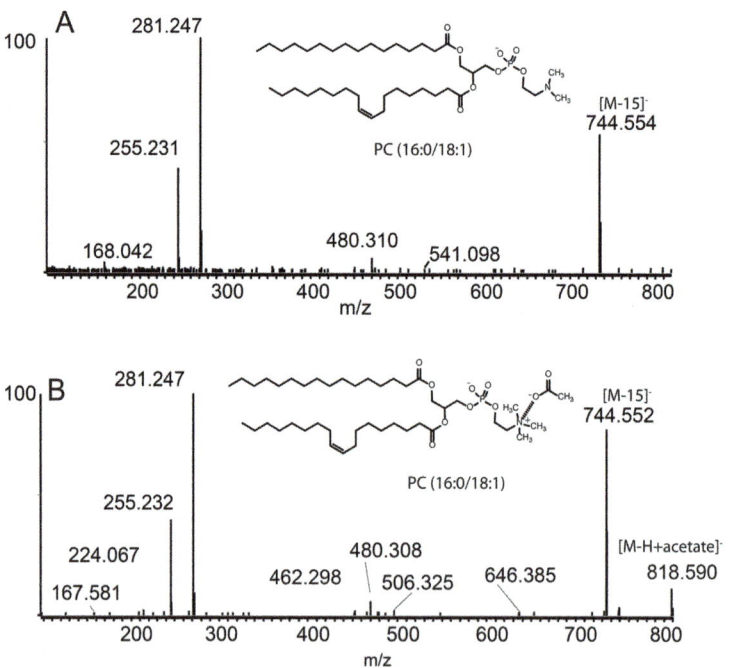

Scheme 5.18

Figure 5.4 Electrospray ionization (negative ions) and tandem mass spectrometry of phosphatidylcholines. (A) Product ions obtained following collisional activation of PC(16 : 0/18 : 1) $[M − 15]^-$ at m/z 744; (B) product ions obtained following collisional activation of PC(16 : 0/18 : 1) $[M − H + acetate]^-$ at m/z 818. These high resolution MS/MS spectra were obtained using a quadrupole time-of-flight mass spectrometer.

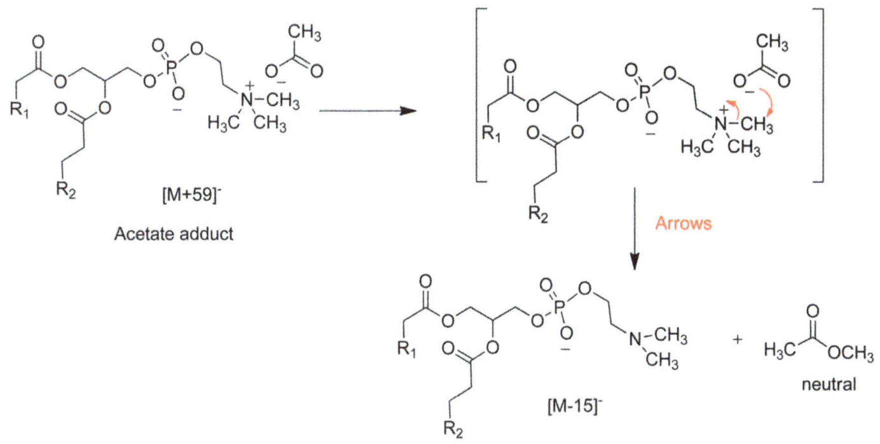

Scheme 5.19

as an acetate, formate, chloride, or other anionic adduct to neutralize the permanent charge of the quaternary ammonium cation, thus rendering a net negative charge on the molecule due to the phosphate anion.[17] The anionic group can drive a demethylation reaction forming, for example, methyl acetate (Scheme 5.19), methyl formate or methyl chloride as neutral molecules, leaving the dimethylamino (uncharged group) on the phospholipid.

One interesting observation was that if one of the fatty acyl side chains contains a carboxylic acid group such as one derived from oxidation of PC lipids by free radical reactions,[18] an abundant $[M - H]^-$ is observed because of the presence of the additional anionic site on the fatty acyl group. These carboxylate anions can in fact demethylate the choline residue, forming a methyl ester *in situ*, which is readily observed following collisional activation as a carboxylate anion with a terminal methyl ester, as shown in Scheme 5.20.[18] Nonetheless, the collisional activation of either the anion adducted species or the $[M - 15]^-$ demethylated PC, yields abundant carboxylate anions as product ions that readily reveal the nature of the fatty acyl groups at both *sn*-1 and *sn*-2 (Scheme 5.3).

There have been various claims about the ratio of the carboxylate anion abundances to each other, suggesting that this information reveals which fatty acid acyl group is at *sn*-1 and which is at *sn*-2. Yet the dominance of the *sn*-2 carboxylate anion over the *sn*-1 appears to be an over simplification, with important exceptions.[19] One of the significant issues is that the resultant carboxylate anions may not have the same stabilities. For example, loss of CO_2 from docosahexaenoate or arachidonate can readily take place (Chapter 1) and these decarboxylated hydrocarbon ions are observed as additional product ions. This is especially true in a tandem quadrupole collision cell, where multiple collisions take place. This process lowers the apparent abundance of these polyunsaturated carboxylate anions and thus the *sn*-2/*sn*-1 carboxylate anion ratio.

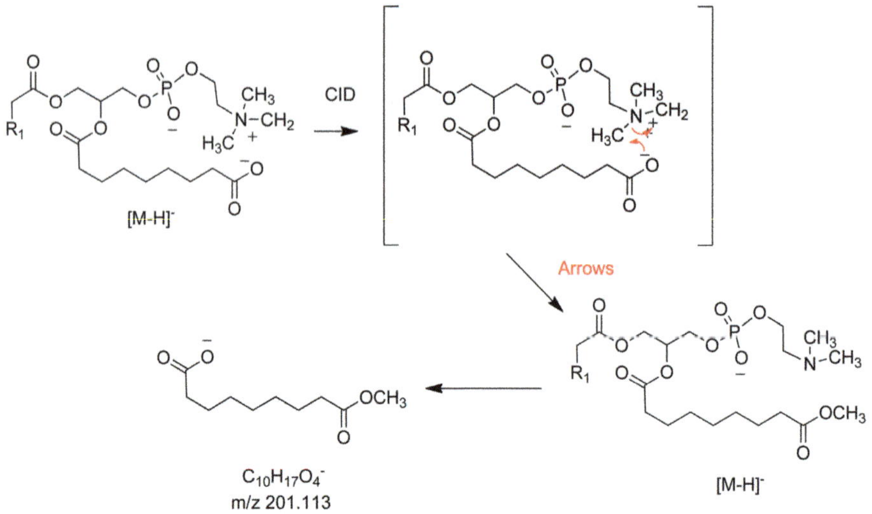

Scheme 5.20

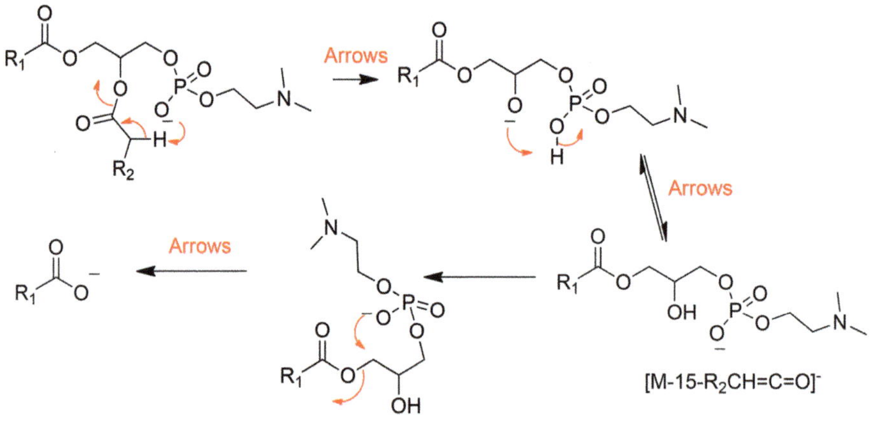

Scheme 5.21

Ions of substantially lower abundance are observed corresponding to the loss of each fatty acyl ester as a free acid $[M - 15\text{-RCOOH}]^-$ as well as ketene $[M - 15\text{-RCH}{=}C{=}O]^-$. While detailed studies of the mechanisms by which these ions are formed have not been reported, they are most likely similar to that reported for PE, where studies have appeared.[20] Interestingly, it is the ratio of the abundances of the ions corresponding to the loss of the free carboxylic acid that best predicts acylation position of the fatty acyl groups at *sn*-1 and *sn*-2.[19] The somewhat more abundant loss of ketene is likely the mechanism adapted from the work of Hsu and Turk[20] shown in Scheme 5.21 and is also an additional mechanism for the formation of carboxylate anions.

The loss of either fatty acyl group as the ion $[M - 15\text{-}RCOOH]^-$ is usually less abundant and suggested to be a charge remote fragmentation of the $[M - 15]^-$ from PC that involves the glycerol backbone protons. This general mechanism (Scheme 5.5) is based on the lack of the loss of RCOOD from a side chain deuterated analog.[10] This mechanism is quite similar to that seen as a loss of free carboxylic acids from glycerol esters (Chapter 4).

The lyso species of plasmenyl, plasmanyl, and 1-acyl-2-lyso-phosphatidyl-cholines have distinctively different product ion spectra following collisional activation of $[M - 15]^-$.[15] As expected, the 1-acyl-2-lyso-PC $[M - 15]^-$ generates a carboxylate anion by the mechanism previously presented (Scheme 5.17). Since the ether PC lipids cannot form carboxylate anions, collisional activation yields different product ions. The vinyl ether at *sn*-1 can undergo an ene-type reaction, likely removing a glycerol backbone proton at the carbon atom *sn*-2 with concerted loss of a neutral aldehyde from the *sn*-1 side chain (Scheme 5.22).

The alkyl ether lyso-PC has very few facile reaction pathways to decompose and relax the excited $[M - 15]^-$ and the dominate product ions observed are loss of the polar head group and loss of water. The loss of water from the *sn*-2 hydroxyl group is surprisingly difficult and not seen for the 1-acyl and 1-vinyl ether lyso-PCs. The mechanism is likely an initial charge remote loss of dimethylaminoethylene to form the phosphatidic acid intermediate. The phosphate anion can then attack the carbon at *sn*-2 which drives the hydroxyl group to form a covalent bond with a proton to form neutral water from the proton of the phosphatidic acid. This reaction could have a concerted mechanism (Scheme 5.23). It is interesting that this reaction sequence and

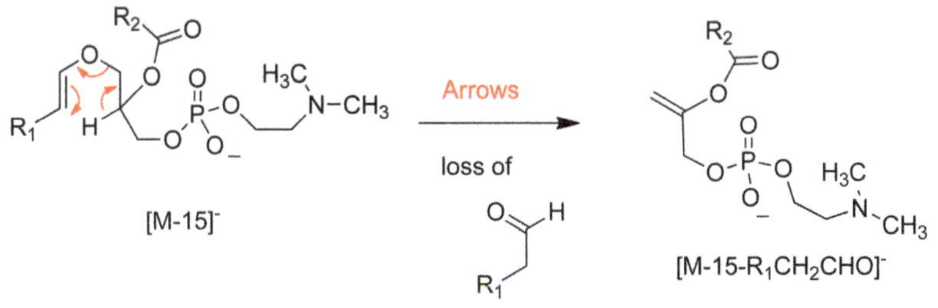

Scheme 5.22

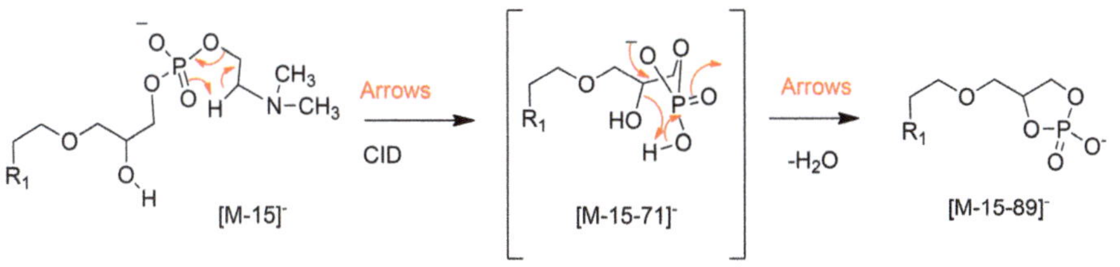

Scheme 5.23

pathway is available to all three lysoPC subspecies, but only the $[M - 15]^-$ of the alkyl ether lysoPC and the vinyl ether plasmalogen PC have abundant ions as the result of collisional activation of the $[M - 15]^-$.[15]

5.6   Glycerophosphoethanolamine Lipids (PE)

Glycerophosphoethanolamine lipids are very common phospholipids containing nitrogen and this lipid class is present in all cells. These lipids are typically more abundant on the inner leaflet of the outer bilayer membrane of the cell due to the action of several enzymes that render this bilayer quite asymmetric as to phospholipid composition between inner and outer leaflets. In addition, the biosynthesis of plasmalogen phospholipids proceeds by way of a PE polar head group. Interestingly, the PE lipids generate more abundant negative ions than positive ions in spite of the fact of having a primary amino moiety.

5.6.1   Positive Ion $[M + H]^+$ PE

The formation of a positive ion by attachment of a proton $[M + H]^+$ or alkali metal ion, $[M + Li]^+$ or $[M + Na]^+$, is more facile for this phospholipid class than more acidic phospholipids such as PG, PI, CL, or even PS (Figure 5.5). The CID of these positive ions have been studied in some detail and mechanisms for product ion formation suggested.[11,20,21]

The most abundant product ion formed by collisional activation of $[M + H]^+$ from PE (Figure 5.5A) is a loss of 141 Da. From high resolution and isotope labeling studies, this ion corresponds to the loss of the polar head group forming a "diglyceride-like" ion. For PE (16 : 0/18 : 1) $[M + H]^+$, which has a molecular weight at m/z 718.538, this diglyceride-like ion appears at m/z 577.519 ($C_{37}H_{69}O_4$). This neutral loss of the polar head group is seen for most phospholipids except PC and has a charge remote mechanism outlined in general Scheme 5.2 and specifically for PE in Scheme 5.24. Note the charge site involves proton attachment to one of the ester moieties and the suggestion that a phosphate anion is making a salt bridge to the protonated primary amine, rendering the polar head group neutral. If the phosphate anion was a site of protonation and the primary amine charged ($R-NH_3^+$), then one would expect to observe m/z 142,[11] but there is no ion observed at this mass (Figure 5.5A).

There are other, rather minor, ions, including acylium ions for each fatty acyl group (m/z 239 and m/z 265), but their abundances do not reveal the specific esterification positions on the glycerol backbone. In general, the collisional activation of $[M + H]^+$ provides rather specific information as to the polar head group being phosphoethanolamine. While some have used this loss of 141 Da as a general approach to measure the relative abundances of all PE species in complex mixtures by constant neutral loss experiments,[22] plasmalogen PE molecular species which can be abundant in some tissues

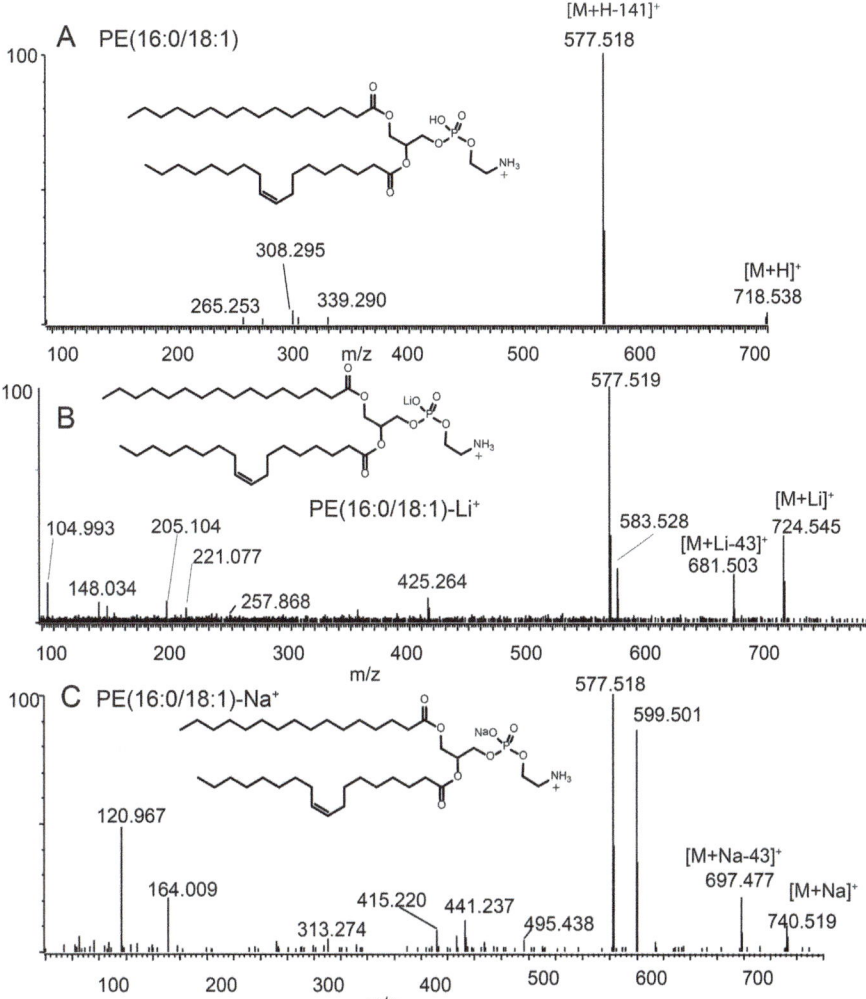

Figure 5.5 Electrospray ionization (positive ions) and tandem mass spectrometry of phosphatidlyethanolamine. (A) Product ions obtained following collisional activation of PE(16 : 0/18 : 1) $[M + H]^+$ at m/z 718; (B) product ions obtained following collisional activation of PE(16 : 0/18 : 1) $[M + Li]^+$ at m/z 724; (C) product ions obtained following collisional activation of PE(16 : 0/18 : 1) $[M + Na]^+$ m/z 740. These high resolution MS/MS spectra were obtained using a quadrupole time-of-flight mass spectrometer.

Scheme 5.24

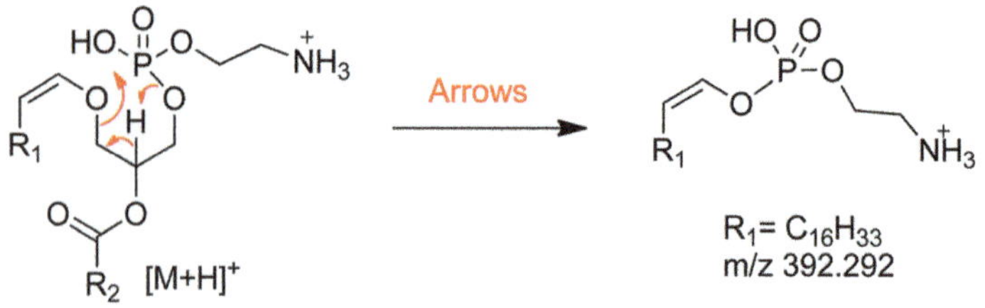

R$_1$= C$_{16}$H$_{33}$
m/z 392.292

Scheme 5.25

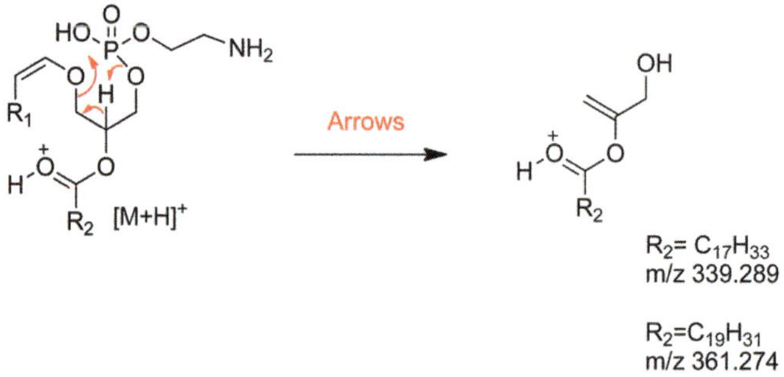

R$_2$= C$_{17}$H$_{33}$
m/z 339.289

R$_2$=C$_{19}$H$_{31}$
m/z 361.274

Scheme 5.26

are observed not to yield this ion as the most abundant product ion.[23] There are two additional ions observed in the product ion spectrum of PE plasmalogens (Figure 5.6) and the mechanisms for formation of these two ions center around unique properties of the vinyl ether group. The product ion characteristic of the *sn*-1 radyl group involves a pair of non-bonded electrons from the vinyl ether oxygen atom attacking the phosphorus atom (Scheme 5.25) with the formation of a new oxygen–phosphorus bond simultaneous to the cleavage of the oxygen–phosphorus bond at *sn*-3, which abstracts a glycerol hydrogen atom at *sn*-2. The loss of the glycerol moiety as a neutral olefin results in the ion unique for each plasmalogen *sn*-1 group. For the PE (p16 : 0/ 20 : 4), PE (p16 : 0/18 : 1), and PE (p16 : 0/22 : 0) this ion is observed at *m/z* 392.

The second unique plasmalogen derived product ion likely involves those population of [M + H]$^+$ ions that have the *sn*-2 ester carbonyl oxygen atom protonated which facilitates loss of the neutral polar head group. This would be the same mechanism as above only the product ion which results has the initial charge site at a different position (Scheme 5.26). The mechanism presented in this scheme is consistent with deuterium labeling studies.[23]

5.6.2 Positive Ion Lithiated PE

The product ions formed following collisional activation of the positively charged monolithiated adduct [M + Li]$^+$ (Figure 5.5B) are similar to those observed for the corresponding phosphatidylcholine lithium adduct. This

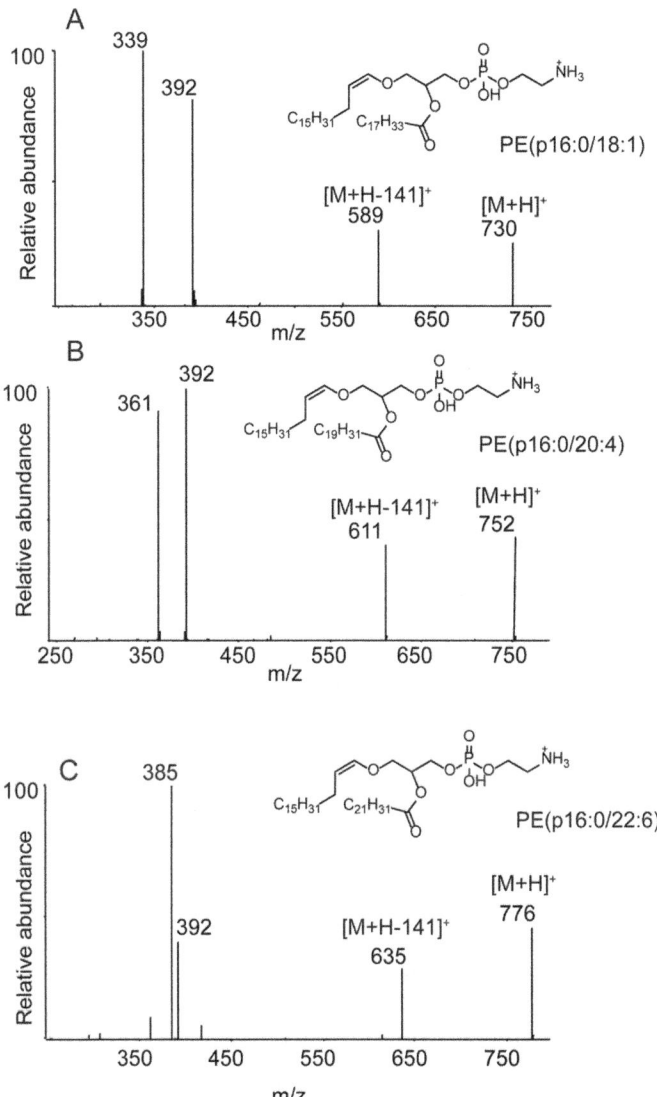

Figure 5.6 Electrospray ionization (positive ions) and tandem mass spectrometry of plasmalogen glycerophosphoethanolamine lipids. (A) Product ions obtained following collisional activation of PE(p16 : 0/18 : 1) [M + H]$^+$ at *m/z* 730; (B) product ions obtained following collisional activation of PE(p16 : 0/20 : 4) [M + H]$^+$ at *m/z* 752; (C) product ions obtained following collisional activation of PE(p16 : 0/20 : 6) [M + H]$^+$ at *m/z* 776. This figure was redrawn from the data presented in ref. 23.

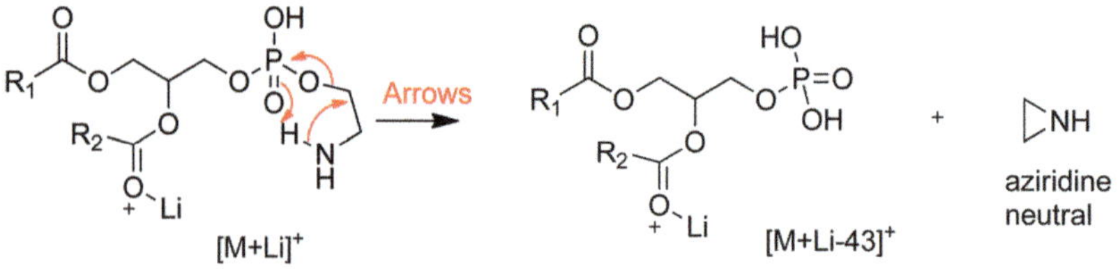

Scheme 5.27

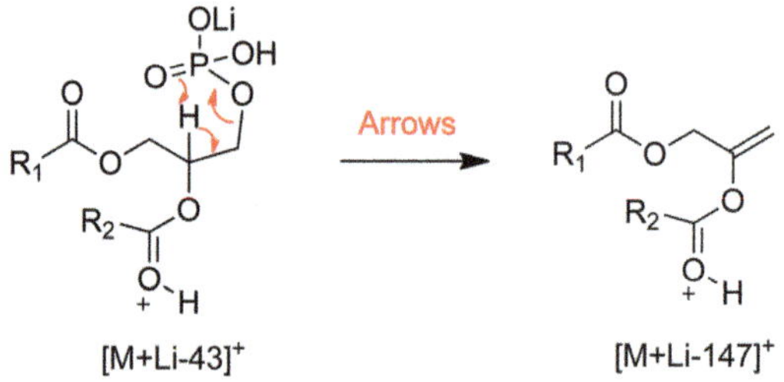

Scheme 5.28

includes loss of the nitrogen-containing moiety in the polar head group of PE which corresponds to the loss of 43 Da as the loss of aziridine illustrated in Scheme 5.27.[21]

The other two abundant product ions correspond to the loss of the entire phosphoethanolamine moiety with and without the lithium ion. When the lithium ion is lost, the charging site is at an ester group as a proton adduct ($[M + Li - 147]^+$) (Scheme 5.28). When the lithium atom is retained in the product ion, the loss observed for the protonated PE is the loss of 141 Da, as described earlier (Scheme 5.22). In the case of the lithium adduct, the charge site corresponds to the lithium ion adducted to one of the ester carbonyls. Interestingly, when there are deuterium atoms on the fatty acyl side chain, these product ions appear as doublets (for example, $[M + Li - 148]^+$ and $[M + Li - 147]^+$), indicating a second mechanism operating for the formation of the phosphoethanolamine losses that include transfer of a fatty acyl proton.[16] Most likely this would be at the 2′ proton as outlined in Scheme 5.29. These mechanisms can also operate for the $[M + Li - 43]^+$ as a precursor ion which would decompose by these pathways. There are other product ions formed, but in general they are fairly low in abundance. The mechanisms responsible for the formation of these low abundant ions have been presented.[21]

Several abundant product ions are observed from the lithium adduct of a plasmalogen PE followed by collisional activation (Figure 5.7).[16,21] This subtle structural alteration opens additional pathways for decomposition as had been previously described. For example, the PE (p18 : 0/18 : 1) lithium

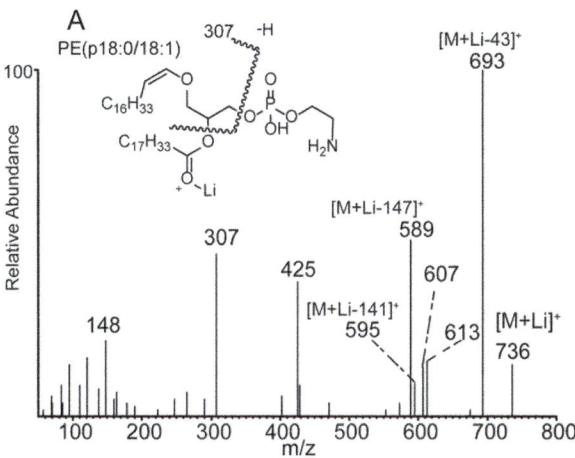

Scheme 5.29

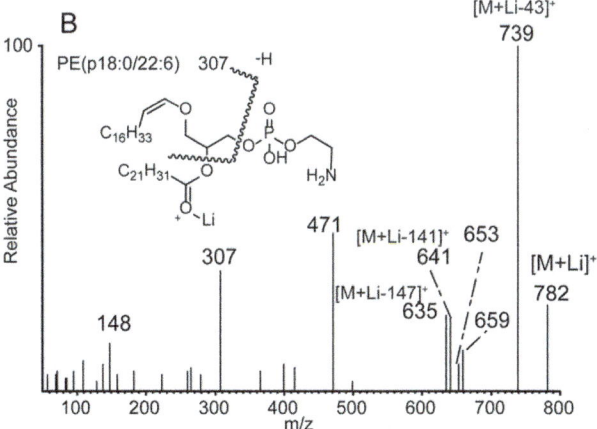

Figure 5.7 Electrospray ionization (positive ions) and tandem mass spectrometry of lithium salts of plasmalogen glycerophosphoethanolamine lipids. (A) Product ions obtained following collisional activation of PE(p18 : 0/18 : 1) [M + Li]⁺ at *m/z* 736; (B) product ions obtained following collisional activation of PE(p18 : 0/22 : 6) [M + Li]⁺ at *m/z* 782. This figure was redrawn from data presented in ref. 21.

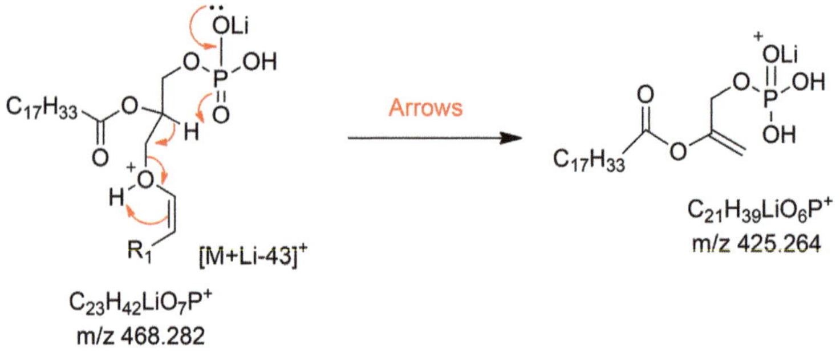

Scheme 5.30

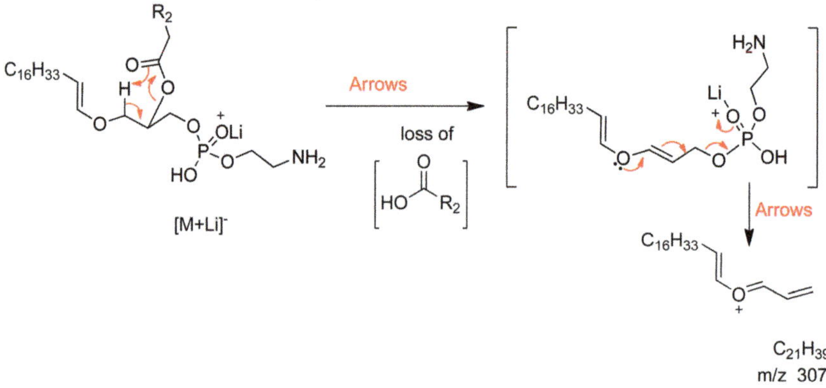

Scheme 5.31

adduct has two abundant ions at m/z 307 and 425. M/z 425 is formed by the loss of 43 Da (aziridine) and the vinyl ether moiety, likely as an aldehyde (Scheme 5.30).

The other abundant ion appears at m/z 307 and corresponds to the loss of the polar head group plus the lithium ion as well as the sn-2 group as a free carboxylic acid. The likely mechanism involves initial loss of the sn-2 carboxylic acid described before, followed by the loss of the lithiated polar head group driven by formation of a conjugated oxonium ion facilitated by the vinyl ether double bond (Scheme 5.31). These two product ions are rather unique to the lithium adducted plasmalogen PE and therefore, quite useful in characterizing the radyl groups at sn-1 and sn-2.

5.6.3 Positive Ions Sodiated PE

The sodium ion adduct of PE is collisionally activated to product ions that are very similar to that observed for the lithiated adduct (Figure 5.5C). The molecular ion loses aziridine to form $[M + Na - 43]^+$ which for

PE (16 : 0/18 : 1) Na$^+$ is seen at m/z 697. This ion, which has the sodium ion adduct localized as a phosphate salt, is the precursor of the abundant ions corresponding to the loss of the *sn*-1 and *sn*-2 fatty acyl groups as free carboxylic acids (as seen in Figure 5.5C at m/z 441 and 415, respectively). The mechanism for these losses was outlined in Scheme 5.19, except for the presence of the sodium adducted ion, which replaces the lithium ion.

The most abundant ions observed in the tandem mass spectra correspond to the loss of the phosphoethanolamine polar head group with and without the Na$^+$ ion adducted to the leaving group. These ions at m/z 577 and 599 are formed by identical mechanisms described in Schemes 5.28 and 5.29, respectively. There are two low mass product ions at m/z 121 and 164 derived from sodiated phosphate $[H_3PO_4Na]^+$ and the sodiated PE polar head group $[C_2H_8PO_4Na]^+$, respectively.

5.6.4 Negative Ion PE

Negative ions generated by electrospray ionization of glycerolphosphatidyl-ethanolamine molecular species are observed as the abundant $[M - H]^-$. The collisional activation of these ions renders product ions commonly observed for all glycerophospholipid $[M - H]^-$ ions (Figure 5.8) and those previously described for the collisional activation of $[M - 15]^-$ from PC.

The most abundant product ions are carboxylate anions corresponding to the *sn*-1 and *sn*-2 fatty acyl group formed by attack in a 5-membered ring or a 6-membered ring by the phosphate anion (Scheme 5.3) for the *sn*-2 and *sn*-1 carboxylate anion formation, respectively. For many mono- and diunsaturated fatty acyl PE molecular species, the most abundant ion observed is derived from the *sn*-2 fatty acyl group (Figure 5.8A) and for many molecular species this reveals the fatty acyl position.[7] However, for polyunsaturated fatty acid acyl groups, especially those containing more than five double bonds such as docosahexaenoate (22 : 6), the carboxylate anion is of lower abundance even though it is at the *sn*-2 position. This in part is due to the further decomposition of this carboxylate anion through the loss of 44 Da (CO_2), forming a hydrocarbon ion ($C_{21}H_{31}$, m/z 283.243) that can have the anionic site delocalized over several double bond systems after rearrangement.[20] With low resolution mass spectrometry, this ion cannot be distinguished from a carboxylate anion derived from the stearate ester ($C_{18}H_{35}O_2$, m/z 283.264), which can lead to misinterpretation of the presence of stearate in the PE (spectra published in ref. 20).

Of considerably less abundance are those ions corresponding to the loss of neutral ketene (R–CH=C=O). The loss of the ketene neutral from the *sn*-2 position has been observed to be in the greatest abundance without any exceptions yet noted.[20] This loss of ketene is the result of the abstraction of the 2′-proton from the *sn*-2 fatty acyl chain with initial formation of an alkoxide anion at *sn*-2 (Schemes 5.4 and 5.32). Due to the reactivity of this oxygen-centered anion, it is likely that a further reaction abstracts the proton from the phosphate hydroxyl group, forming a more stable phosphate anion.

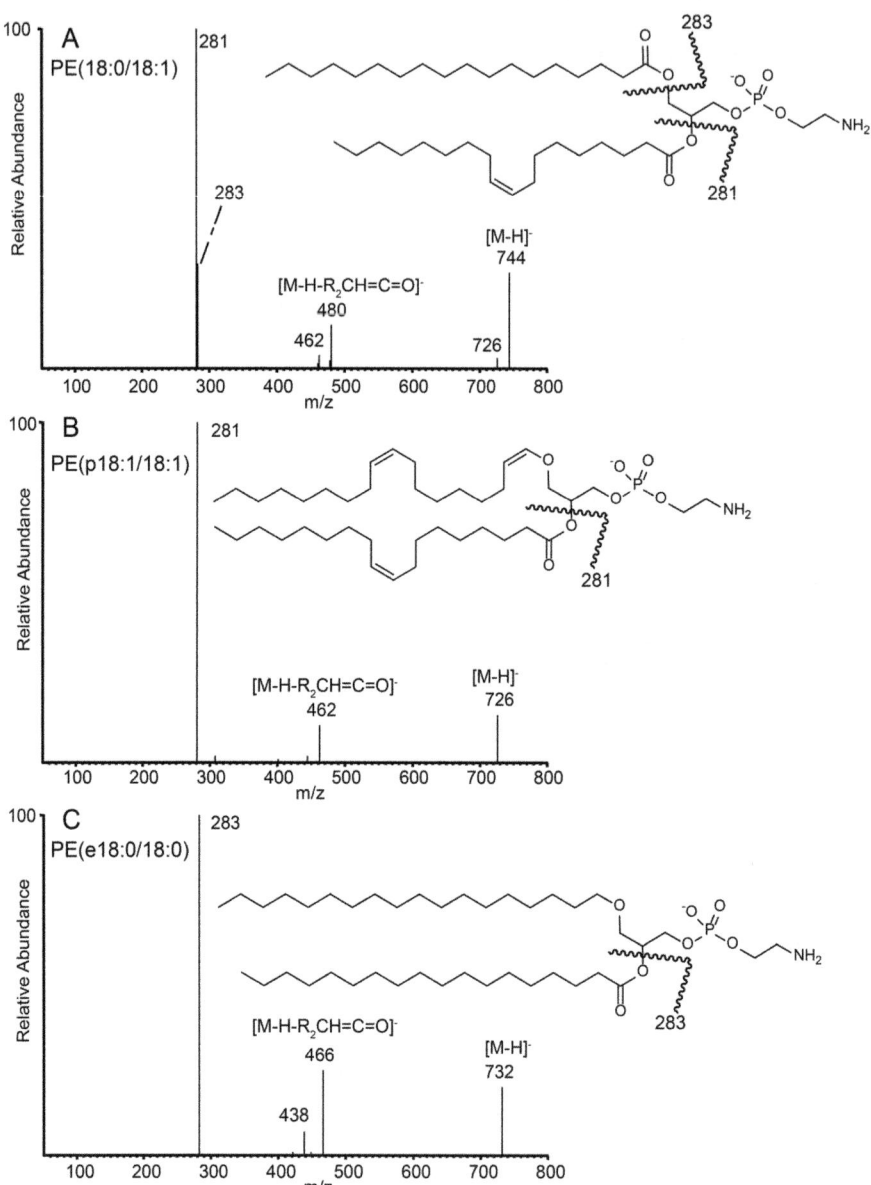

Figure 5.8 Electrospray ionization (negative ions) and tandem mass spectrometry of phosphatidylethanolamine molecular species. (A) Product ions obtained following collisional activation of PE(18 : 0/18 : 1) [M − H]⁻ at *m/z* 744; (B) product ions obtained following collisional activation of PE(p18 : 1/18 : 1) [M − H]⁻ at *m/z* 726; (C) product ions obtained following collisional activation of PE(e16 : 0/18 : 1) [M − H]⁻ at *m/z* 732. This figure was redrawn from data presented in ref. 24.

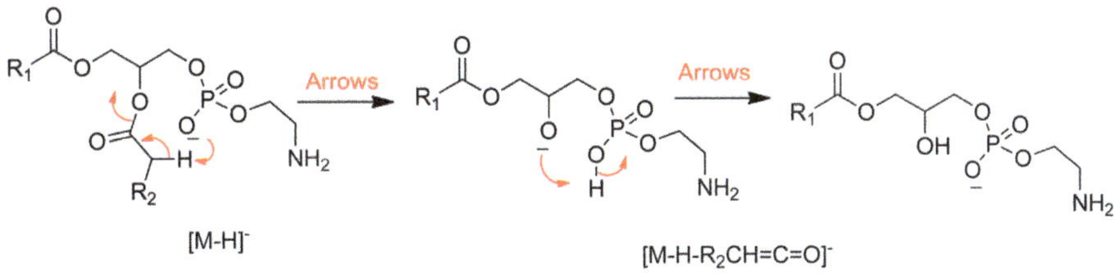

Scheme 5.32

The negative molecular ion from PE plasmalogen (Figure 5.8B) and the O-alkyl PE species (Figure 5.8C) behave identically to that of the diacyl-PE except for the single abundant carboxylate anion being formed. These ether PE lipids can also show the loss of the *sn*-2 fatty acyl groups as ketene.[24]

5.7   Phosphatidylserine (PS)

A third abundant nitrogen containing phospholipid is phosphatidylserine. This phospholipid also contains an α-amino acid moiety and, as such, has an additional functional group on the polar head group, compared to other phospholipids, that can participate in adduct formation during the electrospray process. While PS is most efficiently observed as the negative ion $[M - H]^-$, it is quite possible to form alkali metal adduct ions such as $[M - 2H + Na]^-$ or $[M - 2H + Li]^-$ species in the tandem mass spectrometry of these ionic forms.[25] The $[M - H]^-$ likely is composed of two anionic charges and one cationic site as a result of an internal salt formed by the α-amino acid functionality and the phosphate anion.

Positive ions can also be formed by electrospray ionization leading to the formation of $[M + H]^+$ as well as alkali metal adducted ions corresponding to $[M - H + Na_2]^+$ and even $[M - 2H + 3Na]^-$ species as examples. The tandem mass spectrometry of these species do differ to some extent, but very similar behavior is observed following collisional activation compared to the single metalated ions. However, the product ions observed do suggest the site on which the charge resides.[25]

5.7.1   Positive Ion $[M + H]^+$ PS

The protonated molecular ion $[M + H]^+$ has a behavior identical to that observed following collisional activation of the $[M + H]^+$ from PE. Major product ions are likely formed by the general mechanism Scheme 5.2 previously described as involving abstraction of a proton from the 2′ fatty acyl position, only in this case the charge site is retained on the fatty acyl portion of the molecule. The resultant "diglyceride-like" ion $[M + H - 185]^+$ (Scheme 5.33) reveals the total number of fatty acyl carbon atoms and double bonds, and if measured at high resolution, the exact elemental composition of the

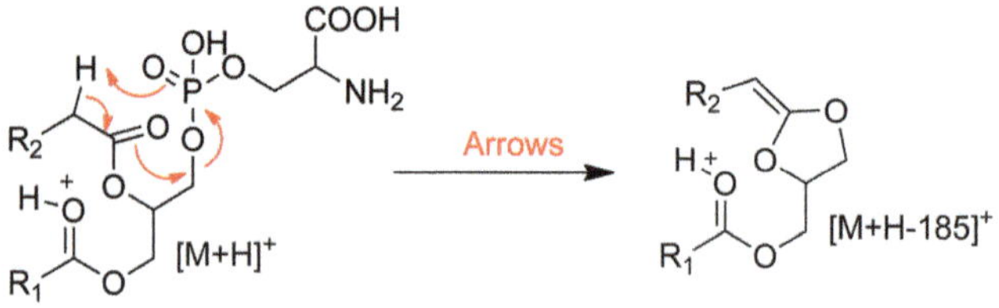

Scheme 5.33

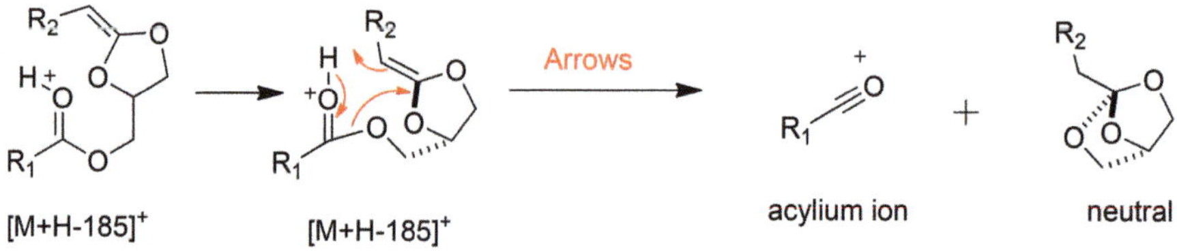

Scheme 5.34

hydrophobic portion of PS (Figure 5.9A). Since PS often appears in nature with polyunsaturated fatty acyl groups, any oxidation of the homoconjugated double bonds on this portion of the molecule can be ascertained through this ion.

In addition to the diglyceride-like ion, there are less abundant product ions corresponding to the fatty acyl groups appearing as acylium ions ($R_1C \equiv O^+$ and $R_2C \equiv O^+$). These can result from fragmentation of the diglyceride-like ion which are best demonstrated using an ion trap instrument where the $[M + H - 185]^+$ ion is isolated, then collisionally activated to yield these acylium product ions.[11] The mechanism of formation of these ions has not been rigorously studied, but one possibility is the involvement of the newly formed double bond in the structure of the proposed $[M + H - 185]^+$ ion. This ion can participate in cleavage of the ester oxygen bond to form a highly stable, caged neutral and stable acylium ion (Scheme 5.34). Insufficient molecular species have been studied to reveal significance of the abundance of these acylium product ions related to the fatty acyl position on the glycerol backbone.

5.7.2 Positive Ion $[M + Li]^+$ PS

The alkali metal adducts of PS have somewhat different behavior to that seen for PC or PE adducts and likely represent an influence of the carboxyl group in the polar head group. The product ions obtained in a tandem quadrupole mass spectrometer after collisional activation of $[M + Li]^+$ of a PS species, reveal two distinct and abundant ion species. One corresponds to the diglyceride-like ions observed as a product ion from PS $[M + H]^+$ seen previously, while the other ion is observed at m/z 192. This latter ion is seen

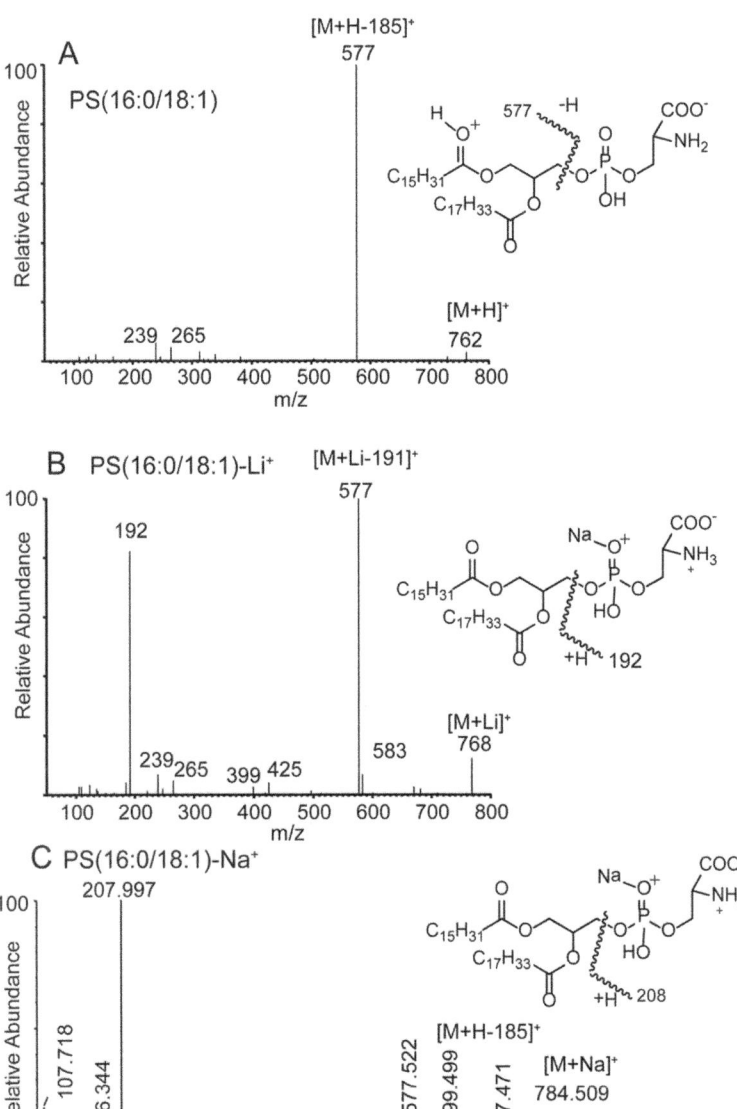

Figure 5.9 Electrospray ionization (positive ions) and tandem mass spectrometry of phosphatidylserines. (A) Product ions obtained following collisional activation of PS(16 : 0/18 : 1) $[M + H]^+$ at m/z 762; (B) product ions obtained following collisional activation of PS(16 : 0/18 : 1) $[M + Li]^+$ at m/z 768. These MS/MS spectra were obtained using a tandem quadrupole mass spectrometer. (C) Product ions obtained following collisional activation of PS(16 : 0/18 : 1) $[M + Na]^+$ at m/z 784. This high resolution MS/MS spectrum was obtained using a quadrupole time-of-flight mass spectrometer.

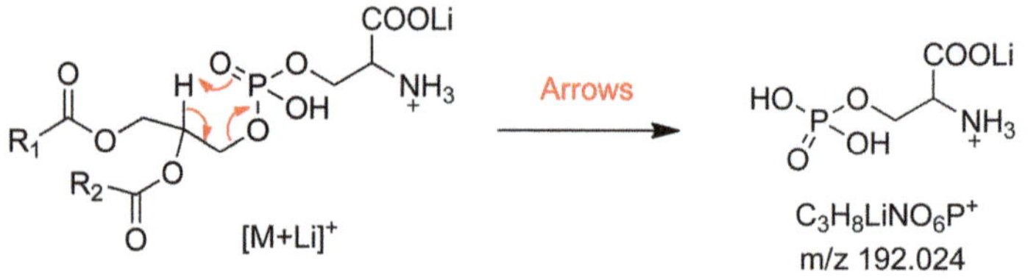

Scheme 5.35

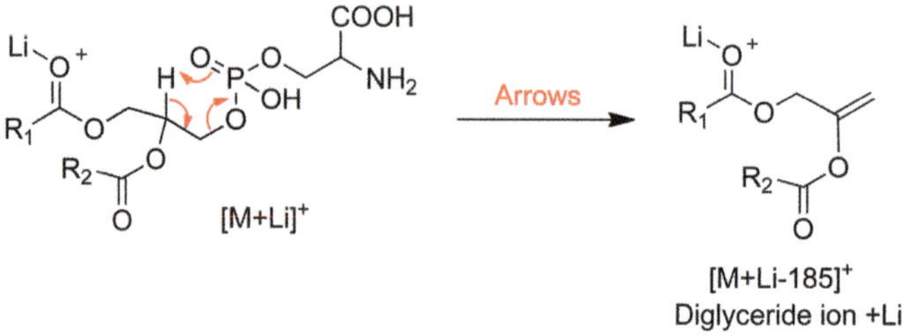

Scheme 5.36

for all $[M + Li]^+$ product ions from PS and corresponds to the lithiated phosphoserine cation that readily forms by an ene-like charge remote mechanism shown in Scheme 5.35, rather than the general mechanism in Scheme 5.2.[25] It is interesting to compare the appearance of this ion with a total absence of the corresponding ion at m/z 185 from collisional activation of the protonated PS $[M + H]^+$. The appearance of this ion renders this a uniquely associated PS product ion.

The major diglyceride ion corresponds to the same product ion illustrated in Scheme 5.32, where the lithium attachment site is either as a carboxylate or phosphate salt and the charge site is located on one of the ester moieties. Thus, this ion corresponds to the loss of the polar head group plus lithium. Another scenario of charge site arrangement would be an internal salt of the α-amino acid and the lithium salt in the phosphate anion. There is a rather minor ion corresponding to an additional 6 Da above the diglyceride ion (observed at m/z 583), which corresponds to the retention of Li^+ on the diglyceride fragment observed. This indicates that some population of the $[M + Li]^+$ involves metalation of one of the two ester functionalities as a residual charge site and the loss of the polar head group does not involve loss of the Li^+ atom (Scheme 5.36).

5.7.3   Positive Ion $[M + Na]^+$ PS

The collisional activation of the sodiated PS positive ion $[M + Na]^+$ is quite similar to that of the lithiated adduct, as expected. There is a prominent ion corresponding to the sodiated phosphoserine ion at m/z 208 (Figure 5.9B) and this ion dominates the mass spectrum (Scheme 5.35 with Na$^+$ instead of Li adduct). Of lower abundance are the diglyceride-like ions (*m/z* 577) and the sodiated diglyceride ion (*m/z* 599), which are most likely formed by the identical mechanisms suggested for the lithiated PS species. There are somewhat more abundant ions corresponding to [M + Na − 87]$^+$ seen at *m/z* 697 in Figure 5.9B that likely involve loss of serine (Scheme 5.37). These ions are also observed after collisional activation of [M + Li]$^+$, but are not as prominent.

5.7.4   Negative Ion [M − H]$^-$ PS

The collision activation of [M − H]$^-$ ion of PS has some unusual features not seen with other phospholipids (Figure 5.10). There is a strikingly abundant ion corresponding to [M − H − 87]$^-$ that involves loss of the serine residue

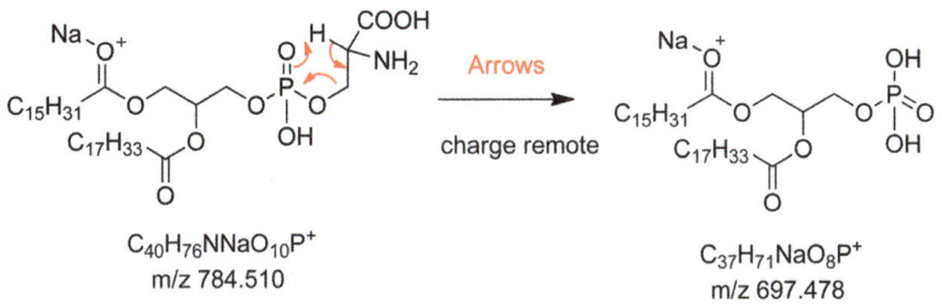

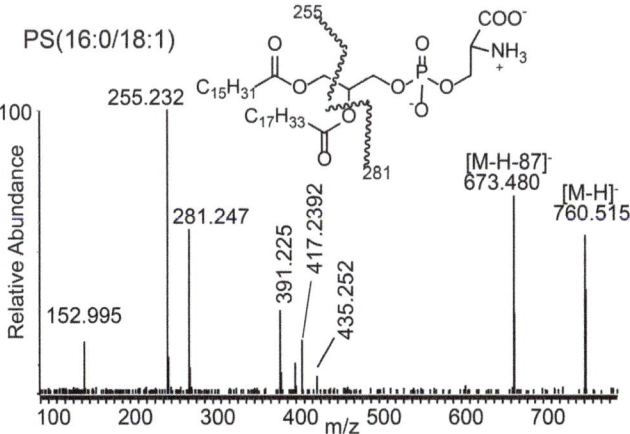

Scheme 5.37

Figure 5.10   Electrospray ionization (negative ions) and tandem mass spectrometry of a phosphatidylserine molecular species. Product ions obtained following collisional activation of PS(16 : 0/18 : 1) [M − H]$^-$ at *m/z* 760. This high resolution MS/MS spectrum was obtained using a quadrupole time-of-flight mass spectrometer.

and formation of an ion identical to that of phosphatidic acid. Furthermore, many of the other product ions appear to be derived from this rather facile $[M - H - 87]^-$ product ion. The mechanism of formation of this ion likely operates by at least two different mechanisms. This has been revealed through deuterium labeling of the exchangeable protons of the α-amino residue, where the loss of the serine involves partial loss of one of these exchangeable protons.[25] One mechanism for the formation of this ion would be the rather facile 6-membered transition state attack of the proton on the α-carbon atom of serine by the phosphoryl carbonyl group (Scheme 5.38). This likely is facilitated because of the acidic nature of this proton and its somewhat weakened bond strength that would enable this reaction to take place. This loss would not involve one of the exchangeable protons. The other mechanism likely involves exchangeable protons and several mechanisms have been proposed,[25] but these involve formation of highly strained 3- and 4-membered ring structures as leaving groups. A much more facile reaction might be the attack of the phosphate anionic site on the protonated carboxylic acid, which can then drive loss of CO_2 and ethyleneamine (Scheme 5.39). As two small neutral species are lost in this mechanism, an exchangeable proton from the serine carboxylic acid group would be retained in the structure of the phosphatidic acid-like product ion. Clearly, multiple mechanisms may result in the formation of this ion, which likely accounts for its very high and unique abundance.

The other abundant ions observed correspond to the carboxylate anions seen in the collisional activation tandem mass spectra of all other $[M - H]^-$ phospholipid species and a loss of ketene from the phosphatidic acid ion $[M - H - 87]^-$. These are observed as follows: $[M - H - 87 - R_1CH=C=O]^-$ at *m/z* 435 and $[M - H - 87 - R_2CH=C=O]^-$ at *m/z* 409. The loss of R_1COOH and R_2COOH from the $[M-H-87]^-$ ion observed at *m/z* 417 and 391,

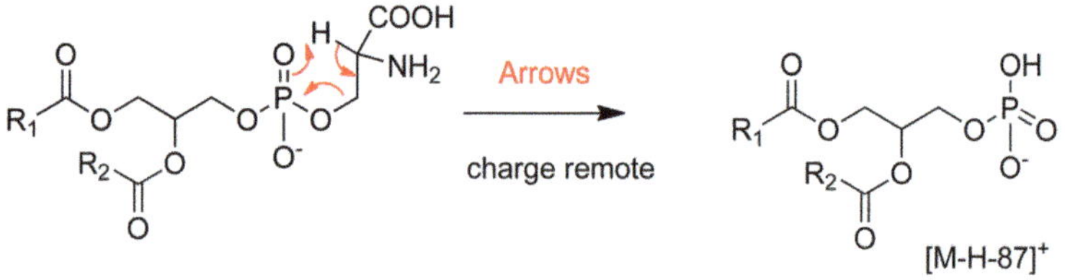

Scheme 5.38

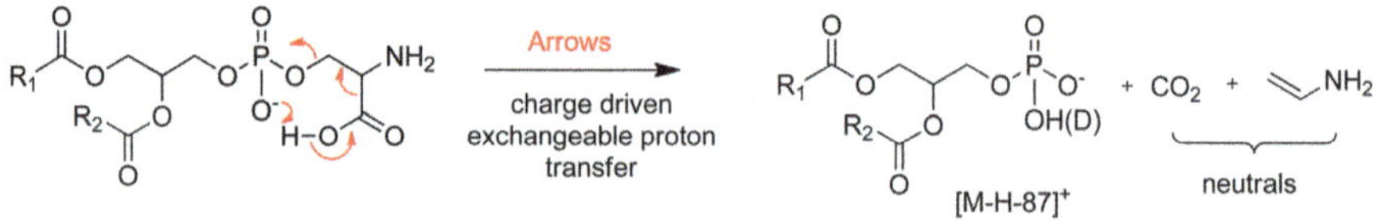

Scheme 5.39

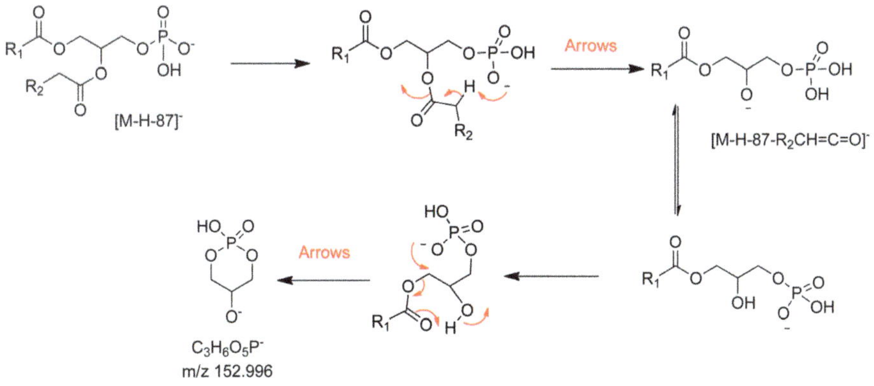

Scheme 5.40

respectively. The formation of these ions has been described previously for collisional activation of $[M - H]^-$ from PE ions and in general by Schemes 5.4 and 5.5, respectively. It is interesting to note that the relative abundance of the carboxylate anions to each other are reversed in order for PS (16 : 0/18 : 1) presented in Figure 5.10 to that seen for PE or PC negative ion derived carboxylate anions considering the *sn*-1 and *sn*-2 positions. However, they are in the same order seen for the collisional activation of phosphatidic acid (to be discussed later in this chapter) where the *sn*-2 fatty acyl carboxylate anion is somewhat lower in abundance. However, the loss of the neutral carboxylic acid ketene moieties do follow the same order of abundance seen for other fatty acids where the abundance of these losses are more abundant from the *sn*-2 position than the corresponding loss of ketene and carboxylic acid from the *sn*-1 position. Thus, these losses are typically employed to ascertain the acylation position of the two fatty acyl groups.[25]

Another abundant ion seen for this and other acidic phospholipids is the ion at *m/z* 153, which is characteristic of glycerophospholipids and is thought to have the structure shown in Scheme 5.40 as a cyclic phosphate anion.[11]

5.8 Phosphatidic Acid (PA)

The simplest glycerophospholipid is phosphatidic acid (PA), which is a phosphomonoester of diacylglycerol. PA is an intermediate in the biosynthesis of all other phospholipids by the Kennedy pathway.[26] Also, recent interest has centered on the role of lyso-PA as a signaling molecule.[27]

5.8.1 Positive Ion $[M + H]^+$ PA

While negative ionization is the most facile mode of ion formation, it is still possible to generate positive $[M + H]^+$ (Figure 5.11), $[M + Li]^+$, and $[M - H + 2Li]^+$ molecular ion species. The collisional activation of PA $[M + H]^+$

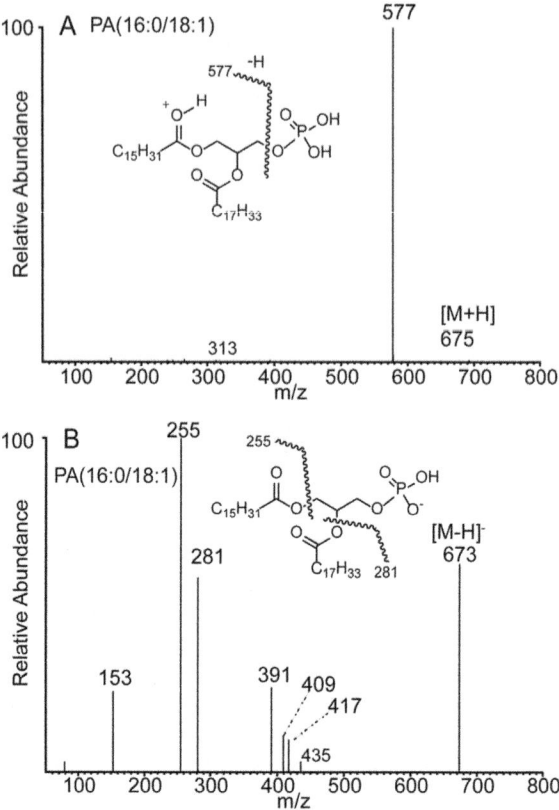

Figure 5.11   Electrospray ionization (positive and negative ions) and tandem mass
spectrometry of a phosphatidic acid molecular species. (A) Product
ions obtained following collisional activation of PA(16 : 0/18 : 1)
[M + H]⁺ at *m/z* 675; (B) product ions obtained following collisional
activation of PA(16 : 0/18 : 1)[M − H]⁻ at *m/z* 673. These MS/MS
spectra were obtained using a tandem quadrupole mass spectrometer.

generates a diglyceride-like ion corresponding to loss of the polar head group
(Scheme 5.2), which in this case is phosphoric acid [M + H − 96]⁺ as well as
acylium ions.[11] Another product ion corresponding to the loss of the *sn*-2
group as neutral ketene is formed from the diglyceride-like ion, perhaps in
a concerted mechanism (Scheme 5.41), but could also be formed from the
intermediate [M + H − 96]⁺ ion.[11]

5.8.2   Negative Ion [M − H]⁻ PA

The tandem mass spectrometry of the [M − H]⁻ generated from PA has been
studied in greater detail compared to the positive molecular ions.[8] Abundant
product ions are observed after collisional activation of this molecular ion
species that provides a great deal of structural information. The ions observed

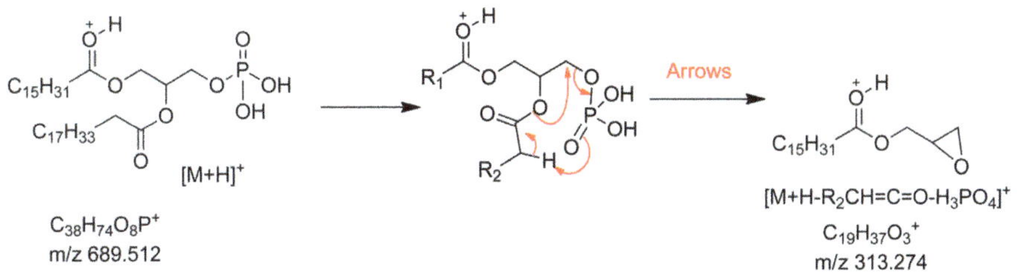

Scheme 5.41

are identical to those found for collisional activation of the $[M - H]^-$ from PE and PS as well as $[M - 15]^-$ from PC, as described previously. These include carboxylate anions $(R_1COO^-$ and $R_2COO^-)$ seen at m/z 255 and 281, loss of both fatty acyl groups as neutral ketene $[M - H - R_{1,2}CH=C=O]^-$ (seen at m/z 435 and 409), loss of both carboxylic acids $[M - H - R_{1,2}COOH]^-$ (seen at m/z 417 and 391), and m/z 153 (Figure 5.11B). However, an interesting feature is that the abundance of the carboxylate anions, relative to each other, is reversed to that typically found following collisional activation of most other phospholipid classes, in this case for PA where R_1COO^- is more abundant than R_2COO^-. This feature has been noted previously, which has led to the suggestion that the mechanism of formation of these ions is different to that observed for the more complex phospholipids with polar head groups containing additional functional groups.[19] The suggestion has been made that the carboxylate anions are derived from decomposition of an intermediate product ion, following from the initial loss of either R_1COOH or R_2COOH.[8] Furthermore, Turk and Hsu have suggested that this loss of the free carboxylic acid is due to the fact that PA is a very acidic phospholipid which facilitates loss of an acid as a gas phase reaction.[8] The structures of these intermediates are seen in Scheme 5.42 for these two losses of the neutral carboxylic acids.

The formation of a bicyclo phosphotriester from either $[M - H - R_{1,2}CH_2COOH]^-$ would result from the loss of the remaining fatty acyl group as a carboxylate anion (Scheme 5.43). Since the formation of the initial loss of RCOOH is observed as a more abundant ion for R_2COOH than R_1COOH, the higher abundant of the precursor ion for the R_1COO^- and R_2COO^- would drive the observed ratio of R_1COO^-/R_2COO^- greater than 1. Molecular models also suggest some steric hindrance in the attack of the phosphate anion at the *sn*-2 carbon atom that perhaps leads to an overall decrease in the release of R_2COO^- from the 6-membered ring phosphodiester intermediate.

5.8.3 Lyso-PA (LPA)

Positive ions corresponding to $[M + H]^+$, $[M + Li]^+$, and $[M + Na]^+$ can be readily formed by electrospray ionization (Figure 5.12A–C). The formation of the monoglyceride ion at m/z 339 following collisional activation of these

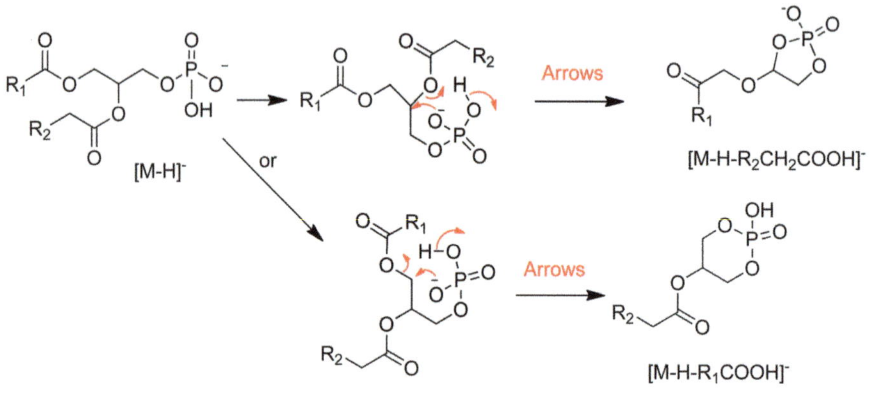

Scheme 5.42

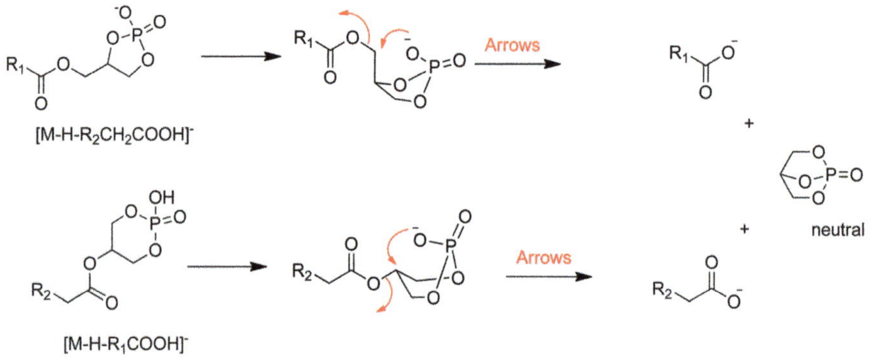

Scheme 5.43

positive ion species is found as a prominent product ion (Scheme 5.2). The product ions formed after collisional activation of LPA $[M − H]^-$ yield the expected carboxylate anion (Scheme 5.3) corresponding to the single fatty acyl group esterified to the glycerol backbone.[28] For LPA(16 : 0) this is observed at m/z 255 (Figure 5.12D). The dominant ion, however, is at m/z 153 seen at low mass-to-charge ratio in many other tandem mass spectra of acidic phospholipids. The mechanism of formation has been outlined in Scheme 5.40 for a *sn-2* lyso type (LPA) ion that was a product ion in the decomposition of the $[M − H − 87]^-$ from PS.

5.9 Phosphatidylinositol (PI)

Phosphatidylinositol is a phosphodiester of diacylglycerol and myoinositol, which renders this an acidic phospholipid with only the phosphate anion as the ionic form observed in either solution or in the gas phase. PI is found in

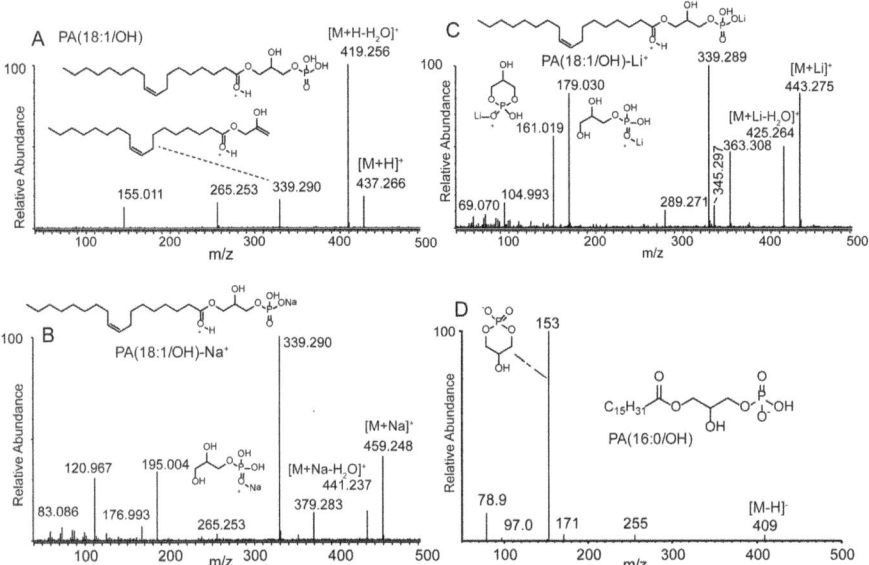

Figure 5.12 Electrospray ionization (positive and negative ions) and tandem mass spectrometry of lysophosphatidic acid molecular species. (A) Product ions obtained following collisional activation of PA(18 : 1/OH) [M + H]⁺ at *m/z* 437; (B) product ions obtained following collisional activation of PA(18 : 1/OH) [M + Na]⁺ at *m/z* 459; (C) product ions obtained following collisional activation of PA(18 : 1/OH) [M + Li]⁺ at *m/z* 443. These high resolution MS/MS spectra were obtained using a quadrupole time-of-flight mass spectrometer. (D) Product ions obtained following collisional activation of PA(16 : 1/OH) [M − H]⁻ at *m/z* 409. This MS/MS spectrum was obtained using a tandem quadrupole mass spectrometer.

most mammalian and higher animal membranes composed of a major molecular species containing arachidonate, PI (18 : 0/20 : 4), many other molecular species are observed but they are not as abundant. In addition, PI can be phosphorylated at several of the free hydroxyl groups of inositol in a surprisingly large number of unique polyphosphoinositides. Singly phosphorylated PI can be PI(3)P, PI(4)P, and PI(5)P which are typically abbreviated PIP as products of enzymatic phosphorylation by the enzyme PI3 kinase.[29] The doubly phosphorylated PI that have been characterized include PI(3,4)P$_2$, PI(3,5)P$_2$, and PI(4,5)P$_2$ and the triply phosphorylated PI as PI(3,4,5)P$_3$. These polyphosphorylated forms are difficult to analyze by mass spectrometry because they are present in very low abundance in cells due to their important role as signaling molecules, as well as being difficult to efficiently extract and they have less than ideal chromatographic properties for LC-MS analysis. This is a very active area of research and advances have been made on several lines to improve the analysis of these molecules by several laboratories.[29,30]

5.9.1   Positive Ion [M + H]⁺ PI

The formation of positive ions from PI during electrospray ionization is possible, but requires addition of acid to the solvent system that is sprayed. Under more typical conditions of electrospray LC-MS, only negative ions $[M - H]^-$ are observed. Nonetheless, the collisional activation of $[M + H]^+$ from PI generates a few significant ions (Figure 5.13A). The most abundant product ion corresponds to the loss of the phosphoinositol head group and formation of the diglyceride-like ion $[M + H - 260]^+$ by general mechanism seen in Schemes 5.2 and 5.44. The other ions corresponds to the formation of acylium ions from the fatty acyl groups (m/z 239 and 265) in Figure 5.13A. The loss of neutral ketene from each fatty acyl groups observed at m/z 313 and 339

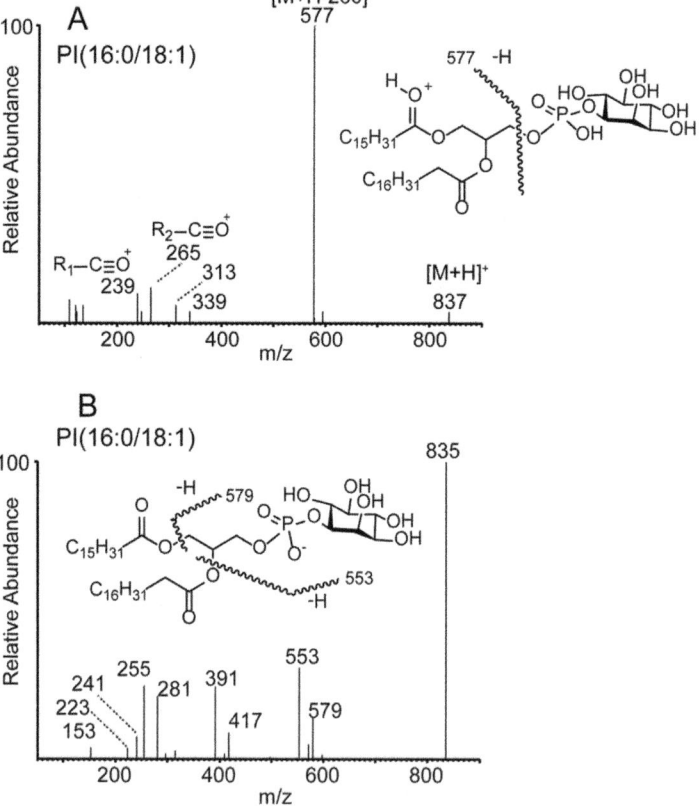

Figure 5.13   Electrospray ionization (positive and negative ions) and tandem mass spectrometry of a phosphatidylinositol molecular species. (A) Product ions obtained following collisional activation of PI(16 : 0/18 : 1) $[M + H]^+$ at m/z 837; (B) product ions obtained following collisional activation of PI(16 : 0/18 : 1) $[M - H]^-$ at m/z 835. These MS/MS spectra were obtained using a tandem quadrupole mass spectrometer.

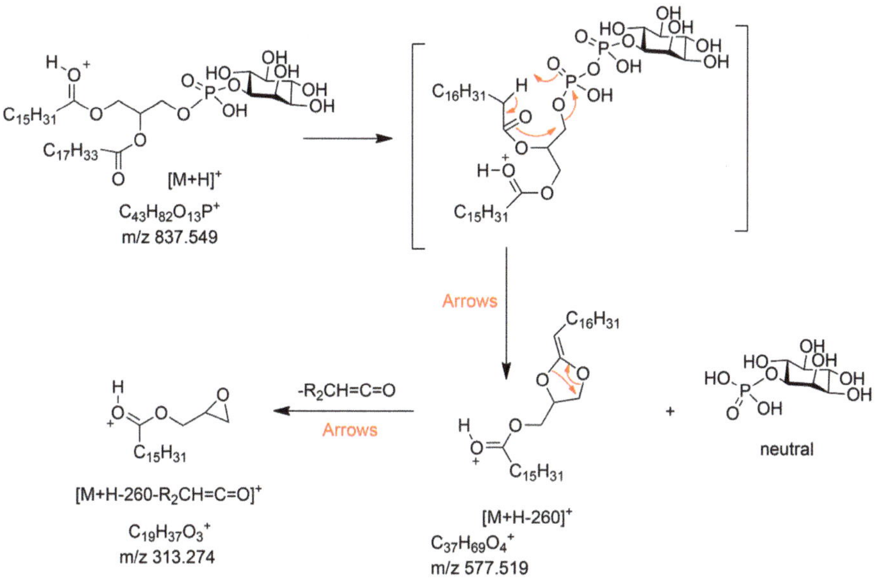

Scheme 5.44

occurs by a mechanism that involves the loss of the inositol head group as presented in Scheme 5.44.

5.9.2 Negative Ion [M − H]⁻ PI

The product ions formed following collisional activation of the negative ions $[M - H]^-$ from PI are quite similar to that seen for the acidic phospholipid PA with a few additional abundant ions unique to the phosphoinositol polar headgroup. The loss of both carboxylate groups as free carboxylate anions likely are a result of several mechanisms operating and have been discussed for all other $[M - H]^-$ ions, and even the $[M - 15]^-$ anion from PC. The mechanism that is most likely responsible for the most significant proportion of the R_1COO^- and R_2COO^- was outlined in Scheme 5.3, as well as Schemes 5.17 and 5.30 where a phosphate anion attacks a glycerol carbon at *sn*-1 or *sn*-2 to yield the respective carboxylate anions. An alternative mechanism for formation of these carboxylate anions involves an initial loss of the free carboxylic acid $[M - H - R_xCOOH]^-$ from the molecular ion, followed by release of $RCOO^-$ species as outlined for PA in Scheme 5.42.

Of noticeable abundance are the ions corresponding to a neutral loss of each fatty acyl group as a free carboxylic acid and ketene. These are present in Figure 5.13B at *m/z* 553, $[M - H - R_2COOH]^-$, *m/z* 579 $[M - H - R_1COOH]^-$, *m/z* 571 $[M - H - R_2CH=C=O]^-$, and *m/z* 597 $[M - H - R_1CHC=C=O]^-$, respectively. In the several cases studied,[16,24,31] the loss of the *sn*-2 free acid and ketene are the most abundant ions observed of this type, suggesting that these ions

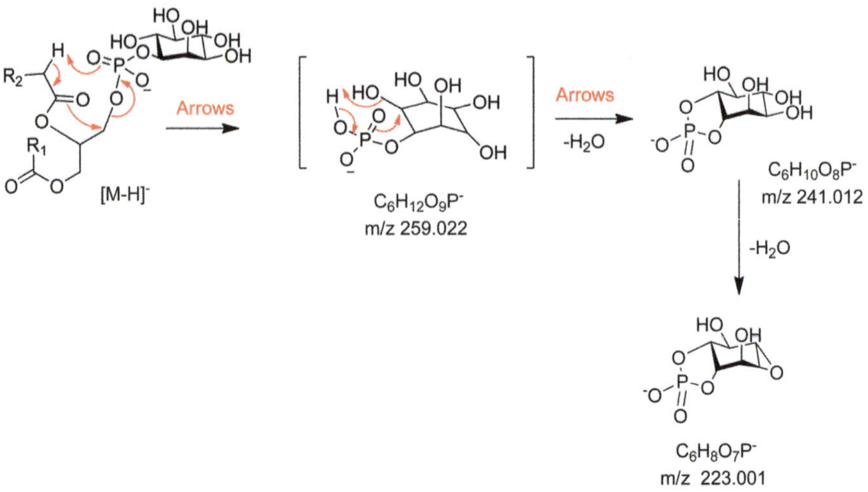

Scheme 5.45

could be indicative of the respective fatty acyl group location on the glycerol backbone. The mechanisms of formation of these ions have been suggested to be identical to those presented previously as Schemes 5.41 and 5.31.

An ion quite characteristic of PI is observed at m/z 241 and corresponds to a cyclic anion of inositol phosphate. The formation of this ion is initiated by remote site fragmentation quite analogous to the mechanism of formation of the positive ion at m/z 184 from PC (Scheme 5.2), only in this case the charge site is on the anion localized at the phosphate headgroup. The mechanism operating for the m/z 241 formation from PI is suggested in Scheme 5.45. After a hydrogen atom rearrangement from the 2′ fatty acyl chain and cleavage of the phosphoester bond, a final loss of water forms the stable cyclic phosphate anion. The ion at m/z 223 (Figure 5.13B) would arise from a further loss of water from m/z 241.

The formation of the ion at m/z 391 (Figure 5.13B) has been suggested to have at least two different pathways of formation, either from the loss of the sn-2 ketene or from loss of the sn-2 carboxylic acid, followed by loss of a neutral inositol epoxide (180 Da).[24,31] A possible mechanism for this is outlined in Scheme 5.46. The ions at m/z 417 and 391 correspond to the loss of the sn-1 group as ketene and carboxylic acid, respectively.

5.9.3 Polyphosphoinositides [M − H]⁻ and [M − 2H]⁼

The phosphoinositide phospholipids can be further phosphorylated to PI(X)P, as indicated at the beginning of this section. The tandem mass spectrometry of the singly charged PI(X)P (negative ion) is quite similar to the PI [M − H]⁻ product ions (Figure 5.14), with formation of carboxylate anions, loss of ketene, as well as free carboxylic acid from the sn-2 position as well as the PI diagnostic

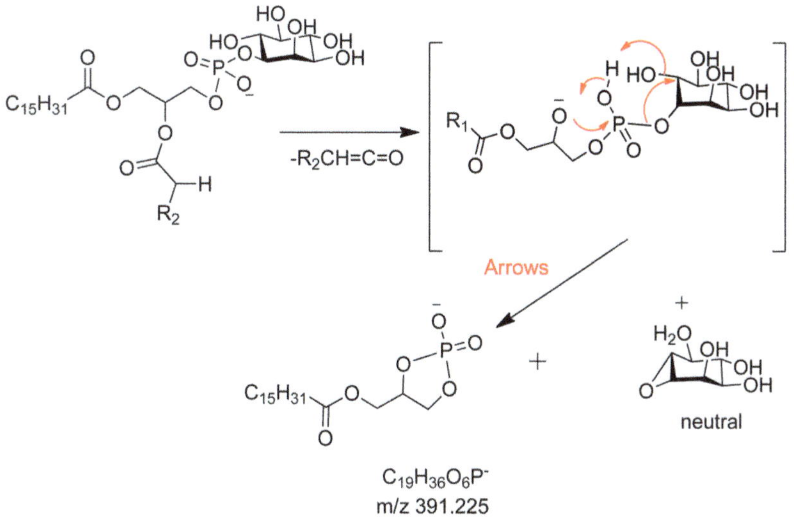

Scheme 5.46

product ions at *m/z* 241, 79, and 97. A prominent ion quite unique to PIP isomers is observed at *m/z* 321, which has the suggested structure of phosphoryl inositol phosphate. This ion has been suggested to form by the loss of the inositol diphosphate headgroup with formation of a neutral diglyceride-like species (Scheme 5.47) shown for PI(4)P.[31]

A unique feature of PIP species is the ability to exist as doubly charged anions since the phosphate residues are sufficiently separated by the rigid inositol ring. Aside from a doubly charged $[M - H]^=$ ion, another prominent doubly charged ion corresponds to the loss of *sn*-2 ketene (*m/z* 325, Figure 5.14) by the mechanisms suggested in Scheme 5.48. The behavior of the positional isomer PI(3)P and of the PI(4)P is quite similar, making it difficult to ascertain a phosphorylation position in the inositol ring by mass spectrometry. It has been suggested, however, that the loss of H_3PO_4 (loss of 98 Da) from $[M - H]^-$ of PI(4)P is more favorable than the isomer PI(3)P, but these can be very minor product ions and very few species have been studied to extend these observations to a more general mechanism.[31]

The more highly phosphorylated PIs only form $[M - H]^-$ and $[M - 2H]^=$ since the additional phosphate moieties are now quite close to each other on the inositol ring, which presents stability problems to localize an additional negative charge. The same general features described for the collisional activation of PIP and PI(2)P operate for these polyphosphoinositides.

5.9.4 Glycophosphoinositides

Even more complex PI molecular species are known to exist in nature, including families of mannose-containing PIs. Such complex phosphatidylmyoinositol mannosides are found in bacterial biochemistry, including

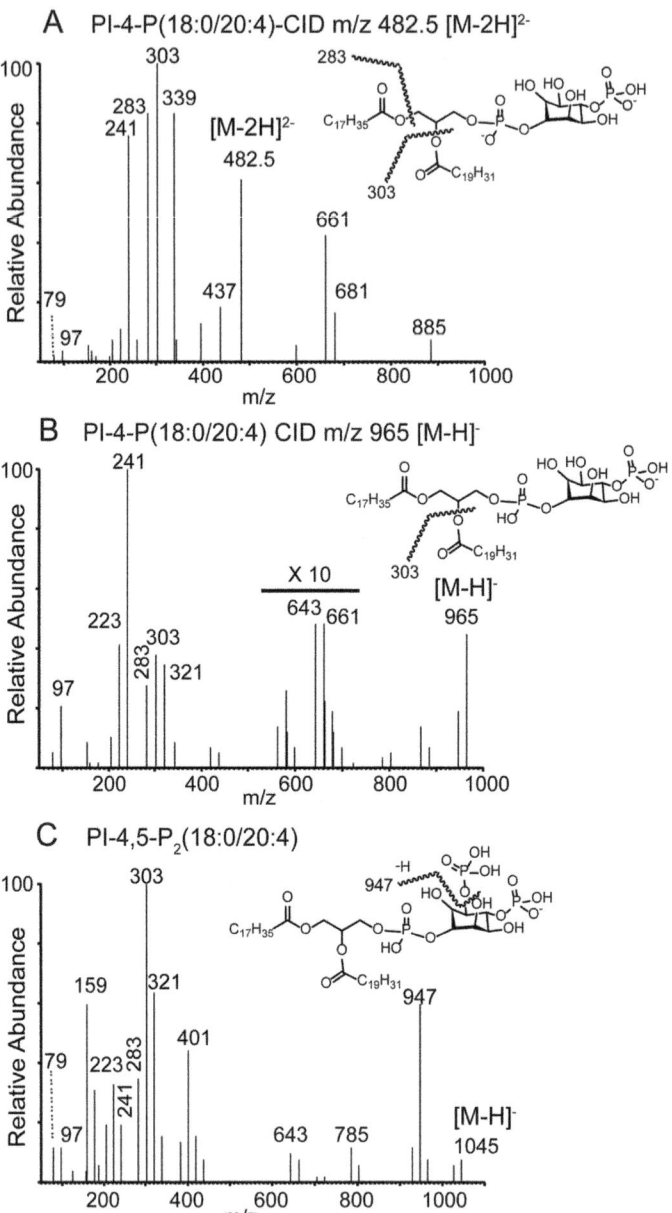

Figure 5.14 Electrospray ionization (negative ions) and tandem mass spectrometry of phosphatidylinositol phosphates. (A) Product ions obtained following collisional activation of the doubly charged ion PI(4)P(18 : 0/20 : 4) [M − 2H]$^{2-}$ at *m/z* 482.5; (B) Product ions obtained following collisional activation of the singly charged ion from PI(4)P (18 : 0/20 : 4) [M − H]$^-$ at *m/z* 965; (C) product ions obtained following collisional activation of PI-4,5-P$_2$(18 : 0/20 : 4) [M − H]$^-$ at *m/z* 1045. This figure was redrawn from data presented in ref. 31.

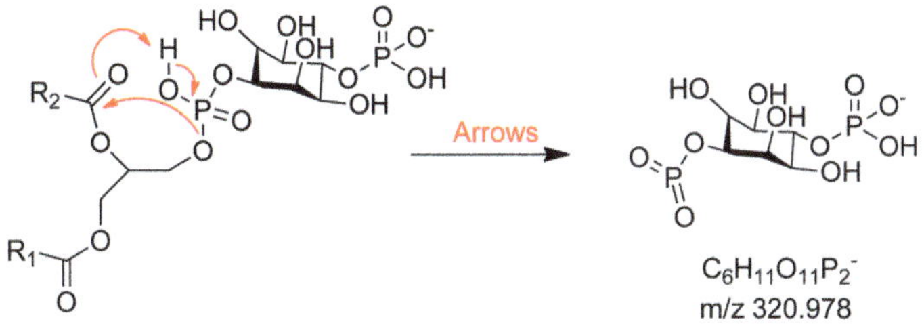

$C_6H_{11}O_{11}P_2^-$
m/z 320.978

Scheme 5.47

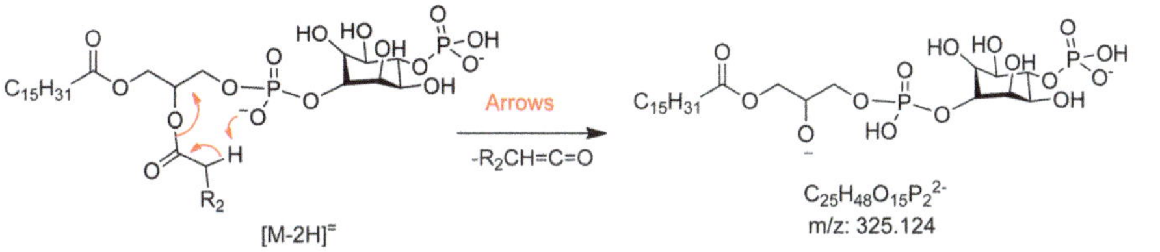

$C_{25}H_{48}O_{15}P_2^{2-}$
m/z: 325.124

Scheme 5.48

Mycobacterium tuberculosis, where PI species with 1–5 mannose sugars are attached. There have been detailed mass spectrometric studies of these unique glycophosphatidylinositides, as well as the identification of sugar-related product ions following collisional activation of $[M - H]^-$.[32]

5.10 Phosphatidylglycerol (PG)

Phosphatidylglycerol is an acidic phospholipid that is primarily analyzed by tandem mass spectrometry as a negative $[M - H]^-$ species after electrospray ionization. This is a very interesting phospholipid from various biochemical and biomedical aspects since it is a significant component of pulmonary surfactant in humans and it is thought to have direct antiviral activity.[33] Under appropriate conditions, positive ions can also be observed and studied.

5.10.1 Positive Ion $[M + H]^+$ PG

The collisional activation of PG $[M + H]^+$ leads to a major ion corresponding to the diglyceride-like structure as $[M + H - 172]^+$ observed at m/z 577 from PG(16 : 0/18 : 1) (Figure 5.15A). Since the site of ionization due to proton attachment is likely at an ester moiety, the mechanism previously discussed for decomposition of such ions is particularly relevant (Scheme 5.2). The minor ions, which can be formed following collisional activation of this positive ion, correspond to acylium ions from *sn*-1 and *sn*-2 fatty acyl positions. These are observed at m/z 239 and 265 for PG (16 : 0/18 : 1) in Figure 5.15A.

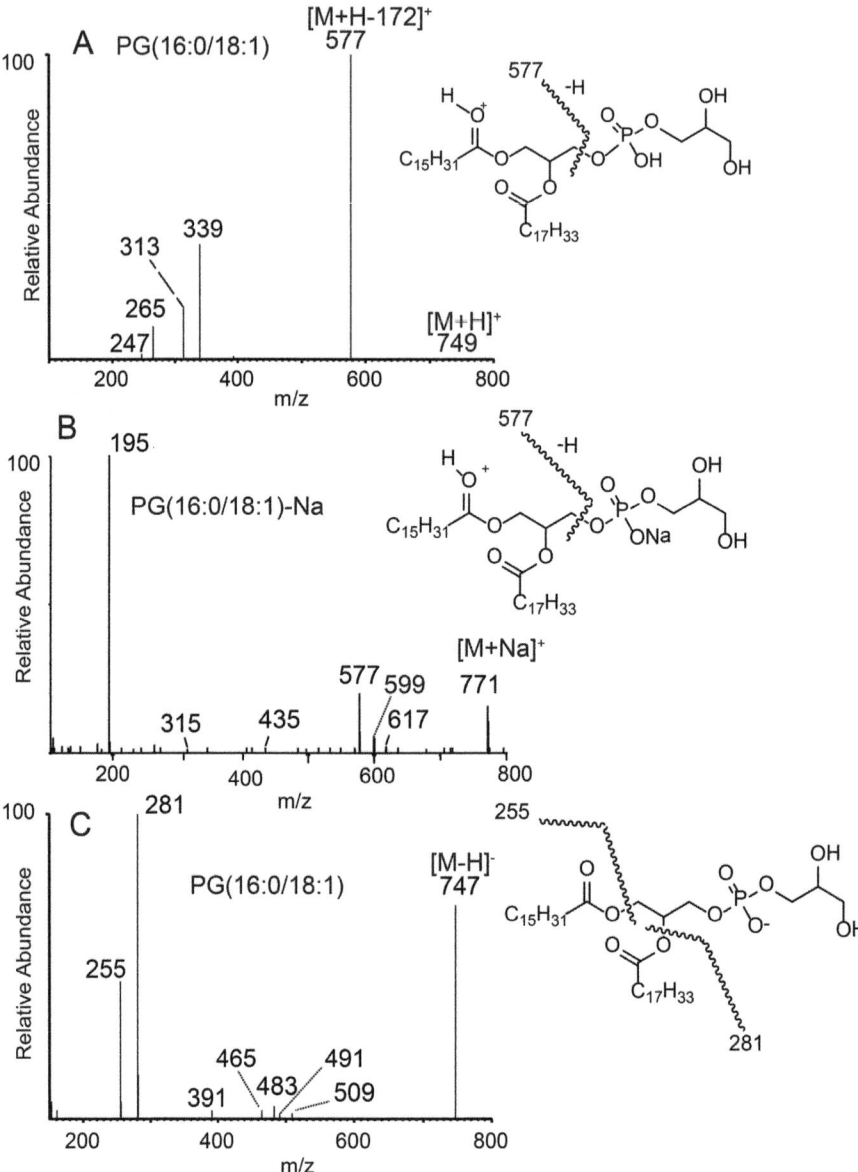

Figure 5.15 Electrospray ionization (positive and negative ions) and tandem mass spectrometry of a phosphatidylglycerol molecular species. (A) Product ions obtained following collisional activation of PG(16 : 0/18 : 1) [M + H]⁺ at *m/z* 749; (B) product ions obtained following collisional activation of PG(16 : 0/18 : 1) [M + Na]⁺ at *m/z* 771; (C) product ions obtained following collisional activation of PG(16 : 0/18 : 1) [M − H]⁻ at *m/z* 747. These MS/MS spectra were obtained using a tandem quadrupole mass spectrometer.

5.10.2 Positive Ion [M + Na]⁺ PG

Alkali attachment ions can also be formed from PG and as exemplified by [M + Na]⁺ for PG (16 : 0/18 : 1) in Figure 5.15B. Collisional activation of this ion leads to two diglyceride-like product ions at m/z 577 and 599 that are formed from the sodium adducts attached to different sites in the molecule. When the sodium ion charges one of the ester moieties as illustrated in the structure in Figure 5.15B, the loss of the polar headgroup appears at m/z 599, corresponding to [M + Na − 172]⁺. The more abundant diglyceride ion corresponds to the sodium ion attaching as a phosphate salt with a proton charging one of the ester groups. The diglyceride ion which forms following collisional activation therefore loses the sodium atom and the observed product ion mass corresponds to [M + Na − 194]⁺. The most abundant product ion corresponds to m/z 195 and arises from the same mechanisms outlined in Scheme 5.2, but the charge site involves a sodium ion attachment site on the phosphoryl oxygen atom (Scheme 5.49). Acylium ions are not very abundant products from this form of the sodium adduct, likely due to the quite facile formation of m/z 195.

5.10.3 Negative Ion [M − H]⁻ PG

The most abundant ions formed following collisional activation of PG [M − H]⁻ molecular anions are the expected carboxylate anions from the *sn*-1 and *sn*-2 fatty acyl esters (Figure 5.15C). The mechanism for the formation of these ions has been presented in generic form (Scheme 5.3). The relative abundance of R_1COO^- to that of R_2COO^- is consistent across several molecular species that have been studied, where the R_2COO^- is more abundant than the R_1COO^-, similar to that formed from [M − 15]⁻ from PC and [M − H]⁻ for PE.[9,11,34]

Less abundant ions correspond to loss of *sn*-1 and *sn*-2 as ketene [M − H − $R_{1,2}CH=C=O$]⁻ observed at m/z 509 and 483, respectively, in Figure 5.15C and free carboxylic acid ([M-H-$R_{1,2}COOH$]⁻) observed at m/z 491 and 465 (Schemes 5.4 and 5.5). It has been suggested that the relative ratio of these ions and the predominance of the loss of ketene over free carboxylic acid, is related to the acidic nature of PG.[9] However, these losses are not always observed as very

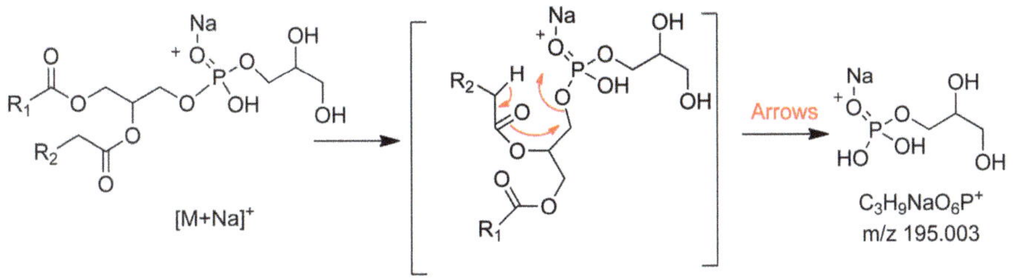

Scheme 5.49

abundant, they are likely a result of collisional activation condition and instrument parameters that effect the yield of product ions as well as the ion transit times (for example, tandem quadrupole *versus* quadrupole time-of-flight, as well as ion mobility mass spectrometers). Nevertheless, the $[M - H - R_2CH=C=O]^-$ and $[M - H - R_2COOH]^-$ appear to dominate in abundance over the corresponding R_1 acyl group losses (Figure 5.15C).

The ions corresponding to the neutral loss of either $R_{1,2}COOH$ from $[M - H]^-$ are interesting in that they arise from at least two different mechanisms. One mechanism is charge-driven and one is a charge-remote.[9] These two mechanisms are illustrated in Schemes 5.50 and 5.51 for loss of the *sn*-2 fatty acyl group. When the polar head group exchangeable hydrogen atoms are labeled with deuterium atoms, these two product ions appear as doublets consistent with two different mechanisms. The ratio of these two ions in the deuterated species can change when collision energy changes, revealing the predominance of one pathway over the other, depending on the internal energy of the ion driving the mechanism.

Two other additional ions are present in the collisional spectra of PG negative ions. One corresponds to the loss of the fatty acyl group at *sn*-2 as a free carboxylic acid along with the polar headgroup glycerol (*m/z* 391). The

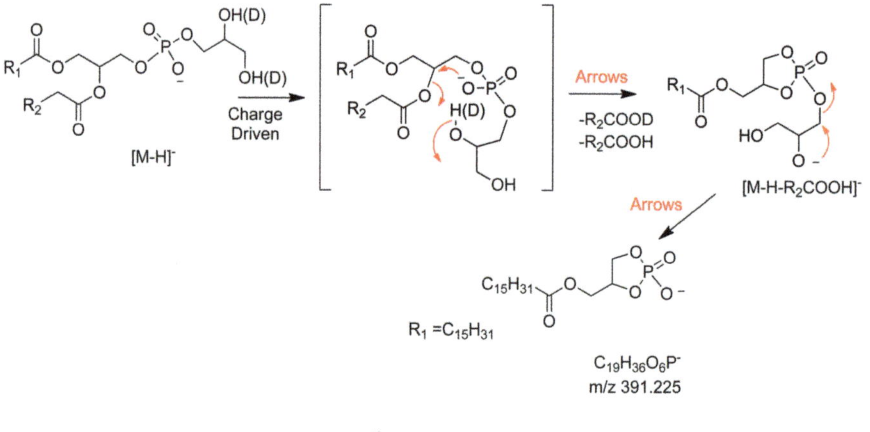

Scheme 5.50

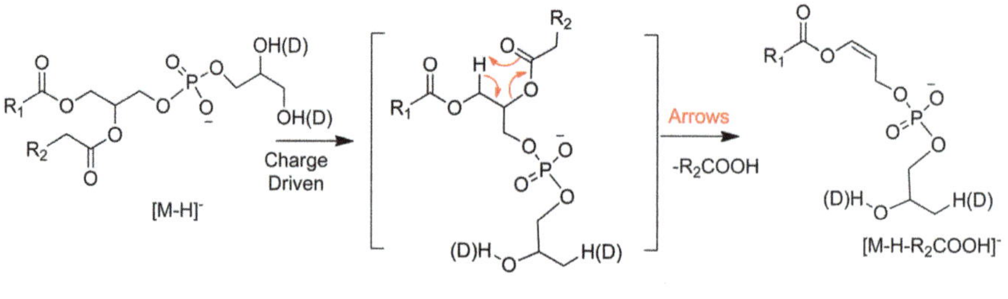

Scheme 5.51

mechanism for this ion has been proposed[9] to involve participation of the glycerol oxygen atom on the polar headgroup and the formation of a cyclic phosphate ester in the charge-driven mechanism (Scheme 5.50). A second reasonably abundant ion is observed at m/z 153, corresponding to the cyclic phosphate anion previously discussed (Scheme 5.42) as a product ion from PA.

5.11 Bis(Monoacyl-Lysophosphatidyl)Glycerol (BMP)

A rather unique phosphatidylglycerol regioisomer is bis(monoacyl lyso-phosphatidyl)glycerol, which has two fatty acyl groups and two free glycerol hydroxyl groups on each glycerol moiety of PG and is often referred to as BMP. This glycerophospholipid is found fairly abundantly in lysosomes and other intracellular structures.[35] Very few detailed studies had been made concerning the collision induced decomposition of the positive or negative ions that can form from this acidic phospholipid. Electrospray ionization generates both abundant negative ion $[M - H]^-$ as well as positive ions corresponding to $[M + H]^+$ and $[M + Na]^+$.

The product ions obtained from collisional activation of the $[M - H]^-$ from BMP (18 : 1/18 : 1) are dominated by the single carboxylate anion at m/z 281 (Figure 5.16A) and likely arise from the mechanisms previously described (Scheme 5.3). This product ion mass spectrum is identical to that of PG(18 : 1/18 : 1) and does not permit assignment of this phospholipid as a BMP or, for that matter, a PG(18 : 1/18 : 1).[36] These carboxylate anions do define the esterified fatty acyl groups; however, no information is available as to which fatty acyl moiety on which glycerol moiety is responsible for this ion. There are also low abundance ions corresponding to $[M - H - RCOOH]^-$ seen at m/z 491 in Figure 5.16A.

Collisional activation of the positive ions from BMP $[M + H]^+$ can be used to differentiate BMP phospholipids from their corresponding isomeric PG phospholipids since the major pathway for acidic phospholipid decomposition is the loss of the polar head group with formation of a diglyceride-like ion, or in this case a monoglyceride ion (Figure 5.16B). The monoglyceride ion mass and the neutral fragment lost define the molecular species of BMP (Scheme 5.52). Another interesting difference between BMP and PG is the loss of one and two molecules of H_2O in the collisional activation of $[M + H]^+$ in BMP (Scheme 5.53).[36] These ions are not observed in the CID of the isomeric PG molecular species. From experiments that labeled the exchangeable protons on BMP, it is evident that at least two separate mechanism are responsible for these losses of water. Possible charge driven mechanisms are presented in Scheme 5.53.

Collisional activation of $[M + Na]^+$ from sodiated BMP yields product ions that include m/z 459 (Figure 5.16C), which is consistent with the polar head group ion in general mechanism (Scheme 5.2) after loss of the diglyceride neutral fragment. In this case the charge is retained on the polar head group as well as the sodium ion.

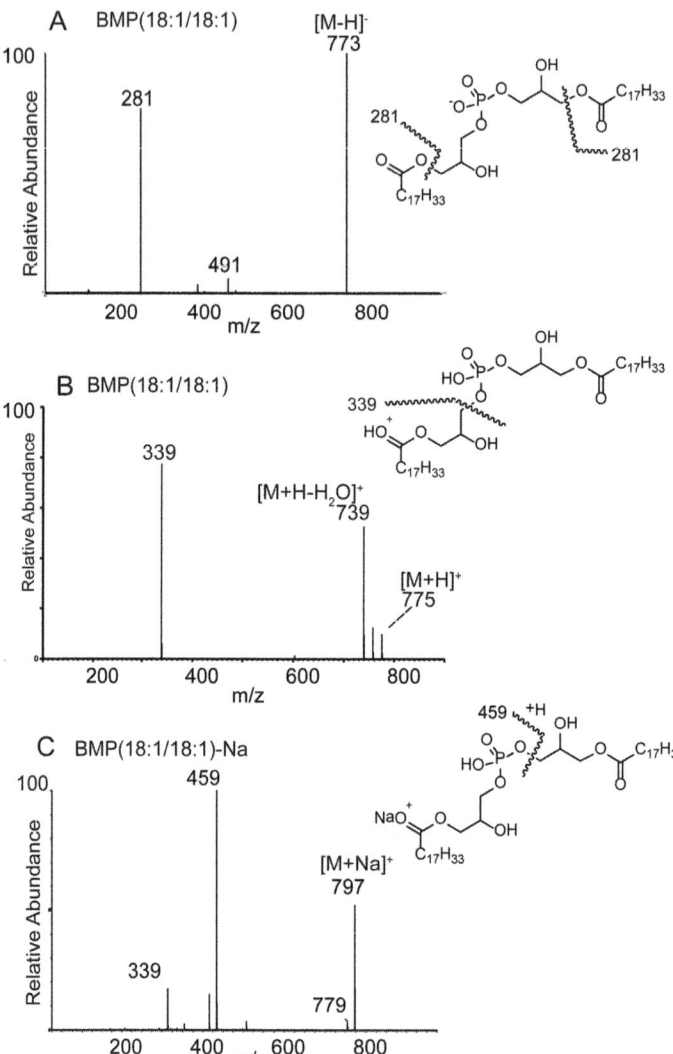

Figure 5.16   Electrospray ionization (negative positive ions) and tandem mass
spectrometry of bis(monoacylglycerol)phosphate. (A) Product ions
obtained following collisional activation of BMP(18 : 1/18 : 1)
[M − H]⁻ at m/z 773; (B) product ions obtained following collisional
activation of BMP(18 : 1/18 : 1) [M + H]⁺ at m/z 775; (C) product ions
obtained following collisional activation of BMP(18 : 1/18 : 1)
[M + Na]⁺ at m/z 797. These MS/MS spectra were obtained using
a tandem quadrupole mass spectrometer.

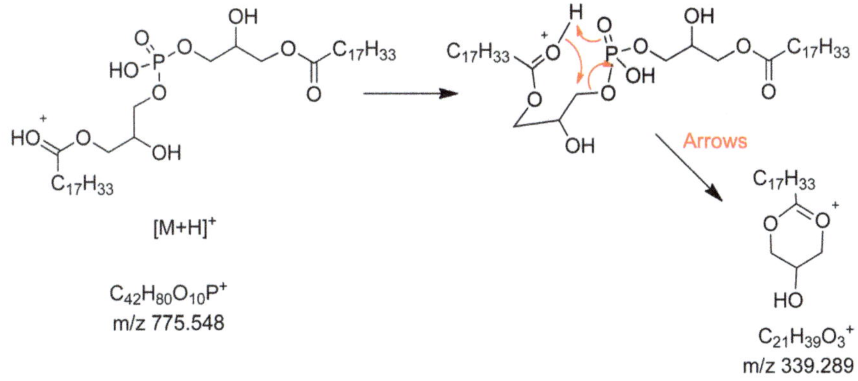

Scheme 5.52

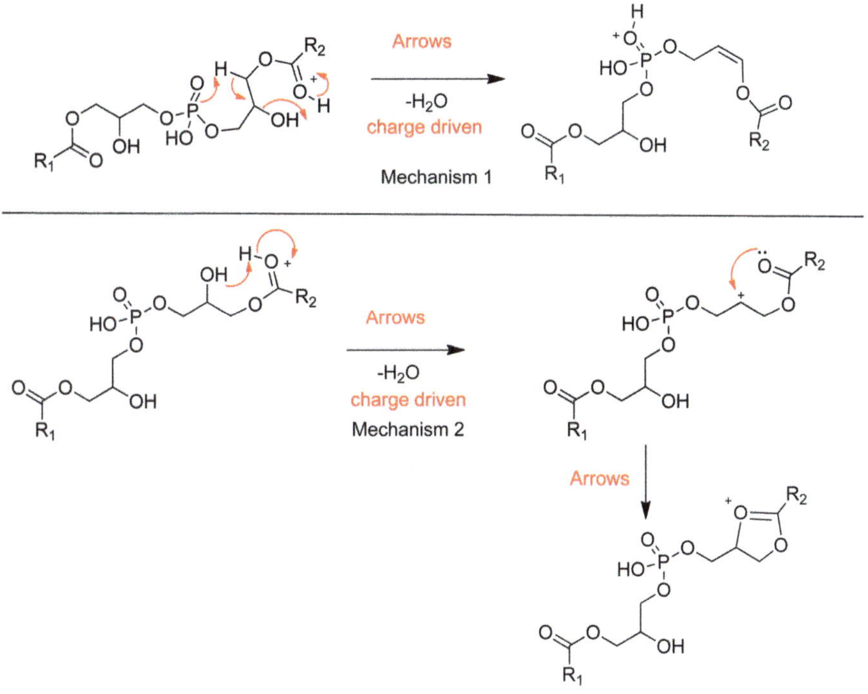

Scheme 5.53

5.12   Cardiolipin

The glycerophospholipids termed cardiolipins (CL) contain two phosphatic acid moieties linked by phosphoester bonds to a common third glycerol moiety. Thus, these lipids have four fatty acyl chains, three glycerol moieties, and two phosphodiester groups, rendering these species the highest molecular weights that are typically encountered for glycerophospholipids. These lipids are found in high abundance in the mitochondria, where they play a central role in the electron transport systems and oxidative phosphorylation.[37] The tandem mass spectrometry of CL has been studied in some detail as both positive and negative ions, however; negative ions as $[M - H]^-$ and $[M - 2H]^=$ are the more abundant electrospray generated ions.

As seen in Figure 5.17A, the doubly charged negative ions are typically more abundant than the singly charged ions and these ions can be readily identified since the carbon-13 isotope containing species differ by only 0.5 Da (Figure 5.17B). When the fatty acyl chains are much larger than the 18-carbon chains in the example (Figure 5.17), the isotopic species containing a single carbon-13 is almost as abundant as the all carbon-12 mono-isotopic species. Furthermore, the exact mass of the molecular ion is found almost one mass unit higher than if one calculated the cardiolipin molecular weight based on integer values for the elements. This could lead to some confusion, making it important to keep track of the exact number of carbon atoms and protons in these species. Singly charged ions $[M - H]^-$ can be observed, as well as singly charged ions that have the addition of sodium atoms $[M - 2H + Na]^-$.

Designation of the cardiolipin exact glycerol atoms will be followed here essentially following the suggestion of Schlaume where the central glycerol atoms of CL are indicated as $1'$, $2'$, and $3'$, and the 1,2-diacyl-*sn*-glycerol-3-phosphoryl groups are designated as pro-S when it is linked to the $1'$ central glycerol carbon and 1,2-diacyl-*sn*-glycerol-3-phosphoryl group linked to the $3'$ will be termed the Pro-R (Scheme 5.54).[38] The ability of CLs to generate doubly charged ions in solution requires somewhat higher pH than expected. In solution, the first phosphate of CL ionizes with a pKa of 2.8, but the remaining phosphate (although identical in all respects to the first phosphate) has its pKa raised to over 7.5.[38] After electrospray ionization, the doubly charged ion $[M - 2H]^=$ can dominate over the $[M - H]^-$ (Figure 5.17A) in the negative ion mode,[39] while in the positive ion mode efforts need to be taken to remove alkali metal ions such as Na^+ or high abundance of both $[M + H]^+$ and $[M + Na]^+$ are observed. In addition to these species, $[M - 2H + 3Na]^+$,[40] $[M - 2H + 3Li]^+$,[41] and $[M - 2H + Na]^+$,[42] have been studied by tandem mass spectrometry and mechanism of product ion formation discussed.

5.12.1   Cardiolipin $[M + H]^+$ Positive Ions

The collisional spectra of $[M + H]^+$ are surprisingly simple, with a major product ion corresponding to each diglyceride-like moiety (termed R and S diglycerides) following the general mechanism seen for other $[M + H]^+$

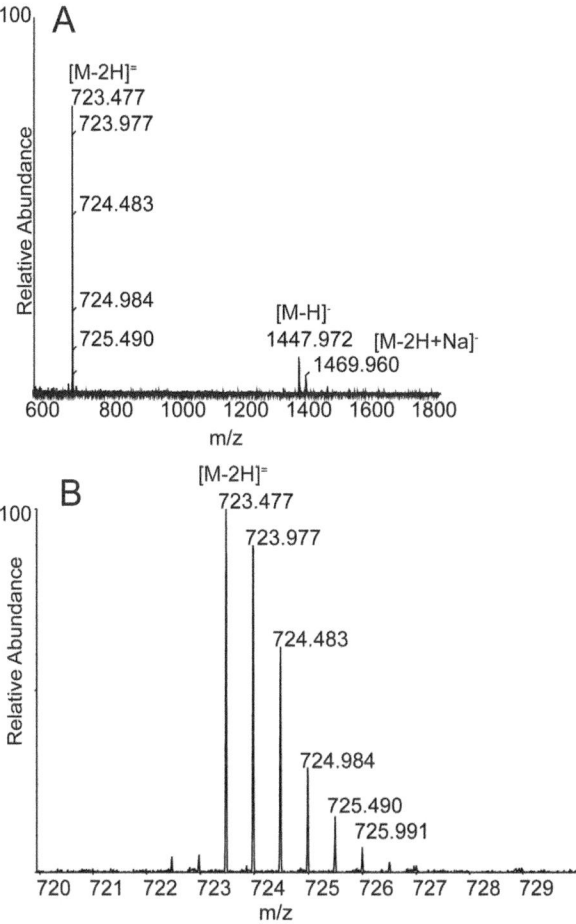

Figure 5.17 Electrospray ionization (negative ions) and tandem mass spectrometry of a cardiolipin molecular species. (A) The electrospray ions obtained from CL(18 : 2/18 : 2/18 : 2/18 : 2) that leads to abundant $[M - 2H]^{2-}$ as doubly charged ions and $[M - H]^-$ as singly charged ions; (B) detail of the doubly charged molecular ion region for CL(18 : 2/18 : 2/18 : 2) $[M - 2H]^{2-}$. These high resolution electrospray mass spectra were obtained using a quadrupole time-of-flight mass spectrometer.

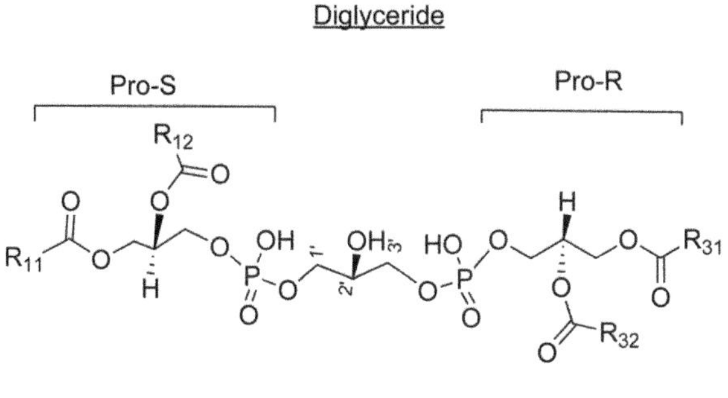

Diglyceride

Scheme 5.54

phospholipid product ions (Scheme 5.2). In the case of CL, the diglyceride-like ion observed corresponds to either the R or S diglyceride retaining the positive charge site. The remaining portion of the molecule is not observed. For example, the diglyceride-like ions from CL(18 : 2/18 : 2/18 : 2/18 : 2) illustrated in Figure 5.18A appear as one ion at *m/z* 599.5 (Scheme 5.55).

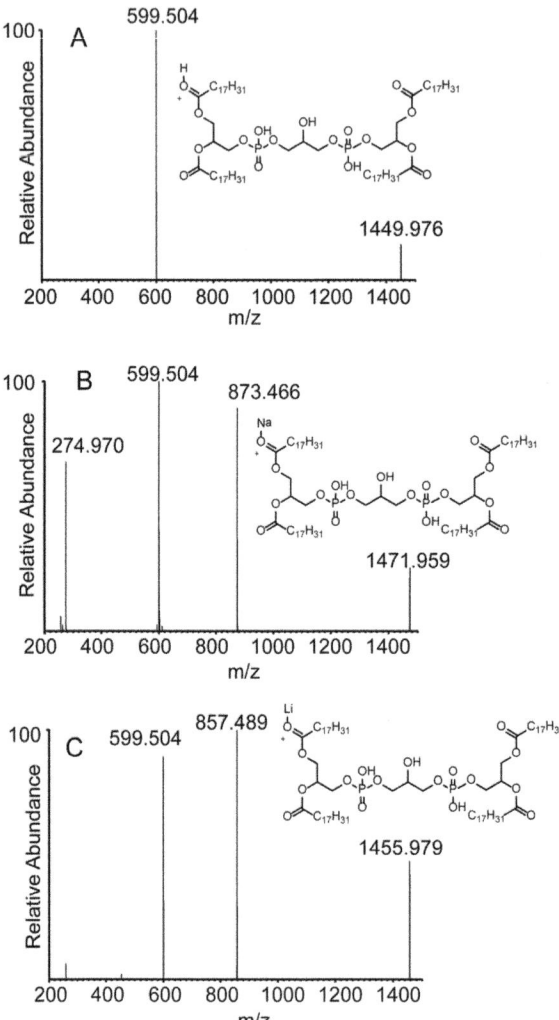

Figure 5.18 Electrospray ionization (positive ions) and tandem mass spectrometry of a cardiolipin molecular species. (A) Product ions obtained following collisional activation of CL(18 : 2/18 : 2/18 : 2/18 : 2) [M + H]$^+$ at *m/z* 1449.9; (B) product ions obtained following collisional activation of CL(18 : 2/18 : 2/18 : 2/18 : 2) [M + Na]$^+$ at *m/z* 1471.9; (C) product ions obtained following collisional activation of CL(18 : 2/18 : 2/18 : 2/18 : 2) [M + Li]$^+$ at *m/z* 1455. These high resolution MS/MS spectra were obtained using a quadrupole time-of-flight mass spectrometer.

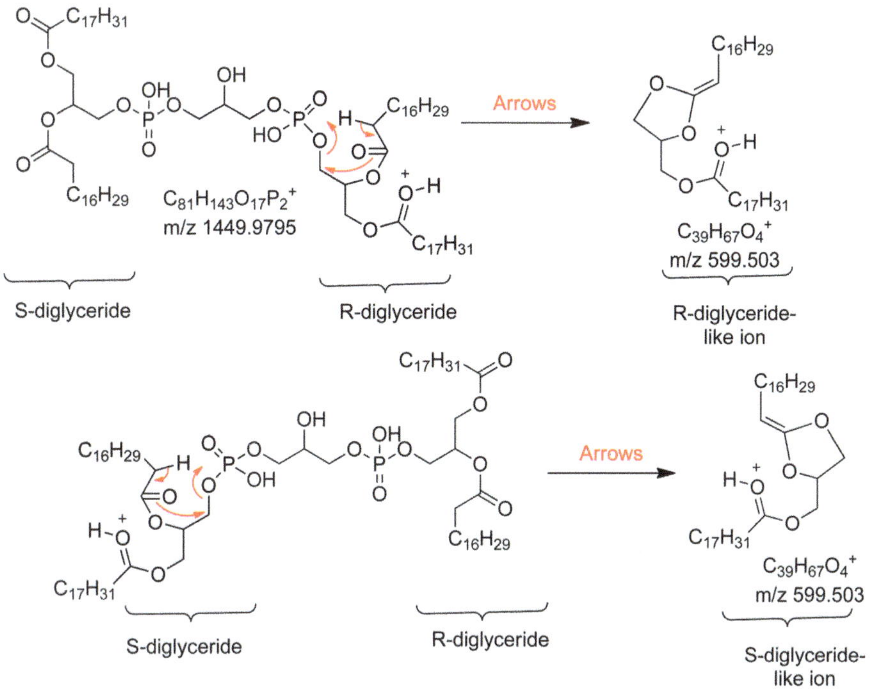

Scheme 5.55

However, both R and S diglycerides are present within this single ion, but they are identical in mass due to identical fatty acyl substituents. With multiple fatty acyl groups the symmetry is destroyed and these diglyceride product ions appear as two separate ions. Interestingly, this CL(18 : 2/18 : 2/18 : 2/18 : 2) is the most abundant cardiolipin molecular species found in the mitochondria from cells isolated from the hearts of mammalian organisms.[43]

The same mechanism suggested by Scheme 5.2 could occur with the phosphodiester distal to either the R or S diglyceride that carries the charging proton, which would lead to an ion at *m/z* 850.4. Such a theoretical ion would be the polar head group ion in this scheme. Yet this ion is not observed, perhaps indicating that decomposition to the diglyceride-like ion which carries the charge (that is observed) is the dominate reaction pathway.

5.12.2 Cardiolipin $[M + Na]^+$ Positive Ions

The behavior of the single alkali metal adduct $[M + Na]^+$ is somewhat more complex than that observed for $[M + H]^+$ in that three ion species are observed even for this very symmetric CL species (Figure 5.18B). The most abundant ion corresponds to the R and S diglyceride product ions, which most likely are formed as described in Scheme 5.55, where the Na^+ is a salt of one of the

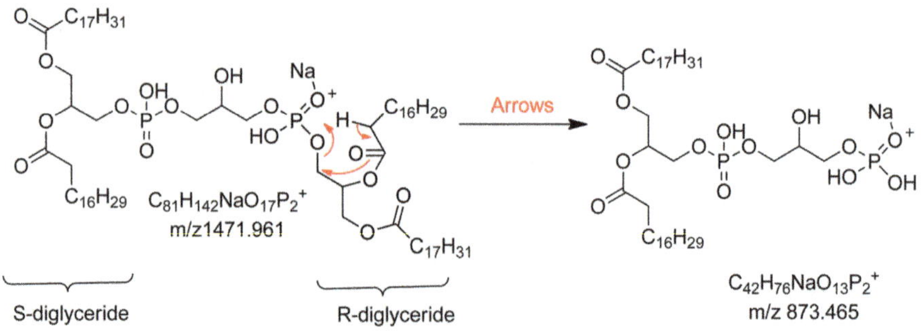

Scheme 5.56

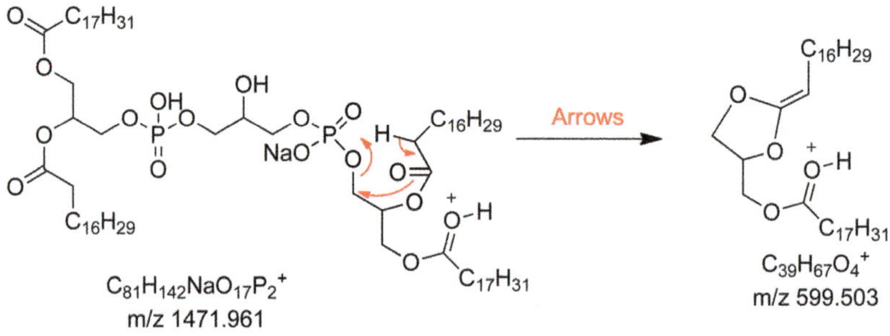

Scheme 5.57

two phosphates and there is a population of proton attachments at the four different ester moieties. The ion corresponding to the neutral species lost in this case could now retain the charge site as illustrated in Scheme 5.56 for the ion at m/z 873 $[M + Na - 559]^+$. Again, this would appear as two ions when the diglyceride R and S are not identical and this would divide the intensities of these two diglyceride ions. The lithiated cardiolipin product ion is observed at m/z 857 (Figure 5.18C). A second ion from the sodiated cardiolipin is derived from a sodium phosphate salt that has a proton charging site on one of the fatty acyl ester groups that would decompose as illustrated in Scheme 5.57 and retain no sodium atom in the product ion.

The third ion appears at m/z 275 and corresponds to the neutral loss of both R-diglyceride and S-diglyceride when one phosphate retains the charge site on the central glycerol moiety, but not on either of the R or S diglycerides (Scheme 5.58). A corresponding lithiated ion is not observed (Figure 5.18C).

5.12.3   Cardiolipin $[M - H]^-$

The collisional activation of $[M - H]^-$ of CL leads to common charge-driven fragment ions previously described, including carboxylate anions by direct attack of either phosphate anion when localized at the S-diglyceride or

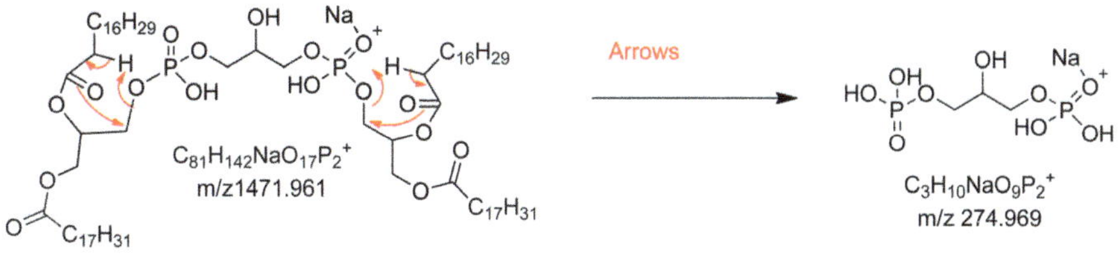

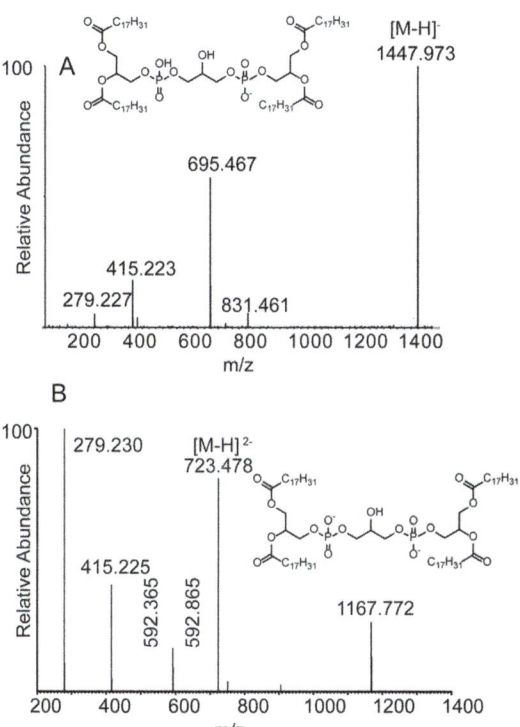

Scheme 5.58

Figure 5.19 Electrospray ionization (negative ions) and tandem mass spectrometry of singly and doubly charged cardiolipin molecular species. (A) Product ions obtained following collisional activation of CL(18 : 2/18 : 2/18 : 2/18 : 2) [M − H]⁻ at *m/z* 1447.9; (B) product ions obtained following collisional activation of the doubly charged CL(18 : 2/18 : 2/18 : 2/18 : 2) [M − H]²⁻ at *m/z* 723.5. These high resolution MS/MS spectra were obtained using a quadrupole time-of-flight mass spectrometer.

R-diglyceride phosphates (Scheme 5.3). These product ions provide direct information as to the fatty acyl groups making up the CL molecular species. In the example (Figure 5.19), a single carboxylate anion is seen, corresponding to the 18 : 2 carboxylate ion at *m/z* 279. For asymmetric cardiolipin species up to four different R–COO⁻ ions will be formed. When CLs are

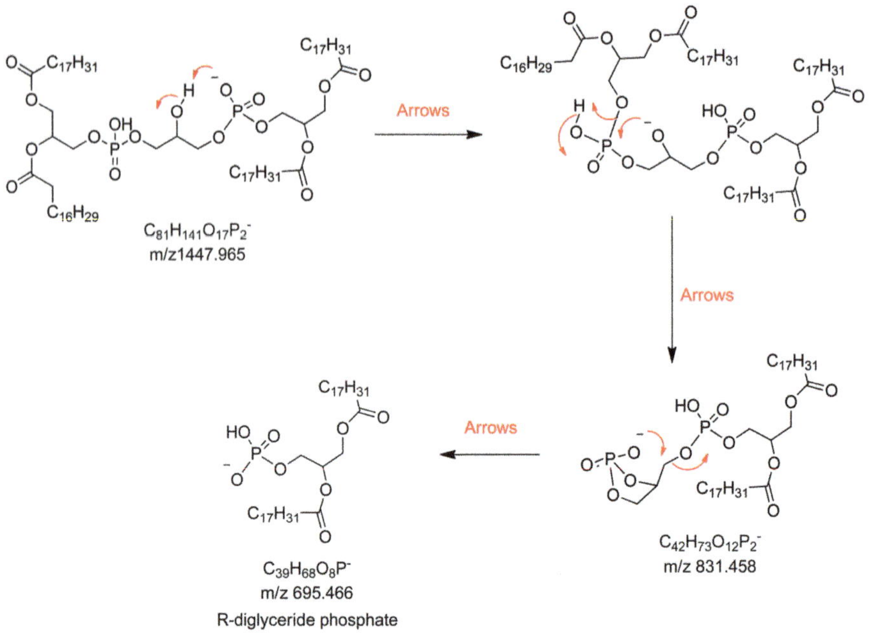

$C_{81}H_{141}O_{17}P_2^-$
m/z 1447.965

Arrows

Arrows

$C_{42}H_{73}O_{12}P_2^-$
m/z 831.458

Arrows

$C_{39}H_{68}O_8P^-$
m/z 695.466
R-diglyceride phosphate

Scheme 5.59

isolated from biological systems, even from simple organisms such as *E. coli*, more than four carboxylate anions have been observed from a single [M − H]⁻ since multiple isobaric molecular species exist with very different fatty acyl compositions. For example, studies of [M − 2H + Na]⁻ from cardiolipins isolated from *E. coli* had up to five or six different isobaric molecular species.[42]

The more abundant product ion from CL collision activation of [M − H]⁻ corresponds to the charge-driven loss of neutral R-diglyceride or S-diglyceride with intermediate formation of the central glycerol hydroxyl group as an alkoxide anion and concomitant formation of a cyclic phosphate anion from the loss of either the R or S-diglyceride neutral.[39] The loss of the S-diglyceride is illustrated in Scheme 5.59, but there appears to be little preference for which neutral diglyceride is lost even with an asymmetric assembly of the fatty acyl group. Two of these types of ions would normally appear, however; for the illustrated, symmetric cardiolipin only one appears at *m/z* 831. This ion further loses a neutral bicyclic phosphate (136 Da) to yield the abundant R-diglyceride phosphate ion (*m/z* 695). These same type of reaction product ions were observed in the tandem mass spectra of PG molecular species.

Either diglyceride phosphate anion can continue to decompose by loss of a fatty acyl group as a RCOOH in the mechanism previously described (Scheme 5.15) to form the ions in Figure 5.19 at *m/z* 415. The loss of the *sn*-2 fatty acyl group gives a more abundant product ion.[39] Again, more complex cardiolipin species yield up to four ions each from these mechanisms. Thus, it is possible by MS/MS of cardiolipins to obtain a fairly detailed set of information concerning fatty acyl groups, but unless one carries out MS of specific ions generated from a naturally occurring CL extract, the complexity of possible product ions presents a fairly ambiguous assignment of the fatty acyl positions, rather than being able to specifically assign whether the R- or S-diglyceride and the corresponding *sn*-1 and *sn*-2 positions have which fatty acyl group.[3] An interesting approach has been suggested to employ an ion trap mass spectrometer to sequentially isolate product ions from $[M - 2H + 3Li]^+$ and up to MS[5] to establish fatty acyl positions and even double bond positions for cardiolipins.[41]

5.12.4 Cardiolipin $[M - 2H]^{2-}$

The presence of the second phosphate in CL leads to an abundant $[M - 2H]^=$ ions for the cardiolipin 18 : 2/18 : 2/18 : 2/18 : 2 (Figure 5.19B). This doubly charged ion is observed at m/z 723.5 Collisional activation of this ion yields product ions at both higher as well as lower mass-to-charge ratios. Those ions at higher m/z are singly charged. A few doubly charged product ions are observed from collisionally activated $[M - 2H]^=$, with the most abundant being loss of either *sn*-2 fatty acyl group as ketene. This is seen in Figure 5.19B at m/z 592.364 $((723.479 - 592.364) \times 2 = 262.230)$ (Scheme 5.60).

There are no ions corresponding to the loss of the fatty acyl groups as free carboxylic acids (loss of R–COOH) likely due to the absence of protons in this doubly charged species that can be abstracted from either phosphate moiety.

The singly charged ions observed as products at higher masses than the ion collisionally activated $[M - 2H]^=$ include m/z 1167, which is a result of the loss of one fatty acyl group as a singly charged carboxylate anion driven by attack of a phosphate anion at one of the glycerol carbon atoms, leaving a second singly charged residue ion as a cyclic phosphate (Scheme 5.61). Interestingly, two negative charged ions result from reaction mechanism.

The other abundant singly charged high mass ion corresponds to the loss of one fatty acyl ketene neutral group from the cyclic phosphate anion as illustrated in Scheme 5.58. This ion appears at m/z 905.495 in Figure 5.19B.

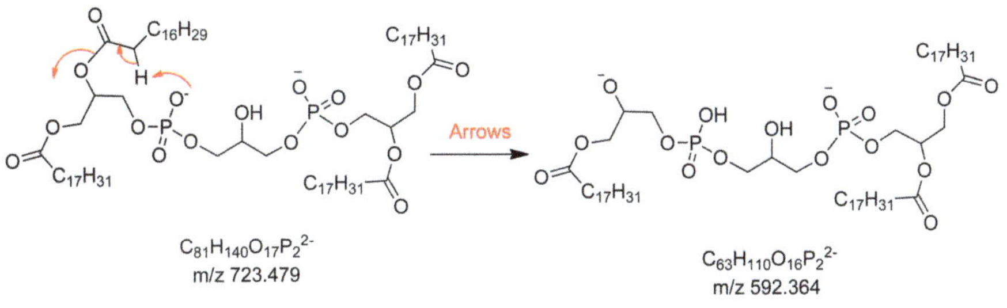

$C_{81}H_{140}O_{17}P_2^{2-}$
m/z 723.479

$C_{63}H_{110}O_{16}P_2^{2-}$
m/z 592.364

Scheme 5.60

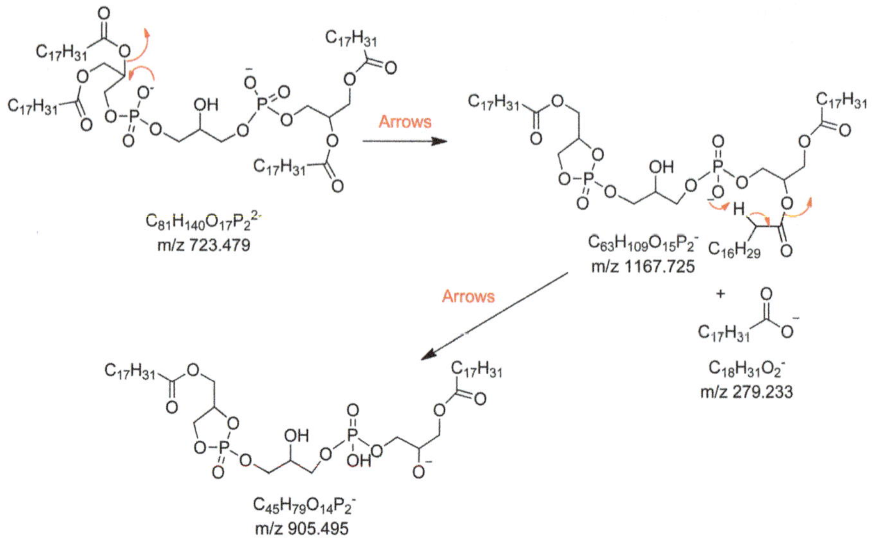

Scheme 5.61

The analysis of natural CL from biological systems is quite challenging. In large part, this is due to the fact that these simple pathways for decomposition of the molecular ion species can operate on any of the four fatty acyl groups, which reduce the net signal abundance for any one product ion when different fatty acyl groups are present. In Figure 5.19, the signals for each ion observed is a sum of four separate reactions, which lead to the identical product ion since all four fatty acyl groups are 18 : 2. The detailed work of Hsu and Turk using both positive and negative ionization to generate product ion information, as well as multiple stage tandem mass spectrometries, should be quite helpful in detailed structural characterization (fatty acyl esterification sites) of complex cardiolipin molecular species.[39–42]

References

1. No authors listed, The nomenclature of lipids (recommendations 1976). IUPAC-IUB Commission on Biochemical Nomenclature, *J. Lipid Res.,* 1978, **19**, 114–128.
2. Y. Koga and H. Morii, Biosynthesis of either-type polar lipids in archaea and evolutionary considerations, *Microbiol. Mol. Biol. Rev.,* 2007, **71**, 97–120.
3. K. Zemski-Berry and R. C. Murphy, Characterization of acrolein-glycerophosphoethanolamine lipid adducts using electrospray mass spectrometry, *Chem. Res. Toxicol.,* 2007, **20**, 1342–1351.
4. F. H. Chilton, III and R. C. Murphy, Fast atom bombardment analysis of arachidonic acid containing phosphatidylcholine molecular species, *Biomed. Mass Spectrom.,* 1986, **13**, 71–76.

5. R. C. Murphy, *Mass Spectrometry of Lipids*, Plenum Press, New York, 1993.

6. P. E. Haroldsen, K. L. Clay and R. C. Murphy, Quantitation of lyso-platelet activating factor molecular species from human neutrophils by mass spectrometry, *J. Lipid Res.*, 1987, **28**, 42–49.

7. E. Hvattum, G. Hagelin and A. Larsen, Study of mechanisms involved in the collision-induced dissociation of carboxylate anions from glycerophospholipids using negative ion electrospray tandem quadrupole mass spectrometry, *Rapid Commun. Mass Spectrom.*, 1998, **12**, 1405–1409.

8. F. F. Hsu and J. Turk, Charge-driven fragmentation processes in diacyl glycerophosphatidic acids upon low-energy collisional activation. A mechanistic proposal, *J. Am. Soc. Mass Spectrom.*, 2000, **11**, 797–803.

9. F. F. Hsu and J. Turk, Studies of phosphatidylglycerol with triple quadrupole tandem mass spectrometry with electrospray ionization: Fragmentation processes and structural characterization, *J. Am. Soc. Mass Spectrom.*, 2001, **12**, 1036–1043.

10. F. F. Hsu and J. Turk, Electrospray ionization/tandem quadrupole mass spectrometric studies on phosphatidylcholines: the fragmentation processes, *J. Am. Soc. Mass Spectrom.*, 2003, **14**, 352–363.

11. R. C. Murphy and P. H. Axelsen, Mass spectrometric analysis of long-chain lipids, *Mass Spectrom. Rev.*, 2011, **30**, 579–599.

12. J. Castro-Perez, T. P. Roddy, N. M. Nibbering, V. Shah, D. G. McLaren, S. Previs, A. B. Attygalle, K. Herath, Z. Chen, S. P. Wang, L. Mitnaul, B. K. Hubbard, R. J. Vreeken, D. G. Johns and T. Hankemeier, Localization of fatty acyl and double bond positions in phosphatidylcholines using a dual stage CID fragmentation coupled with ion mobility mass spectrometry, *J. Am. Soc. Mass Spectrom.*, 2011, **22**, 1552–1567.

13. F. F. Hsu, J. Turk, A. K. Thukkani, M. C. Messner, K. R. Wildsmith and D. A. Ford, Characterization of alkylacyl, alk-1-enylacyl and lyso subclasses of glycerophosphocholine by tandem quadrupole mass spectrometry with electrospray ionization, *J. Mass Spectrom.*, 2003, **38**, 752–763.

14. A. N. Hunt, G. T. Clark, G. S. Attard and A. D. Postle, Highly saturated endonuclear phosphatidylcholine is synthesized in situ and colocated with CDP-choline pathway enzymes, *J. Biol. Chem.*, 2001, **276**, 8492–8499.

15. N. Khaselev and R. C. Murphy, Electrospray ionization mass spectrometry of lysoglycerophosphocholine lipid subclasses, *J. Am. Soc. Mass Spectrom.*, 2000, **11**, 283–291.

16. F. F. Hsu and J. Turk, Electrospray ionization with low-energy collisionally activated dissociation tandem mass spectrometry of glycerophospholipids: mechanisms of fragmentation and structural characterization, *J. Chromatogr. B: Anal. Technol. Biomed. Life Sci.*, 2009, **877**, 2673–2695.

17. K. Kayganich and R. C. Murphy, Molecular species analysis of arachidonate containing glycerophosphocholines by tandem mass spectrometry, *J. Am. Soc. Mass Spectrom.*, 1991, **2**, 45–54.

18. K. A. Kayganich-Harrison and R. C. Murphy, Characterization of chain-shortened oxidized glycerophosphocholine lipids using fast atom bombardment and tandem mass spectrometry, *Anal. Biochem.*, 1994, **221**, 16–24.

19. Z. H. Huang, D. A. Gage and C. C. Sweeley, Characterization of Diacylglycerylphosphocholine Molecular Species by FAB-CAD-MS/MS: A General Method Not Sensitive to the Nature of the Fatty Acyl Groups, *J. Am. Soc. Mass Spectrom.*, 1992, **3**, 71–78.

20. F. F. Hsu and J. Turk, Charge-remote and charge-driven fragmentation processes in diacyl glycerophosphoethanolamine upon low-energy collisional activation: a mechanistic proposal, *J. Am. Soc. Mass Spectrom.*, 2000, **11**, 892–899.

21. F. F. Hsu and J. Turk, Characterization of phosphatidylethanolamine as a lithiated adduct by triple quadrupole tandem mass spectrometry with electrospray ionization, *J. Mass Spectrom.*, 2000, **35**, 595–606.

22. X. Han, K. Yang and R. W. Gross, Multi-dimensional mass spectrometry-based shotgun lipidomics and novel strategies for lipidomic analyses, *Mass Spectrom. Rev.*, 2012, **31**, 134–178.

23. K. A. Zemski Berry and R. C. Murphy, Electrospray ionization tandem mass spectrometry of glycerophosphoethanolamine plasmalogen phospholipids, *J. Am. Soc. Mass Spectrom.*, 2004, **15**, 1499–1508.

24. F. F. Hsu and J. Turk, Differentiation of 1-O-alk-1'-enyl-2-acyl and 1-O-alkyl-2-acyl glycerophospholipids by multiple-stage linear ion-trap mass spectrometry with electrospray ionization, *J. Am. Soc. Mass Spectrom.*, 2007, **18**, 2065–2073.

25. F. F. Hsu and J. Turk, Studies on phosphatidylserine by tandem quadrupole and multiple stage quadrupole ion-trap mass spectrometry with electrospray ionization: structural characterization and the fragmentation processes, *J. Am. Soc. Mass Spectrom.*, 2005, **16**, 1510–1522.

26. F. Gibellini and T. K. Smith, The Kennedy pathway–De novo synthesis of phosphatidylethanolamine and phosphatidylcholine, *IUBMB Life*, 2010, **62**, 414–428.

27. V. A. Blaho and T. Hla, Regulation of mammalian physiology, development, and disease by the sphingosine 1-phosphate and lysophosphatidic acid receptors, *Chem. Rev.*, 2011, **111**, 6299–6320.

28. A. Triebl, M. Trötzmüller, A. Eberl, P. Hanel, J. Hartler and H. C. Köfeler, Quantitation of phosphatidic acid and lysophosphatidic acid molecular species using hydrophilic interaction liquid chromatography coupled to electrospray ionization high resolution mass spectrometry, *J. Chromatogr. A*, 2014, **1347**, 104–110.

29. M. J. Wakelam and J. Clark, Methods for analyzing phosphoinositides using mass spectrometry, *Biochim. Biophys. Acta*, 2011, **1811**, 758–762.

30. S. B. Milne, P. T. Ivanova, D. DeCamp, R. C. Hsueh and H. A. Brown, A targeted mass spectrometric analysis of phosphatidylinositol phosphate species, *J. Lipid Res.*, 2005, **46**, 1796–1802.

31. F. F. Hsu and J. Turk, Characterization of phosphatidylinositol, phosphatidylinositol-4-phosphate, and phosphatidylinositol-4,5.bisphosphate by electrospray ionization tandem mass spectrometry: a mechanistic study, *J. Am. Soc. Mass Spectrom.*, 2000, **11**, 986–999.

32. F. F. Hsu, J. Turk, R. M. Owens, E. R. Rhoades and D. G. Russell, Structural characterization of phosphatidyl-myo-inositol mannosides from Mycobacterium bovis Bacillus Calmette Guerin by multiple-stage quadrupole ion-trap mass spectrometry with electrospray ionization. II. Monoacyl- and diacyl-PIMs, *J. Am. Soc. Mass Spectrom.*, 2007, **18**, 479–492.

33. M. Numata, Y. Nagashima, M. L. Moore, K. Z. Berry, M. Chan, P. Kandasamy, R. S. Peebles, Jr, R. C. Murphy and D. R. Voelker, Phosphatidylglycerol provides short-term prophylaxis against respiratory syncytial virus infection, *J. Lipid Res.*, 2013, **54**, 2133–2143.

34. M. Pulfer and R. C. Murphy, Electrospray mass spectrometry of phospholipids, *Mass Spectrom. Rev.*, 2003, **22**, 332–364.

35. M. Scherer and G. Schmitz, Metabolism, function and mass spectrometric analysis of bis(monoacylglycero)phosphate and cardiolipin, *Chem. Phys. Lipids*, 2011, **164**, 556–562.

36. M. Scherer, G. Schmitz and G. Liebisch, Simultaneous quantification of cardiolipin, bis(monoacylglycero)phosphate and their precursors by hydrophilic interaction LC—MS/MS including correction of isotopic overlap, *Anal. Chem.*, 2010, **82**, 8794–8799.

37. G. Paradies, V. Paradies, V. De Benedictis, F. M. Ruggiero and G. Petrosillo, Functional role of cardiolipin in mitochondrial bioenergetics, *Biochim. Biophys. Acta*, 2014, **1837**, 408–417.

38. M. Schlame, Cardiolipin synthesis for the assembly of bacterial and mitochondrial membranes, *J. Lipid Res.*, 2008, **49**, 1607–1620.

39. F. F. Hsu, J. Turk, E. R. Rhoades, D. G. Russell, Y. Shi and E. A. Groisman, Structural characterization of cardiolipin by tandem quadrupole and multiple-stage quadrupole ion-trap mass spectrometry with electrospray ionization, *J. Am. Soc. Mass Spectrom.*, 2005, **16**, 491–504.

40. F. F. Hsu and J. Turk, Characterization of cardiolipin as the sodiated ions by positive-ion electrospray ionization with multiple stage quadrupole ion-trap mass spectrometry, *J. Am. Soc. Mass Spectrom.*, 2006, **17**, 1146–1157.

41. F. F. Hsu and J. Turk, Toward total structural analysis of cardiolipins: multiple-stage linear ion-trap mass spectrometry on the $[M - 2H + 3Li]^+$ ions, *J. Am. Soc. Mass Spectrom.*, 2010, **21**, 1863–1869.

42. F. F. Hsu and J. Turk, Characterization of cardiolipin from *Escherichia coli* by electrospray ionization with multiple stage quadrupole ion-trap mass spectrometric analysis of $[M - 2H + Na]^-$ ions, *J. Am. Soc. Mass Spectrom.*, 2006, **17**, 420–429.

43. G. C. Sparagna, C. A. Johnson, S. A. McCune, R. L. Moore and R. C. Murphy, Quantitation of cardiolipin molecular species in spontaneously hypertensive heart failure rats using electrospray ionization mass spectrometry, *J. Lipid Res.*, 2005, **46**, 1196–1204.

CHAPTER 6

Sphingolipids (SP)

Sphingolipids are a major family of lipids that contain a long-chain base, such as sphingosine or sphinganine as a common structural unit. This family includes some of the highest molecular weight and most complex lipids found in the animal and plant kingdoms. The long-chain base originates from a fatty acyl CoA ester, predominately palmitoyl-CoA, which condenses with an amino acid, typically serine.[1] After several intermediate steps, the long-chain base which is formed (Scheme 6.1) serves as a scaffold for modification of the amino group by fatty acid acylation to form ceramides. Subsequent modification of the primary alcohol at the first carbon atom of the long-chain base by phosphorylation forms the signaling molecule ceramide-1-phosphate (termed here as C1P), attachment of phosphocholine to this primary alcohol yields sphingomyelin, glycation with simple sugars such as galactose, glucose, lactose, as well as more complex carbohydrates, yields the glycosphingolipids. Other modifications of the glycosphingolipids include incorporation of quite acidic species, neuraminic acid (gangliosides) and sulfuric acid esters found in sulfatides (Scheme 6.1).

Within this diverse family are also many members that have originated from slightly different long-chain bases and a large number of fatty acyl groups that collectively constitute a large number of molecular species found in cells. One interesting feature is the occurrence of, not only medium chain acyl species found typically in phospholipids (16 to 20 acyl carbon atoms), but also long-chain fatty acyl substituents up to 24-carbon atoms and α-hydroxy fatty acyl substituents. Our understanding of the complexity is still emerging.[2] Many trivial names are used to describe these sphingolipids, as these have been discovered over the past century and these names include ceramides, sulfatides, sphingomyelin, globosides, psychosine, and gangliosides.

New Developments in Mass Spectrometry No. 4
Tandem Mass Spectrometry of Lipids: Molecular Analysis of Complex Lipids
By Robert C Murphy
© Robert C Murphy 2015
Published by the Royal Society of Chemistry, www.rsc.org

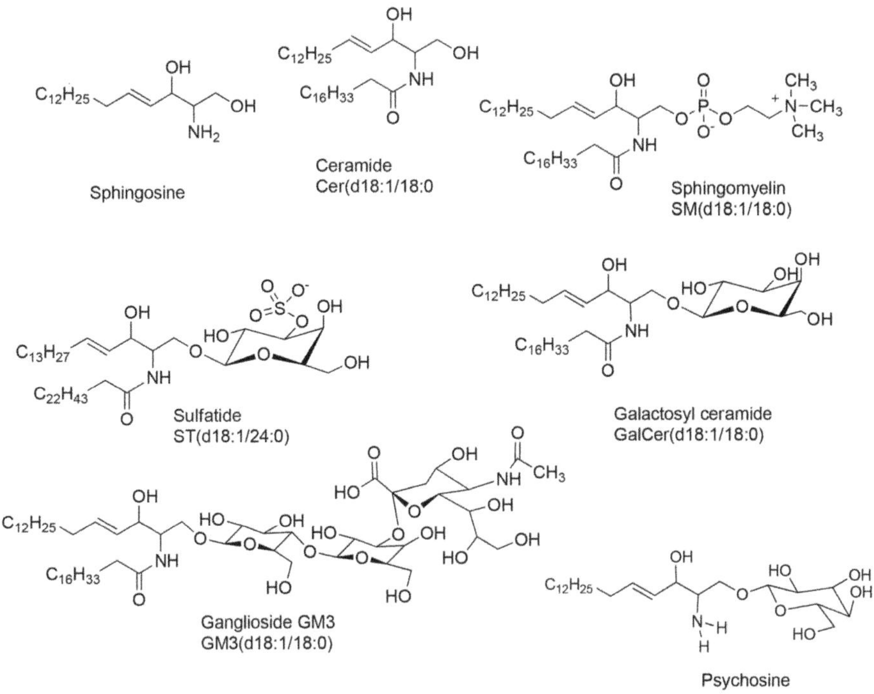

Scheme 6.1

Specific abbreviations are in common use that provide information as to the structural diversity of these molecular species and the modifications that have taken place to the long-chain base. Some are presented in Table 6.1 for various sphingolipid subclasses, with examples where the sphingolipid base are abbreviated as to the number of hydroxyl groups ("d" for 2 and "t" for 3), and the carbon chain length of the long-chain base followed by the number of double bonds. An example is sphingosine, which can be abbreviated So(d18 : 1). When the nitrogen atom of a sphingolipid is an amide (fatty acyl group), the abbreviation includes the number of the N-acyl carbon atoms, and following the colon, the total number of double bonds in the fatty acyl chain. For example, a ceramide molecular species would be abbreviated Cer(d18 : 1/24 : 1). When an α-hydroxyl group is present in the N-acyl substituent, the letter "h" is included such as Cer(d18 : 1/h24 : 1)

6.1 Sphingosine, Sphinganine, and Sphingoid Long Chain Bases

The basic building block of sphingolipids is the long-chain base (LCB) that results from the condensation of a fatty acyl-CoA ester with an amino acid.[1] There have not been mechanistic studies of the product ions formed by collisional activation of these LCBs, but tandem mass spectra have been

Table 6.1 Abbreviation and shorthand notations used in this book for sphingolipid molecular species.

Long chain base	Abbreviation	N-substituent	C-1 substituent	Molecular species notation[a]
Sphinganine (dihydrosphingosine)	Sa	H	H	Sa(d18 : 1)
Sphingosine	So	H	H	So(d18 : 0)
Phytosphingosine	PS	H	H	PS(t18 : 0)
Sphingosine-1-phosphate	S1P	H	PO_3	S1P(d18 : 1)
Ceramide	Cer	Fatty acyl	H	Cer(d18 : 1/24 : 1)
		α-hydroxy fatty acyl[b]		Cer(d18 : 1/h24 : 0)
Sphingomyelin	SM	Fatty acyl	Choline	SM(d8:0/16 : 0)
Glycosphingolipid				
Psychosine		H	Galactose	GalSo
Cerebrosides	[glycan]Cer	Fatty acyl	Galactose	GalCer(d18 : 1/24 : 0)
		α-hydroxy fatty acyl		GalCer(d18 : 1/h24 : 0)
Sulfatides	ST	Fatty acyl	Sulfate	ST(d18 : 0/24 : 1)
		α-hydroxy fatty acyl	Sulfate	ST(d18 : 0/h24 : 1)
Globosides	[glycans]Cer	Fatty acyl	Neutral sugar(s)	LacCer(d18 : 0/24 : 1)
		α-hydroxy fatty acyl		LacCer(d18 : 0/h24 : 1)
Gangliosides	GM3[c]	Fatty acyl	Glycan + NANAs[d]	aNeu5Ac(2-3)bDGalp(1-4)b-DGlcp(1-1)Cer(d18 : 1/18 : 0)

[a]d = dihydroxy, t = trihydroxy long chain base (total carbon atoms/total double bonds).
[b]h = α-hydroxy fatty acyl.
[c]G = ganglioside, M = monoNANA, D = two NANA, T = three NANA, Q = four NANA.
[d]NANA = N-acetyl neuraminic acid (Neu).

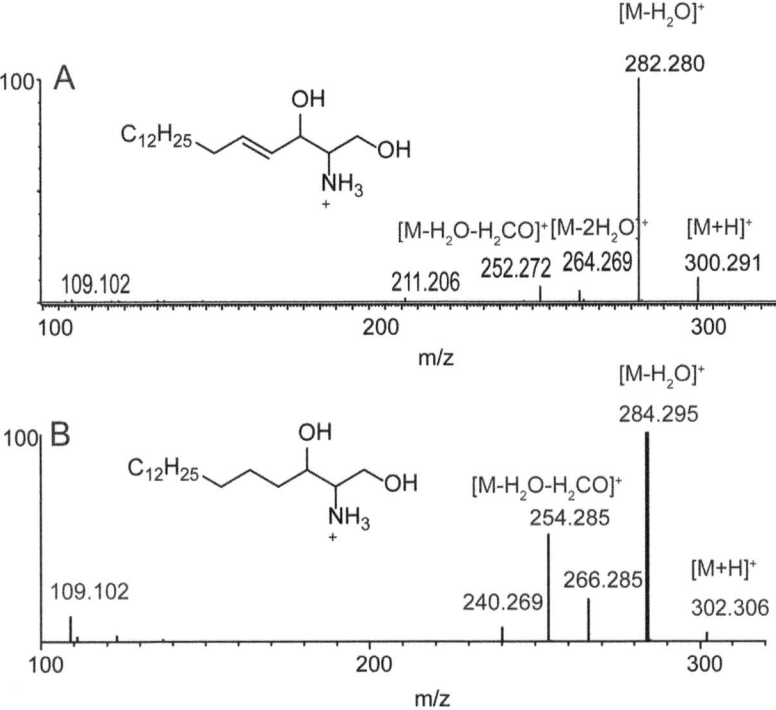

Figure 6.1 Electrospray ionization (positive ions) and tandem mass spectrometry of long chain bases. (A) Product ions following collisional activation of sphingosine [M + H]⁺ at *m/z* 300; (B) product ions obtained following collisional activation of sphinganine [M + H]⁺ at *m/z* 302. These high resolution MS/MS spectra were obtained using a quadrupole time-of-flight mass spectrometer.

published of the product ions formed after collisional activation of the [M + H]⁺ from sphingosine.[2,3] Recent interest has been focused on the signaling molecule sphingosine-1-phosphate (S1P), which is generated from the phosphorylation of sphingosine and forms abundant negative ions by electrospray ionization.

6.1.1 Long Chain Base [M + H]⁺ Positive Ions

Sphingosine, as a highly basic amine, readily forms positive ions by electrospray ionization to yield [M + H]⁺. Collisional activation of this long-chain base yields product ions corresponding to the loss of one (*m/z* 282) and two molecules (*m/z* 264) of water from the primary and secondary carbinol moieties, as seen in Figure 6.1A.

Another product ion corresponds to the loss of water and formaldehyde. This is likely a charge remote rearrangement driven by a newly formed double bond generated from the dehydration of the C-3 secondary alcohol with

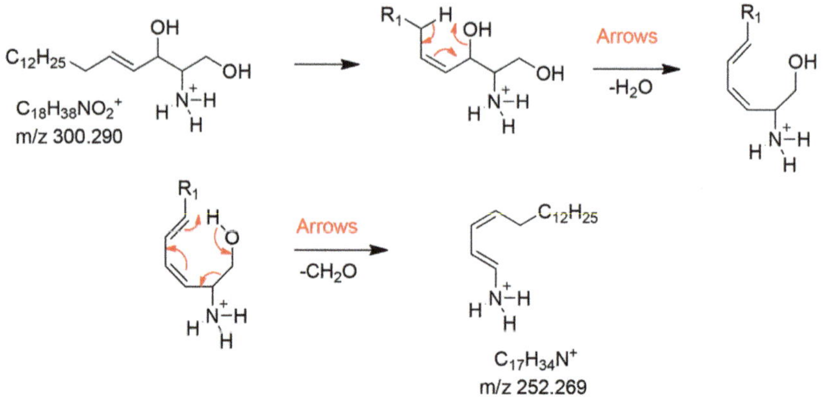

Scheme 6.2

appearance of the conjugated diene (Scheme 6.2). If the conjugated double bond system rearranges after CID to the *cis/trans* geometry, this places the furthest trans double bond in a favorable position for an ene-type reaction seen previously for fatty acids and eicosanoids (Chapters 1 and 2). This reaction then leads to abstraction of the proton on the oxygen atom of the remaining alcohol by the carbon chain during cleavage of the C-1,2 bond while formaldehyde is lost as a small neutral species (Scheme 6.2). This ion appears at m/z 252 for sphingosine and m/z 254 for sphinganine (Figure 6.1).

Low abundance negative ions $[M - H]^-$ can be generated for sphingosine and collision activation leads primarily to the loss of neutral formaldehyde (Figure 6.2A) corresponding to carbon bond cleavage between C-1 and C-2. The charge site is likely an alkoxide anion on C-1, perhaps transferring the charge site to a C-3 alkoxide anion. The absence of the double bond in the long-chain base of sphinganine changes this favorable loss of formaldehyde to that of loss of methanol (32 Da) and formation of the ion at m/z 268 (Figure 6.2B). This subtle structural change suggests participation of the double bond in driving the loss of formaldehyde from sphingosine. Without the assistance of the double bond and with the negative ion charge site at a C-3 alkoxide, the loss of methanol can involve the C-1 carbinol removing a proton from the free amine to form an imine product ion. The lack of other functional groups reduces pathways for favorable decomposition of this negative ion.

6.1.2 Sphingosine-1-Phosphate (S1P)

Sphingosine-1-phosphate (S1P) is the sphingolipid generated by phosphorylation of sphingosine. Studies by many investigators, including extensive work by Spiegel and coworkers have established this sphingolipid as an important biologically active lipid mediator and as such, the focus of many studies engaging quantitative analysis by tandem mass spectrometry.[4-6] S1P

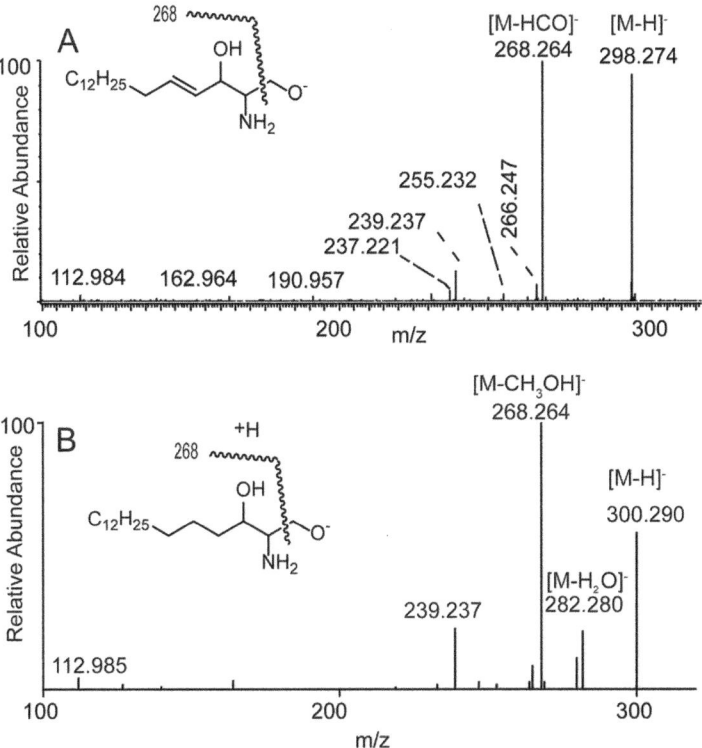

Figure 6.2 Electrospray ionization (negative ions) and tandem mass spectrometry of long chain bases. (A) Product ions obtained following collisional activation of sphingosine $[M - H]^+$ at m/z 298; (B) product ions obtained following collisional activation of sphinganine $[M - H]^-$ at m/z 300. These high resolution MS/MS spectra were obtained using a quadrupole time-of-flight mass spectrometer.

can yield both $[M + H]^+$ (Figure 6.3A) and $[M - H]^-$ (Figure 6.3C) molecule ion species at m/z 380 and 378, respectively, as well as a sodium adduct $[M + Na]^+$ at m/z 402 (Figure 6.3B). Collisional activation of $[M + H]^+$ yields a major product ion at m/z 264, which corresponds to the loss of neutral phosphoric acid plus loss of water (loss of 116 Da), likely by the pathway indicated in Scheme 6.3.

The sodiated adduct predominantly loses 98 Da (m/z 304, Figure 6.3B), suggesting the alkali metal could be located at the C-3 oxygen atom for that specific ion decomposition reaction since the elemental composition of m/z 304.261 is $C_{18}H_{35}NONa$, indicating a loss of H_3PO_4. Clearly, there is a population of S1P ions that have the alkali metal associated to the phosphate moiety, as evidenced by the abundant $[NaH_3PO_4]^+$ ion at m/z 121. Collisional activation of the $[M - H]^-$ from S1P is very simple (Figure 6.3C) with the formation of PO_3^- (m/z 79) and $H_2PO_4^-$ (m/z 97) as the only product ions.[6]

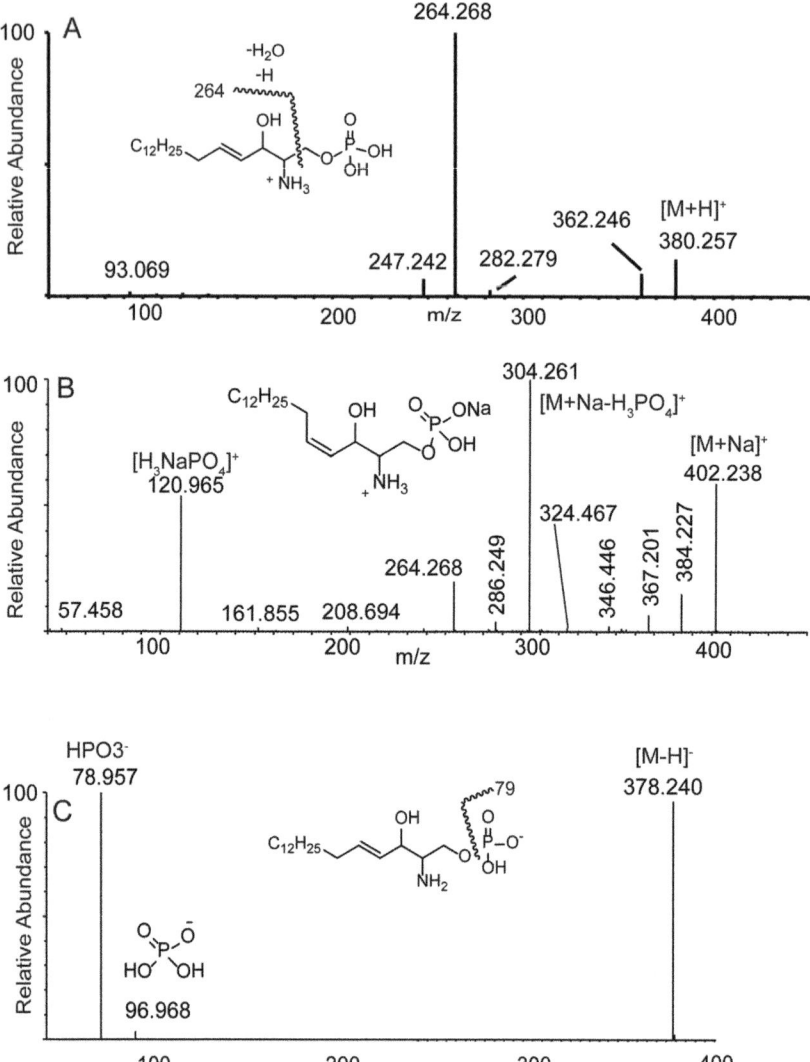

Figure 6.3 Electrospray ionization (positive and negative ions) and tandem mass spectrometry of sphingosine-1-phosphate (S1P). (A) Product ions obtained following collisional activation of sphingosine-1-phosphate [M + H]⁺ at m/z 380; (B) product ions obtained following collisional activation of the sodium salt of sphingosine-1-phosphate [M + Na]⁺ at m/z 402; (C) product ions obtained following collisional activation of the sphingosine-1-phosphate anion [M − H]⁺ at m/z 378. These high resolution MS/MS spectra were obtained using a quadrupole time-of-flight mass spectrometer.

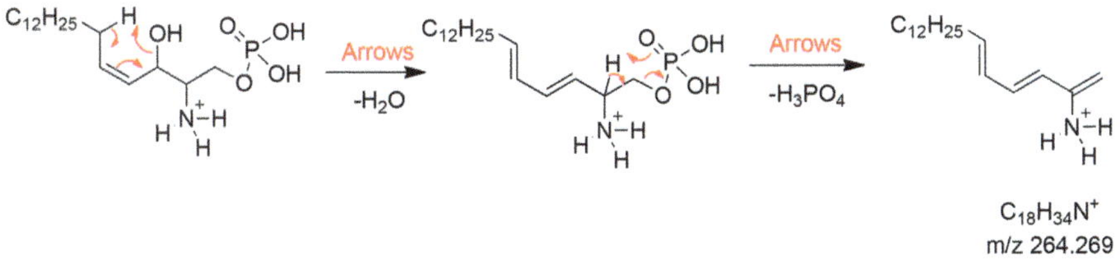

$C_{18}H_{34}N^+$
m/z 264.269

Scheme 6.3

6.2   Ceramides

Ceramides contain the basic chemical structure common to all higher molecular weight sphingolipids in that a long-chain base forms the general framework defining these molecules. However, these lipids are N-fatty acyl amides of the long-chain base and are not only biochemical intermediates for more complex sphingolipid biosynthesis, but they are also quite bioactive molecules in their own right and thus worthy of study by tandem mass spectrometry. The ceramides form both positive and negative ions rather easily (Figures 6.4–6.6) and the site of charge localization can be quite diverse for both positive and negative ions. For positive ions the typical site considered is protonation of the nitrogen atom of the amide group. However, clearly with double bonds in the acyl groups, some fragmentation would suggest a population of ions where the associated proton is in the form of a π-adduct with a fatty acyl double bond. The multiple sites for negative charge location help in understanding certain product ion formation.

While the diversity of the ceramide based sphingolipids appears daunting, when considering tandem mass spectrometry and the collisional activation of molecular ion species, both positive $[M + H]^+$ and negative $[M - H]^-$ yield product ions by a common set of pathways. These pathways have led to the use of a shorthand nomenclature or abbreviations for the types of ions observed (Scheme 6.4) as first suggested by Costello.[7] However, shorthand abbreviations do not reveal mechanistic pathways by which these ions arise. While not discussed in detail here, the nomenclature employed for glycosphingolipids include specific cleavage reactions of the oligosaccharide regions in these more complex lipids. This ion chemistry is not related to the mechanisms of lipid ion decomposition and is a separate study of glycomic tandem mass spectrometry.[8,9]

6.2.1   Ceramide $[M + H]^+$ Positive Ions

The most abundant product ion formed following collisional activation of $[M + H]^+$ from dihydroxy ceramides (*e.g.* d18 : 1) corresponds to cleavage of the amide bond and loss of one or two water molecules. These have been abbreviated to N-ions from the suggestions presented by Costello as well as Adams and Ann (Scheme 6.4) where N′ and N″ correspond to the indicated

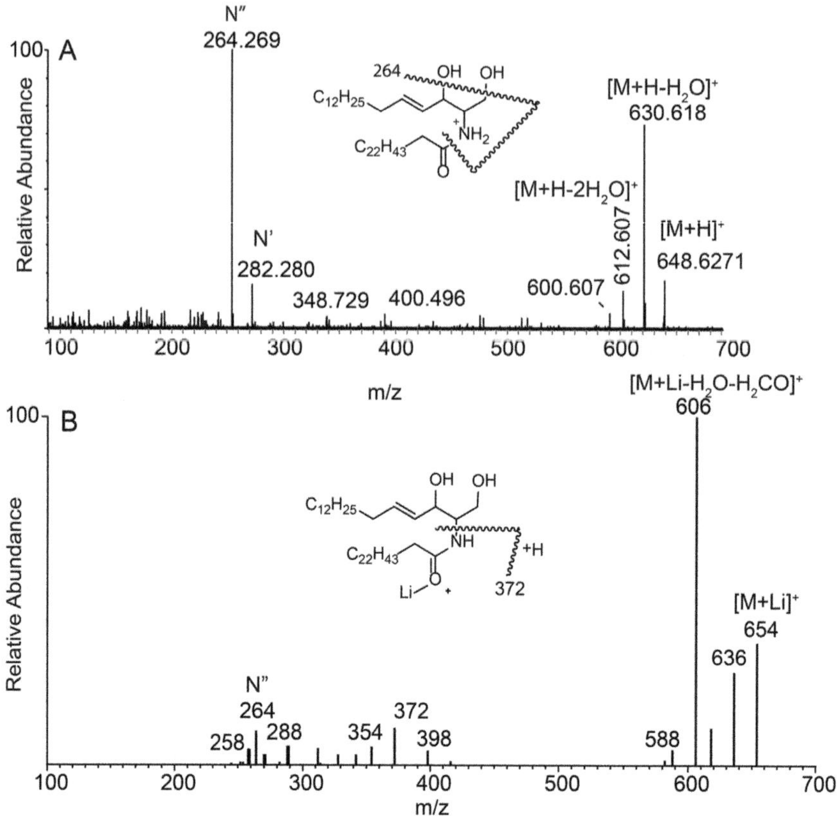

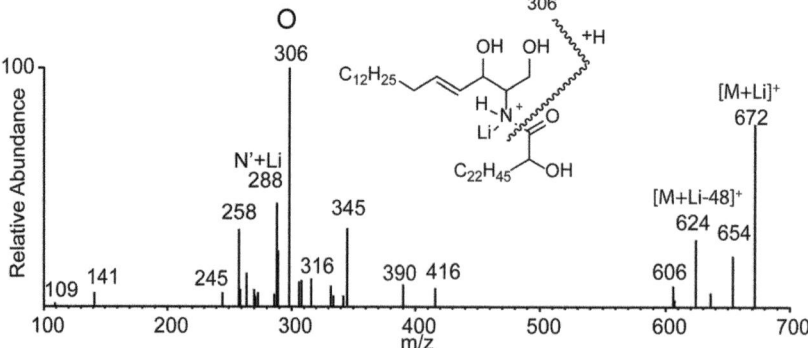

Figure 6.4 Electrospray ionization (positive ions) and tandem mass spectrometry of a ceramide. (A) Product ions obtained following collisional activation of the Cer(d18 : 1/24 : 1) [M + H]⁺ at *m/z* 648. These high resolution MS/MS spectra were obtained using a quadrupole time-of-flight mass spectrometer. (B) Product ions obtained following collisional activation of Cer(d18 : 1/24 : 1) [M + Li]⁺ at *m/z* 654. This figure was redrawn from data presented in ref. 13.

Figure 6.5 Electrospray ionization (positive ions) and tandem mass spectrometry of α-hydroxy fatty acyl sphingosine ceramide. Product ions obtained following collisional activation of Cer(d18 : 1/h24 : 0) [M + Li]⁺ at *m/z* 672. This figure was redrawn from data presented in ref. 13.

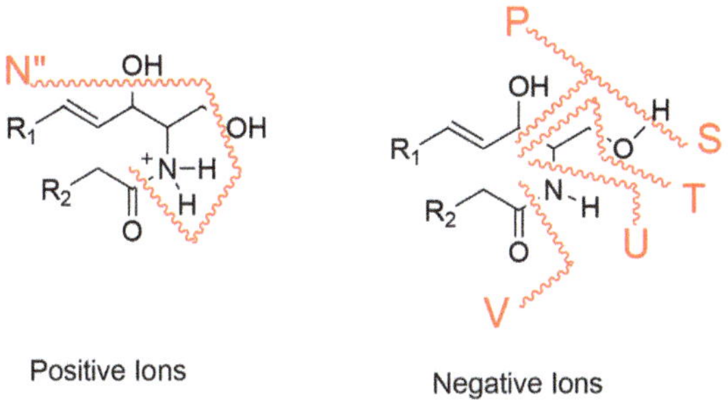

Scheme 6.4

cleavage sites with additional loss of one and two neutral water molecules.[7,10] These ions provide direct information of the nature of the long-chain base, but no information, except by mass difference from the molecular ion, of the nature of the fatty acyl group. These ions have also been referred to as O, O', and O'' by some investigators; however, a more specific use of the "O" shorthand will be reserved for the sphingolipids containing the α-hydroxy fatty acyl amide.[11,12] In addition to these ions, ceramides yield abundant ions close in mass to the molecular ion that are due to the loss of one or two neutral water molecules from the $[M + H]^+$ as seen in Figure 6.4A.

The mechanism for the formation of the N'' and N' has not been studied in detail, but likely follows pathways outlined in Scheme 6.5. This fragmentation pathway is initiated by a charge remote loss of water from the third carbon of sphingosine leading to a conjugated diene. The hydroxyl group on carbon-1 of sphingosine could remove a proton from the 2' carbon atom of the fatty acyl group leading to the loss of water as well as the N-acyl group as ketene. This suggestion comes from the tandem mass spectrum of Cer(d18 : 1/d$_{31}$ − 16 : 0), which is shifted to m/z 265 compared to the nondeuterated Cer(d18 : 1/16 : 0) which has the major N'' ion at m/z 264.[13] This mechanism (Scheme 6.5) is quite similar to that which was seen for the formation of specific ions of phospholipids involving the loss of ketene. A different mechanism has been suggested involving the abstraction of the 2'-proton directly by the amide nitrogen, followed by loss of ketene and formation of an aziridine like structure for this ion.[13]

The formation of the N' ion does not involve an initial loss of water (m/z 282, Figure 6.4), but proceeds to the loss of ketene and water as the last step (Scheme 6.5). These N'' and N' ions are typically used for ceramide analysis by quantitative mass spectrometry (multiple reaction monitoring, MRM) assays.[14,15] For ceramides with the sphingosine backbone, N'' appears at m/z 264 and for sphinganine at m/z 266, and these major N'' product ions are extremely useful in the structural characterization of ceramides.

Another ion of interest corresponds to the loss of 48 Da from the $[M + H]^+$. This ion is most likely the loss of a formaldehyde neutral molecule (H_2CO, 30 Da) and water (18 Da). A mechanism that would explain the formation of

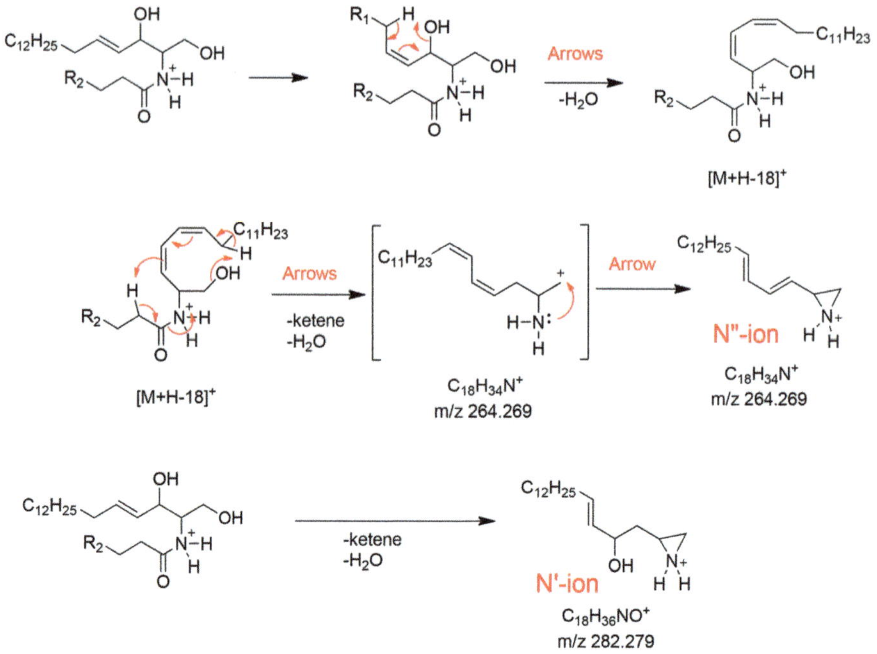

Scheme 6.5

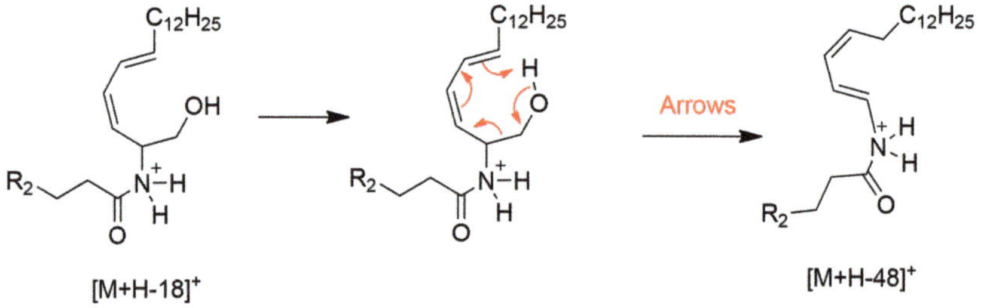

Scheme 6.6

this ion involves the dehydrated intermediate proposed in Scheme 6.5, where the terminal carbon atom of the conjugated diene abstracts a proton from the C-1 alcohol of sphingosine, driving the loss of formaldehyde (Scheme 6.6). This mechanism would also isomerize the diene structure in a position that would delocalize the positive charge located on the amide nitrogen atom and was proposed for sphingosine as Scheme 6.2.

6.2.2 Ceramide [M + Li]⁺ Positive Ions

The first studies of the collisional activation of $[M + Li]^+$ were carried out in a high energy instrument and led to the reports concerning the shorthand abbreviations used for product ions generated following collisional

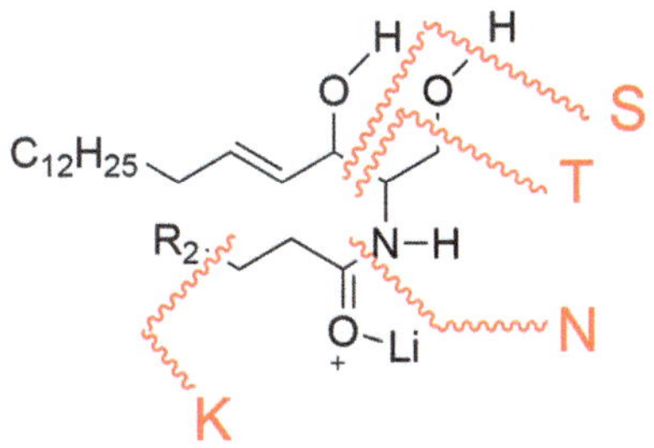

Scheme 6.7

activation of ceramides.[16,17] The product ions formed under these very energetic conditions are presented in the shorthand form in Scheme 6.7. However, some of the prominent cleavage reactions, such as the one indicated by K, are only found following high energy collisions and have not been reported in those tandem mass spectra from instruments such as a tandem quadrupole or ion trap, which employ much lower energy collision activation of ions.

Activation of ceramide $[M + Li]^+$ by low energy instruments have been reported (Figure 6.4B) and mechanisms of ion formation studied using various deuterium labeled analogs.[18] Ions derived from the collision induced decomposition of glycosphingolipid $[M + Li]^+$ yield very similar product ions when compared to product ions from ceramide $[M + H]^+$ decomposition, including major ions corresponding to N'' ions and related structures retaining the lithium cation.[10,18,19] A very abundant product ion corresponding to the loss of 48 Da, most likely by the mechanism presented in Scheme 6.6, was reported as the most abundant ion formed when the ceramide-like ion was formed in the ion source, selected by a first quadrupole then collisionally activated in the second quadrupole-collision cell. This ion appeared at m/z 606 from Cer(d18 : 1/24 : 1) (Figure 6.4B)’ which directly formed the N'' ion seen at m/z 264.[13] Another ion observed was m/z 288 and was proposed to have Li^+ and a hydroxyl group still retained in the structure. An ion not found following CID of Cer$[M + H]^+$ was observed at m/z 372 for the lithiated adduct and Hsu and Turk suggested cleavage of the carbon–nitrogen bond of sphingosine as an important step in the formation of this ion based on retention of an exchangeable deuterium atom.[13] Another possible mechanism, also consistent with the retention of an exchangeable deuterium atom, is presented in Scheme 6.8 which involves attack on the C-1 carbinol proton by the N-acyl carbonyl group that also is a site for Li^+ attachment.

A fairly common ceramide in biological systems is one in which the N-fatty acyl group has an α-hydroxy substituent. These N-α-hydroxy acyl ceramides such as Cer(d18 : 1/h24 : 0) in Figure 6.5 have a significantly altered tandem mass spectrum from the nonhydroxylated lithium adducts. Abundant product ions quite commonly found in the CID spectra from ceramide

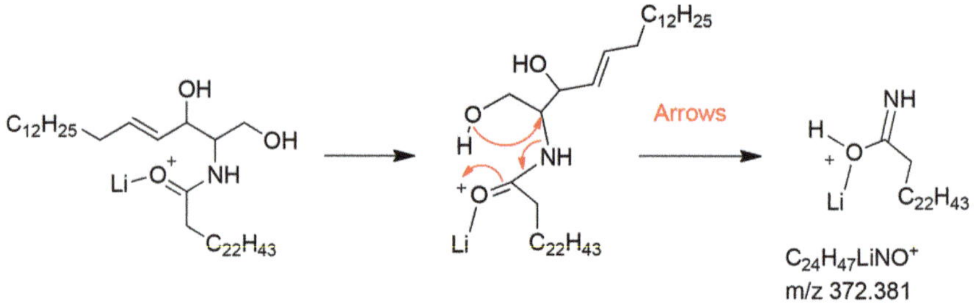

Scheme 6.8

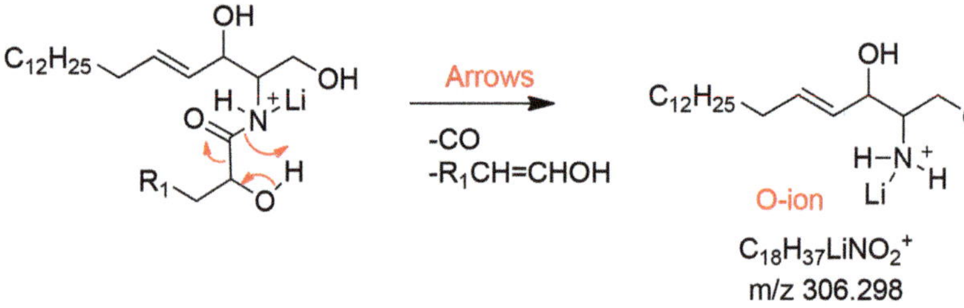

Scheme 6.9

$[M + Li]^+$ discussed above corresponding to $[M + Li - H_2O]^+$ and $[M + Li - H_2O - CH_2O]^+$ (Figure 6.5), are also observed at m/z 654 and 624, respectively for this α-hydroxy ceramide. A distinctive feature of the N-α-hydroxy fatty acyl group is the "O-type" ion observed at m/z 306. The mechanism proposed by Hsu and Turk[13] for this "O-type" cleavage is presented in Scheme 6.9 and involves formation of neutral carbon monoxide and the fatty acyl chain as an enolized aldehyde. The favorable cleavage of the amide bond indicated by the "O-type" fragmentation is a defining feature of the N-α-hydroxy fatty acyl ceramides.

This same mechanism for loss of neutral carbon monoxide and the α-hydroxyl acyl group, as an aldehyde, as a charge remote reaction leads to a different observed ion when the charge site is located on the fatty acyl portion of the ceramide. The ion at m/z 345 would now result (Scheme 6.10) and supports the concept that the $[M + Li]^+$ is in fact a population of structures with different sites at which the lithium cation is attached to the ceramide structure.

Two ions suggested to be derived from the abundant m/z 306 appear at m/z 288 and 258, which would be the loss of H_2O and $H_2O + HCHO$, respectively. The mechanism for the formation of these product ions would follow the reaction previously suggested where an ene-type reaction would lead to the loss of water as well as driving the formation of neutral formaldehyde and the ion at m/z 258 (Scheme 6.11).

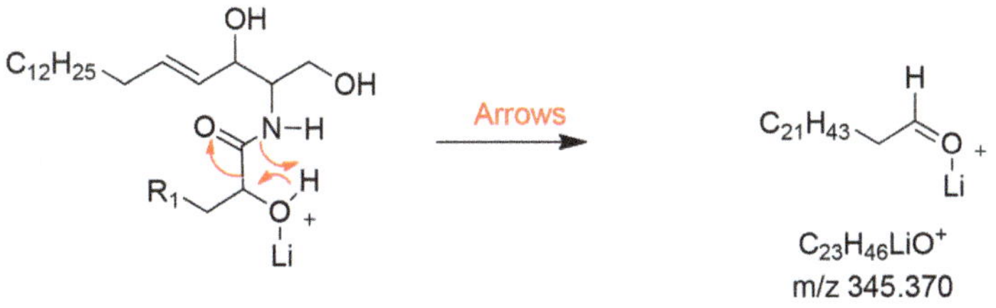

Scheme 6.10

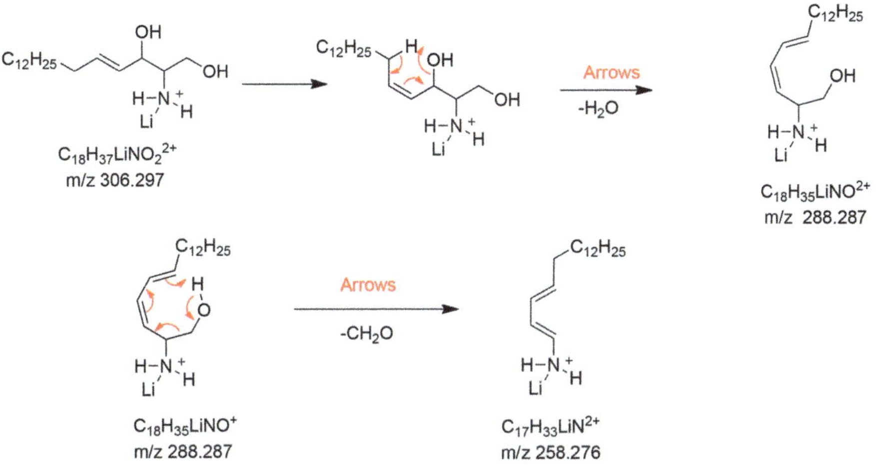

Scheme 6.11

6.2.3 Ceramide [M − H]⁻ Negative Ions

Product ions found following collisional activation of the ceramide $[M - H]^-$ yield complimentary information to that obtained from the collisional activation of the $[M + H]^+$ in that ions corresponding to the fatty acyl amide group are directly observed (Figure 6.6). The most abundant product ion (T) m/z 390 (Figure 6.6A) for Cer(d18 : 1/24 : 1 and m/z 280 for Cer (d18 : 1/16 : 0) (Figure 6.6B) correspond to the cleavage of the long-chain base adjacent to the sphingosine 3-hydroxyl group, likely driven by the indicated charge site as an alkoxide anion followed by the loss of water (Scheme 6.12).[19] When deuterium atoms were only present in the fatty acyl chain for Cer(d18 : 0/ D$_{31}$16 : 0), this ion is split into a doublet, seen at m/z 310 (major) and 322 (minor) (Figure 6.6C). This T-ion should be shifted by 31 Da to m/z 311 but m/z 310 is more abundant, indicating loss of a deuterium from the fatty acyl chain in the mechanism of formation. However, another mechanism must also operate, which has been previously outlined and is shown in Scheme 6.13.[7,18] This mechanism does not lead to the loss of a proton from the fatty acyl chain, but rather a proton (exchangeable) on the amide nitrogen atom.

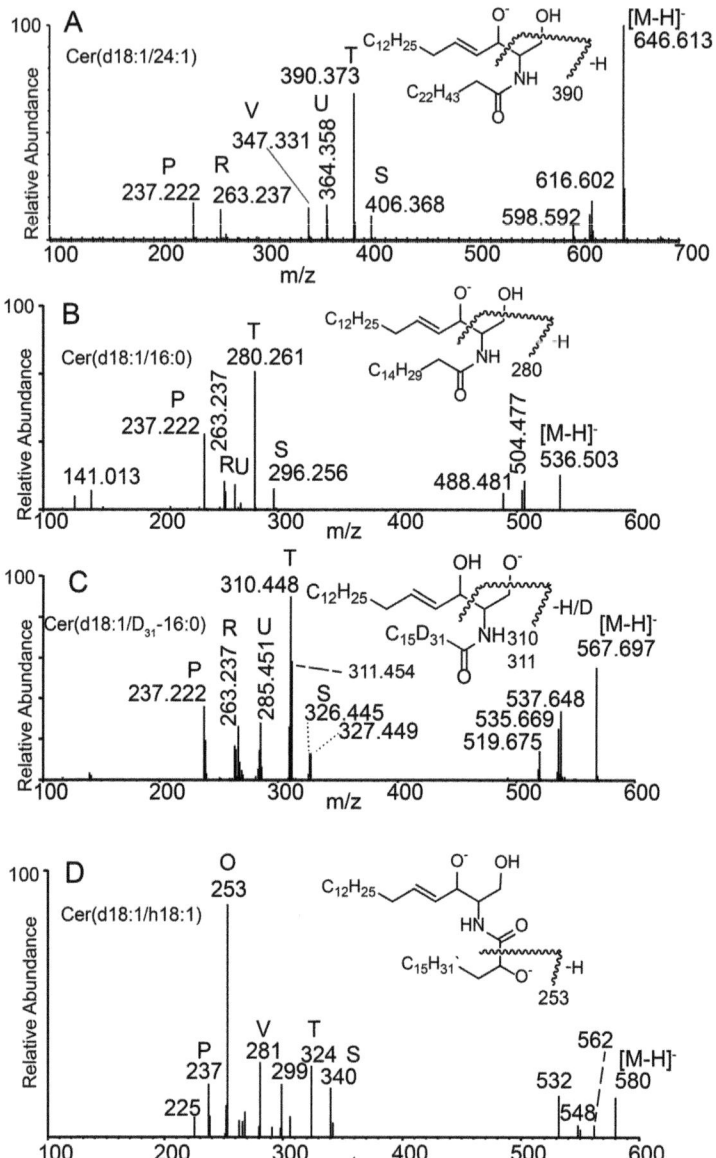

Figure 6.6 Electrospray ionization (negative ions) and tandem mass spectrometry of ceramides. (A) Product ions obtained following collisional activation of Cer(d18 : 1/24 : 1) [M − H]⁻ at *m/z* 646; (B) product ions obtained following collisional activation of Cer(d18 : 1/16 : 0) [M − H]⁻ at *m/z* 536; (C) product ions obtained following collisional activation of deuterated Cer(d18 : 1/D$_{31}$ − 16 : 0) [M − H]⁻ at *m/z* 567. These high resolution MS/MS spectra were obtained using a quadrupole time-of-flight mass spectrometer. (D) Product ions obtained following collisional activation of an α-hydroxy fatty acyl Cer(d18 : 1/h18 : 1) [M − H]⁻ at *m/z* 580. This figure was redrawn from data presented in ref. 18.

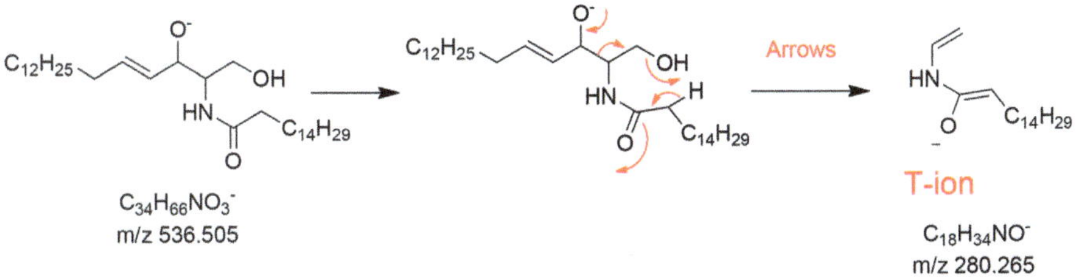

Scheme 6.12

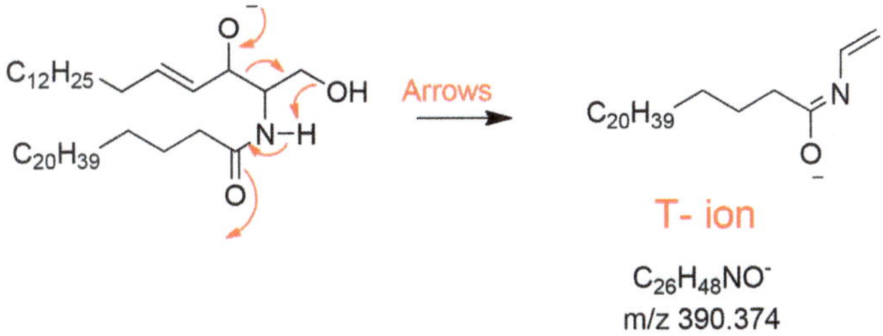

Scheme 6.13

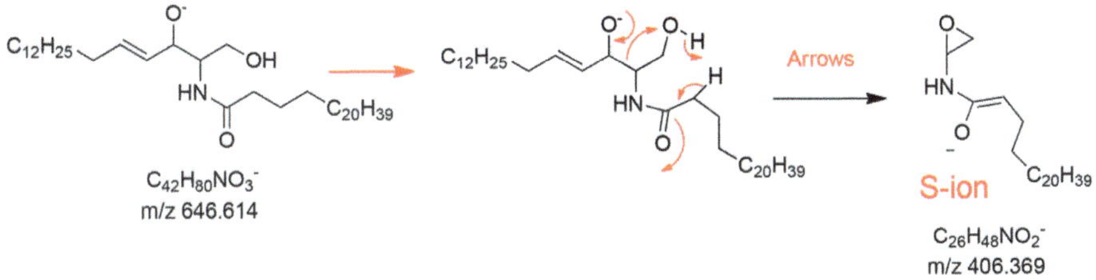

Scheme 6.14

A related ion, the S-ion, is the result of the loss of H_2 rather than H_2O and the formation of an epoxide product ion. This ion is also split into a doublet with the deuterated species suggesting at least two mechanisms.[18] One mechanism would be consistent with a transfer of the 2′ proton from the fatty acyl group during the H_2 formation and subsequent loss of the majority of the sphingolipid base as a neutral aldehyde species (Scheme 6.14). The second mechanism could involve the charge site initially residing at a C-1 alkoxide anion and a remote site loss of H_2 involving the proton from the amide nitrogen atom and the same neutral aldehyde from the LCB.[10,16] Both mechanisms appear to be operating since the S-ion from Cer(d18 : 1/D$_{31}$16 : 0) is split into a doublet.

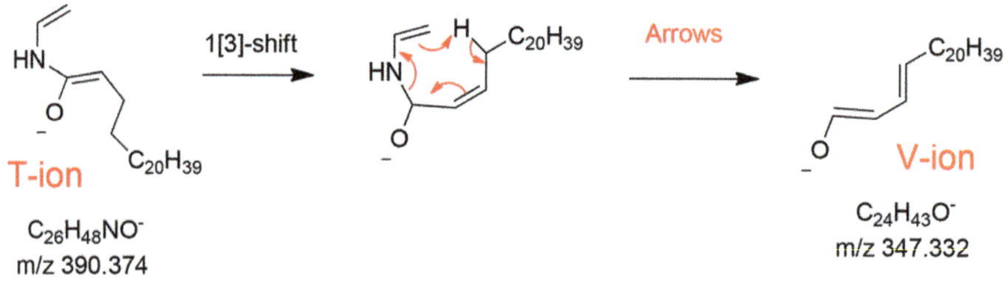

Scheme 6.15

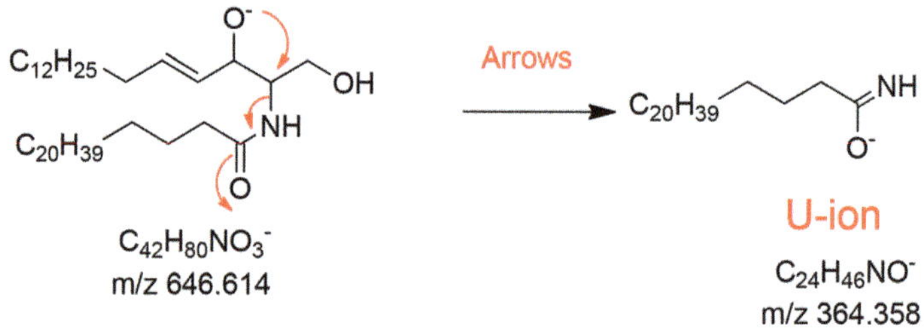

Scheme 6.16

The V and U-ions at *m/z* 347 and 364 (Figure 6.6A) also provide direct information about the N-acyl group in the ceramide. The formation of the V-ion involves the loss of an additional N-acyl proton and could be a product of further decomposition of the abundant T-ion. Following a 1[3]-sigmatropic shift of the double bond in the fatty acyl chain, abstraction of a proton by the terminal olefin of the T-ion (Figure 6.6A) would render a rather stable, doubly conjugated alkoxide anion (Scheme 6.15).

The U-type ion could arise from the ceramide [M − H]⁻ by direct attack of an alkoxide anion either at C-1 or C-3 on the second carbon of the long-chain base to form a neutral epoxide, cleaving the amide bond, and form an alkoxide anion derived from the fatty acyl portion of the molecule (Scheme 6.16).

The remaining common ions formed following collisional activation of [M − H]⁻ are called P and R-ions.[20] These ions are derived from the long-chain base and in the case of Figure 6.6A from sphingosine, the P-ion appears at *m/z* 237 and the R-ion at *m/z* 263. These product ions are likely the result of direct decomposition of [M − H]⁻ driven by the loss of water and a charge driven (Scheme 6.17) and a charge remote reaction (Scheme 6.18), respectively. As a result, very stable conjugated alkoxide anions are formed.

The negative product ions observed for ceramides with the N-α-hydroxy fatty acyl group (Figure 6.6D) are very similar to those observed for the simple N-fatty acyl ceramides. The ion mechanisms described for P,U,V, T, and S ions appear to be operational as described above. The most abundant

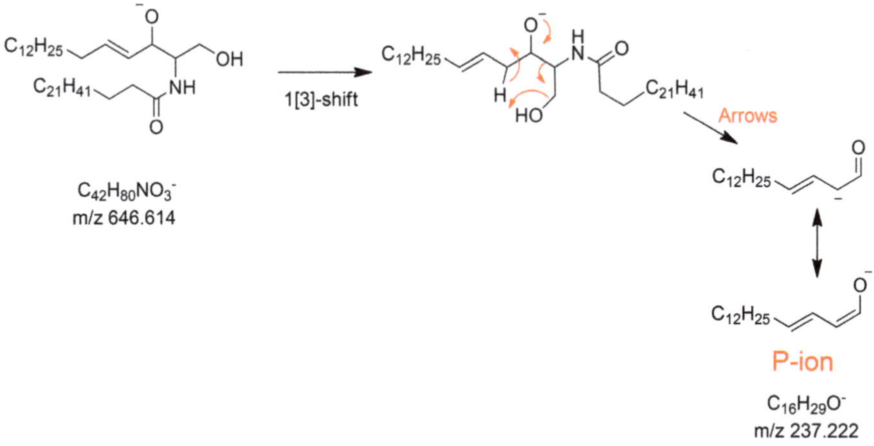

Scheme 6.17

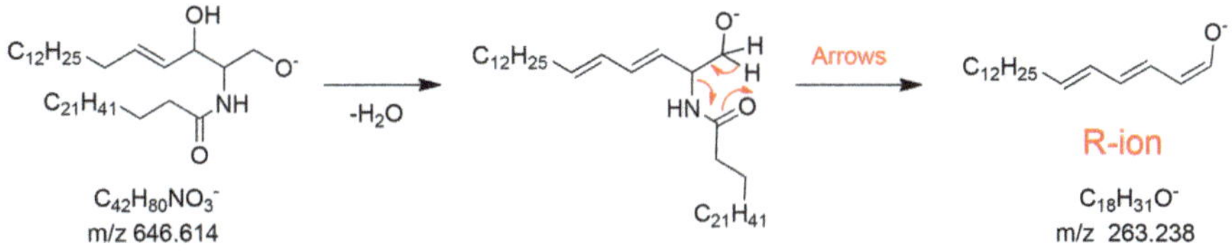

Scheme 6.18

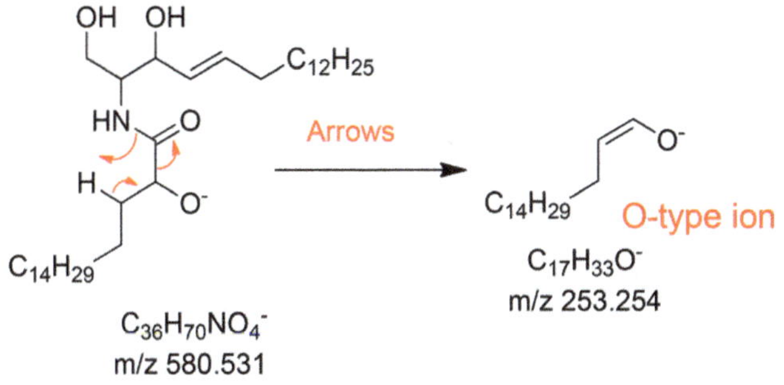

Scheme 6.19

product ion observed for the Cer(d18 : 1/h18 : 0) species is at *m/z* 253 and most likely is an O-type ion (described for positive ion α-hydroxy-fatty acyl ceramides) from cleavage of the amide bond where charge is localized on the N-α-hydroxy-fatty acyl moiety. Perhaps a slightly different mechanism for this bond cleavage involves the amide nitrogen atom removing the hydrogen atom adjacent to the ionized alkoxide ion on the fatty acyl amide chain, followed by loss of carbon monoxide, as previously suggested for the O-type ion (Scheme 6.19).

6.2.4   Ceramide-1-Phosphate (C1P)

The phosphorylation of the ceramide primary hydroxyl group at C-1 of the long-chain base leads to the formation of ceramide-1-phosphate (C1P). This sphingolipid has very interesting biological activity as a signaling molecule, making this an excellent target for mass spectrometric quantitation by tandem mass spectrometry of extracts of biological tissues and fluids.[21] This molecule readily forms negative ions $[M - H]^-$, but it is also possible to generate positive ions using an electrospray solvent system at lower pH (Figure 6.7A).

The structural advantage of the collisional activation of $[M + H]^+$ of C1P is that the very characteristic N''-type rearrangement ion (m/z 264) is formed indicative of the long-chain base. In addition there is an abundant loss of the phosphate head group as a charge remote reaction (Scheme 6.20). The losses of water and various phosphate neutrals from the $[M + H]^+$ do not reveal the nature of the C1P in terms of the long-chain base or N-acyl group, but the mass difference between the $[M + H]^+$ and the N'' ion can be used to establish the N-fatty acyl chain length and total double bonds.

The tandem mass spectrometry of $[M - H]^-$ from C1P (Figure 6.7B) indicates clearly that this molecule is a monophosphoester of a sphingolipid with only abundant product ions at m/z 79 (PO_3^-) and 97 ($H_2PO_4^-$). Very little structural information is revealed by the product ions; however, the $[M - H]^-$ ion is formed in high abundance by electrospray ionization, making it a good ion for sensitive detection of these lipid substances.

The most abundant, high mass product ion from collisional activation of $[M - H]^-$ corresponds to cleavage of the amide group, and transfer of a proton from the fatty acyl chain likely abstracted at the 2'-carbon atom by the phosphate anion (Scheme 6.21) in a charge driven reaction. This product (m/z 378) is rather unique to this phosphate ester.

6.3   Sphingomyelin

Another sphingolipid that is also a phospholipid (as are C1P and S1P) is the compound termed sphingomyelin. This molecule is formed by the esterification of phosphocholine to the hydroxyl group at C-1 of a ceramide, rendering an ionic phosphodiester, due to the quaternary trimethylamino group of choline. Sphingomyelins are very abundant sphingolipids found predominantly in cellular plasma membranes and are thought to play critical roles in membrane stability, packing, and protein stabilization in lipid rafts.[23]

6.3.1   Sphingomyelin $[M + H]^+$ and $[M + Li]^+$ Positive Ions

The dominant ion formed following electrospray ionization of SM is, as expected, the $[M + H]^+$. Collisional activation of $[M + H]^+$ from sphingomyelin appears for all molecular species at m/z 184 (Figure 6.8A). This is the same

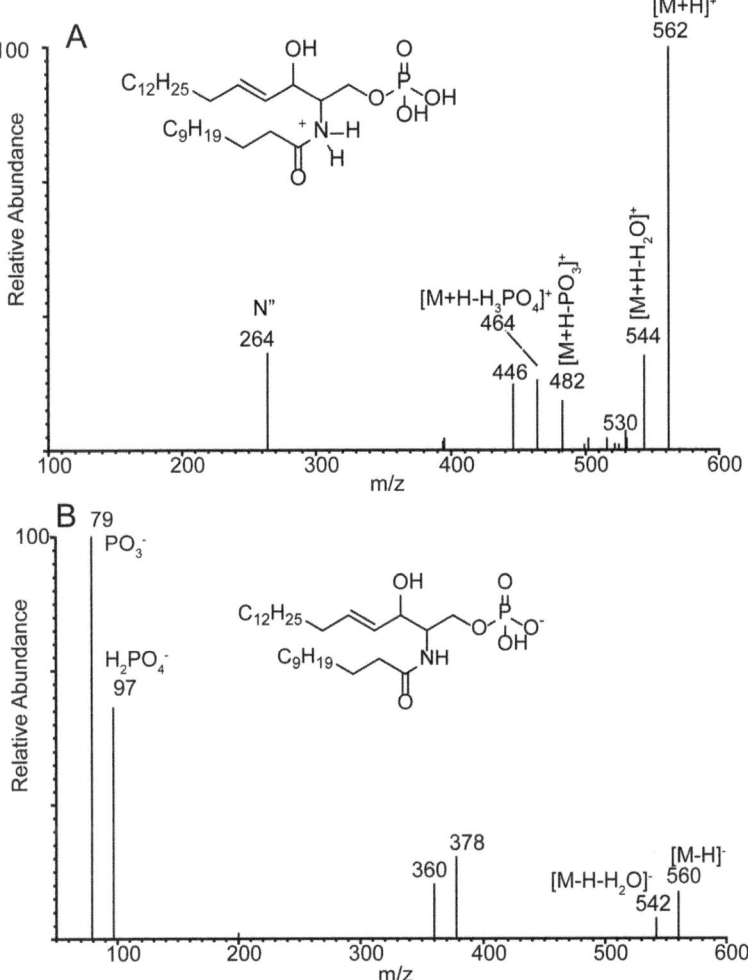

Figure 6.7 Electrospray ionization (positive and negative ions) and tandem mass spectrometry of ceramide-1-phosphate (C1P). (A) Product ions obtained following collisional activation of C1P(d18 : 1/12 : 0) [M + H]⁺ at *m/z* 562; (B) product ions obtained following collisional activation of C1P(d18 : 1/12 : 0) [M − H]⁻ at *m/z* 560. These MS/MS spectra were obtained using a tandem quadrupole mass spectrometer.

product ion [(OH)₂PO–OCH₂CH₂–N⁺(CH₃)₃] seen for glycerophosphocholines as the result of a proton rearrangement and cleavage of the phosphate sphingosine ester bond. However, the formation of this highly abundant product ion does not involve a mechanism removing a 2′ proton from the N-fatty acyl chain. When the N-acyl chain contains only deuterium atoms and is collisionally activated, *m/z* 184 does not shift in mass. However, if the

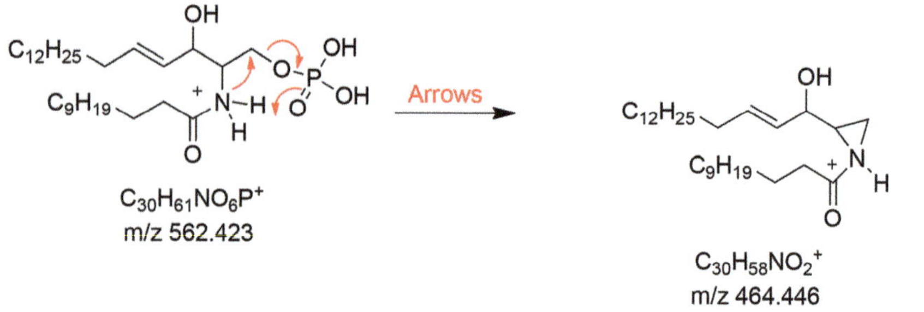

C30H61NO6P+
m/z 562.423

C30H58NO2+
m/z 464.446

Scheme 6.20

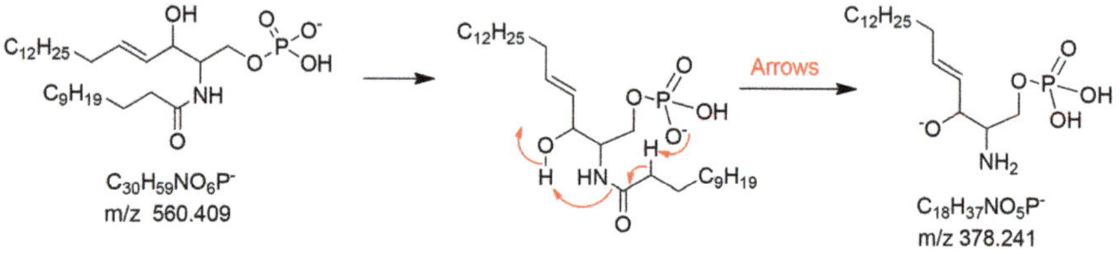

C30H59NO6P−
m/z 560.409

C18H37NO5P−
m/z 378.241

Scheme 6.21

exchangeable protons on the heteroatoms of sphingomyelin are converted to deuterium atoms by H/D exchange (Figure 6.8B), this ion now shifts to m/z 186. One likely mechanism is presented in Scheme 6.22, which involves transfer of a proton from the secondary carbinol of sphingosine to the product ion being formed. Another mechanism (not shown) could involve the proton on the amide nitrogen atom, and would be similar to that of Scheme 6.19 except for the localization of the proton abstracted by the phosphate moiety.

The collisional activation of SM positive ions as lithiated or sodiated species has been studied in some detail by Hsu and Turk.[22] While the ion at m/z 184 is not present, there are two abundant high mass ions corresponding to the loss of the polar head group (phosphocholine) and phosphocholine plus formaldehyde. Both of these product ions retain the alkali metal ions.

6.3.2 Sphingomyelin [M − H]⁻ Negative Ions

Negative ions from SM can also be obtained and collisional activation yields a number of product ions. Negative ions derived from sphingomyelin can be observed corresponding to [M − 15]⁻ (Figure 6.8C) or an adduct such as [M + acetate]⁻ or [M + Cl]⁻ just as a phosphatidylcholine (Chapter 5). These latter negative ion adduct species decompose to product ions corresponding to [M − 15]⁻ by loss of a choline methyl group and the formation of methyl acetate or methyl chloride neutral, respectively.[24] The collisional activation of [M − 15]⁻ yields a major product ion at m/z 168 corresponding to the N-dimethylaminoethylphosphate anion and some HPO_3^- at m/z 79.

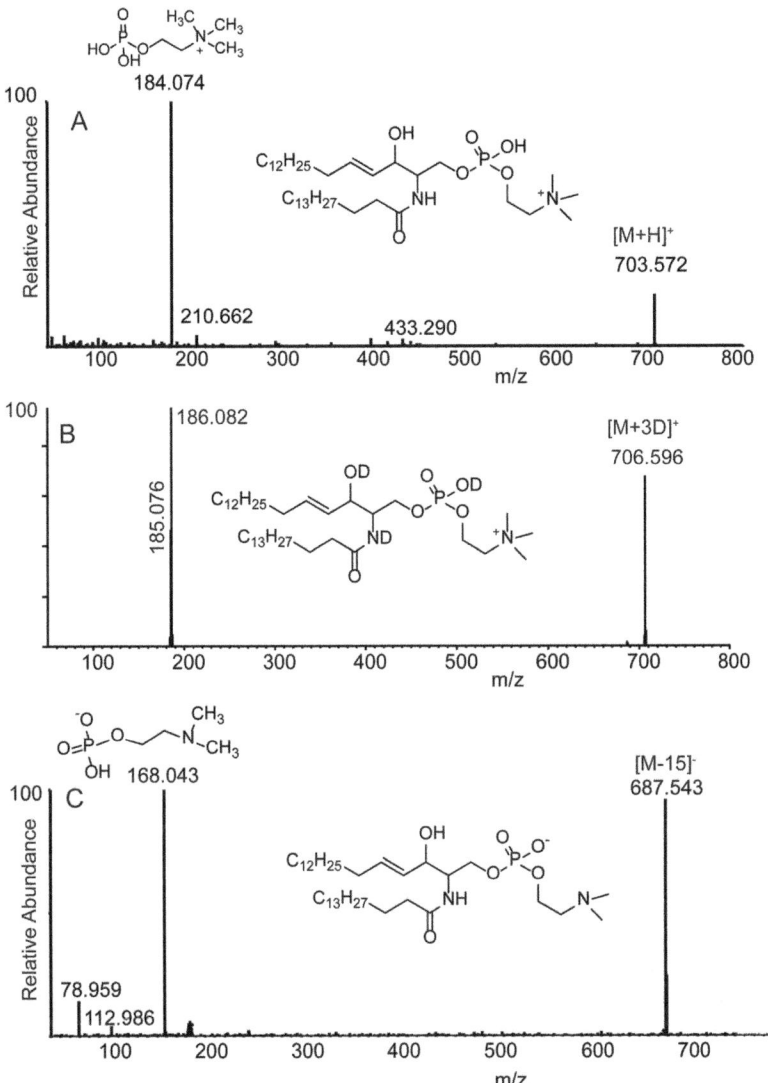

Figure 6.8 Electrospray ionization (positive and negative ions) and tandem mass spectrometry of sphingomyelins (SM). (A) Product ions obtained following collisional activation of SM(d18 : 1/16 : 0) [M + H]$^+$ at *m/z* 703; (B) product ions obtained following collisional activation of the H/D-exchanged SM(d18 : 1/16 : 0) [M + D]$^+$ at *m/z* 760; (C) product ions obtained following collisional activation of the SM(d18 : 1/16 : 0) [M − 15]$^-$ at *m/z* 687. These high resolution MS/MS spectra were obtained using a quadrupole time-of-flight mass spectrometer.

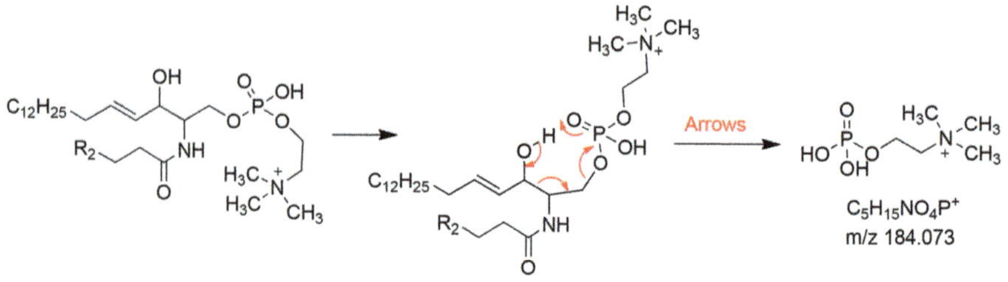

Scheme 6.22

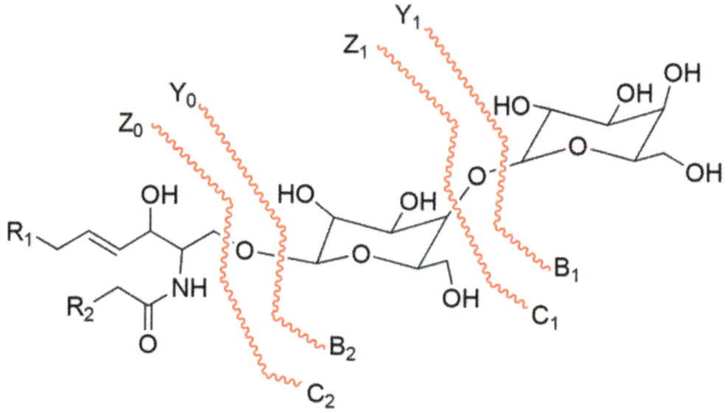

Scheme 6.23

6.4 Glycosphingolipids

The complexity of glycosphingolipids range from fairly simple addition of a monosaccharide at the C-1 alcohol group of the long-chain base to exceedingly complex polysaccharides attached at this position of ceramides with very diverse sugar units and various linkage sites for the glycosidic bonds. The structural challenge presented by these glycosphingolipids for compounds such as gangliosides are enormous. Glycosphingolipids will be divided into two major groups, the first being the neutral family that contain no ionized or charged groups in the carbohydrate portion of the molecule. These molecules include cerebrosides like GalCer(d18 : 1/16 : 0) and LacCer(d18 : 1/16 : 0). The second family group contains structural moieties in the carbohydrate region that are highly ionized in aqueous solution, such as sulfate esters (sulfatides) and sialic acids such as gangliosides.

The challenges of analysis of these glycosphingolipids include assignment of carbohydrate linkage, sugar identification, and structural characterization of the ceramide lipid portion of the molecule. Carbohydrate related ions which can be observed for linkage information are summarized in Scheme 6.23. Mass spectrometry is an essential tool for the characterization of the lipid portion of the molecule, but in most cases a single mass spectrometric technique is insufficient to establish all of the structural features of these molecules. Information from other spectroscopic tools, such as NMR spectroscopy and even microchemical derivatization and enzymatic hydrolysis are typically used for final structural assignments. In the remaining portion of this chapter we will deal with relatively simple glycosphingolipid species to illustrate information that can be obtained concerning the lipid portion of the molecule. A very helpful tool to follow biochemical synthesis and understand the structural complexity of molecular species within this family has been prepared by Merrill, and it is called SphingoMap.[25]

6.4.1 Psychosine $[M + H]^+$ and $[M - H]^-$

The simplest glycosphingolipid is psychosine which has a galactose residue attached to the C-1 carbon atom of sphingosine leaving a free amino group of the long-chain base. This molecule generates abundant $[M + H]^+$ ions by electrospray ionization and collisional activation yields an ion of low abundance for the loss of H_2O, but a very abundant ion corresponding to the loss of galactose.[26] For GalSo (d18 : 1) the $[M + H]^+$ ion appears at m/z 462 and loss of galactose at m/z 282 (loss of neutral 180 Da). This loss of galactose ion can be classified as a Z_0 ion and is formed by a transfer of proton from sphingosine moiety, likely in a charge remote mechanism (Figure 6.9). From molecular models and analogy to various cerebrosides with deuterium atoms on all exchangeable protons, this would likely involve the anomeric oxygen atom removing one proton from the amino group of psychosine forming the stable aziridine ring similar to the N-type ions of ceramides (Scheme 6.24). This loss of the sugar as a neutral species is a common fragment for all glycosphingolipids.

A related mechanism would yield the Y_0 ions involving a hydroxyl proton being transferred to the anomeric oxygen of the glycosidic bond forming a bicyclic sugar neutral and the ion at m/z 300. Other hydroxyl groups of the sugar could also be involved in the mechanism rather than the specific one indicated, but in summation the results are identical with the formation of a neutral, bicyclic sugar moiety (Scheme 6.25). When the charge site is localized on the oligosaccharide portion of the molecule, the bicyclic sugar residue would retain the charge as a B-ion.

Collisional activation of psychosine $[M - H]^-$ yields negative product ions dominated by charge localized on the galactose portion of the molecule and various cross ring fragment ions including $C_4H_5O_3$ (m/z 101), $C_5H_5O_3$ (m/z 113), and the intact galactose anion that has one double bond at m/z 161. When the charge site is on the sphingosine C-3 hydroxy group, the cleavage of the anomeric carbon–oxygen bond with a proton transfer (such as the Y_0 ions in Scheme 6.25) would result in the ion at m/z 298 as a Y_0-type ion.

6.4.2 Neutral Glycosphingolipids $[M + H]^+$

Many sphingolipids contain one or two sugars attached to the C-1 oxygen atom of the sphingosine base and also have the amino group of the long-chain base acylated with fatty acyl or α-hydroxy fatty acyl groups. These lipids

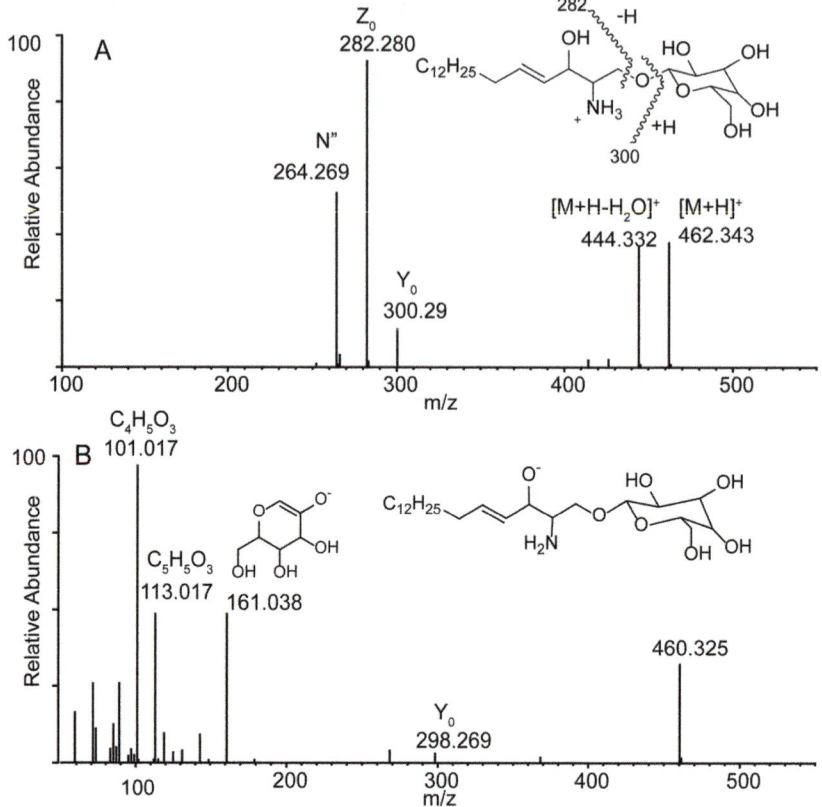

Figure 6.9 Electrospray ionization (positive and negative ions) and tandem mass spectrometry of a psychosine. (A) Product ions obtained following collisional activation of GalSo(d18 : 1) $[M + H]^+$ at *m/z* 462; (B) product ions obtained following collisional activation of the anion from GalSo(d18 : 1) $[M - H]^-$ at *m/z* 460. These high resolution MS/MS spectra were obtained using a quadrupole time-of-flight mass spectrometer.

Scheme 6.24

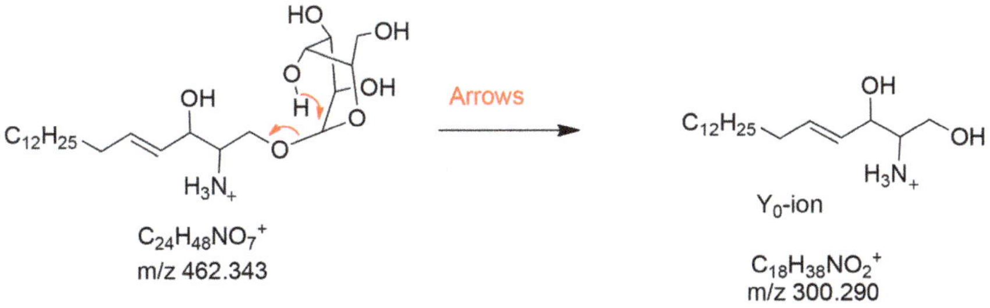

Scheme 6.25

are very abundant in the brain and include compounds given the trivial name of cerebrosides. As expected for ceramides, these sphingolipid species form either positive or negative ions, but most detailed mass spectrometric studies have involved collisional activation of the $[M + Li]^+$ adduct ions of the neutral glycosphingolipids.

The collisional activation of $[M + H]^+$ from GlcCer(d18 : 1/12 : 0) (Figure 6.10A) is fairly similar to that seen for ceramides, with the addition of reactions that lead to cleavage of the glycosidic bond that are competitive with pathways of ceramide cleavage to release the internal energy of the collisionally activated ion. These ions are observed for most all neutral sphingolipids, including this synthetic cerebroside. Glycosidic bond cleavage results in formation of Y_0-type ion due to the loss of the entire sugar residue in a mechanism discussed for psychosine (Scheme 6.25). As a charge site remote reaction, cleavage at the adjacent carbon–oxygen bond to the sugar would lead to a Z_0-ion with formation of an aziridine like structure (Scheme 6.26). This ion can also lose water from the sphingosine C-3 group to yield the ion $[Z_0-H_2O]^+$ by the ene- type reaction following a 1[3]-sigmatropic shift of the $\Delta^{4,5}$-double bond.

Another major product ion following CID of neutral glycosphingolipid appears at *m/z* 264 and corresponds to the N''-type ion for which a mechanism was outlined for ceramides in Scheme 6.5. However, the mechanism (Scheme 6.27) is a variant to that previously suggested.[13] The loss of the neutral sugar residue would result from the abstraction of the proton from the C'-2 position of the N-acyl group. Following the loss of the fatty acyl ketene, a stabilized N''-type ion would result at *m/z* 264.

The slightly more complex lactosyl ceramide has an additional sugar as the disaccharide galactosyl-glucose (lactose) attached to the sphingoid base, which introduces additional glycosidic bond fragmentation (Figure 6.10B). A difference at high mass is the appearance of minor cleavage ions of galactose and formation of a Z_1-type ion at *m/z* 700 for the species LacCer(d18 : 1/16 : 0). The Z_0-type ion (*m/z* 520) and the N''-type ion (*m/z* 264) are observed just as in GlcCer(d18 : 1/16 : 0).

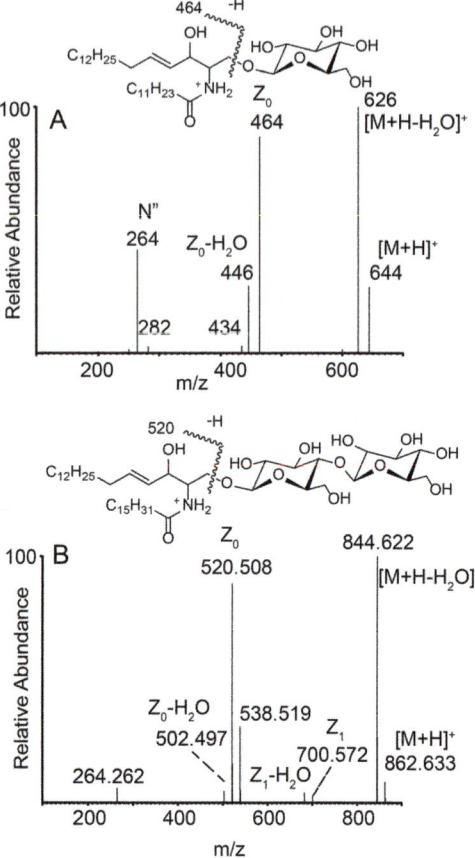

Figure 6.10 Electrospray ionization (positive ions) and tandem mass spectrometry of a cerebroside and globoside. (A) Product ions obtained following collisional activation of GalCer(d18 : 1/12 : 0) $[M + H]^+$ at m/z 640. This MS/MS spectra was obtained using a tandem quadruple mass spectrometer. (B) Product ions obtained following collisional activation of LacCer(d18 : 1/16 : 0) $[M + H]^+$ at m/z 862. This high resolution MS/MS spectrum was obtained using a quadrupole time-of flight mass spectrometer.

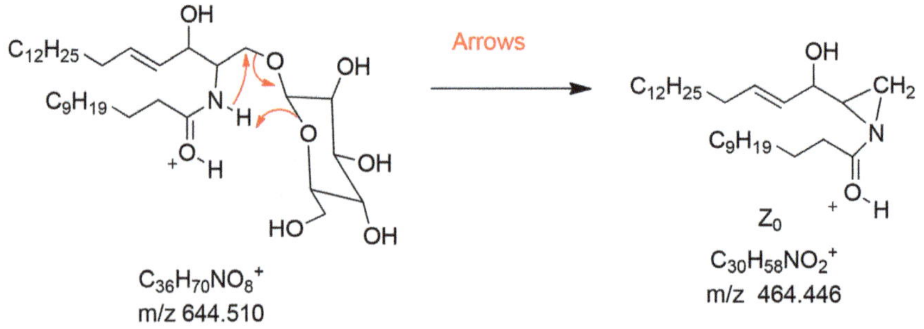

Scheme 6.26

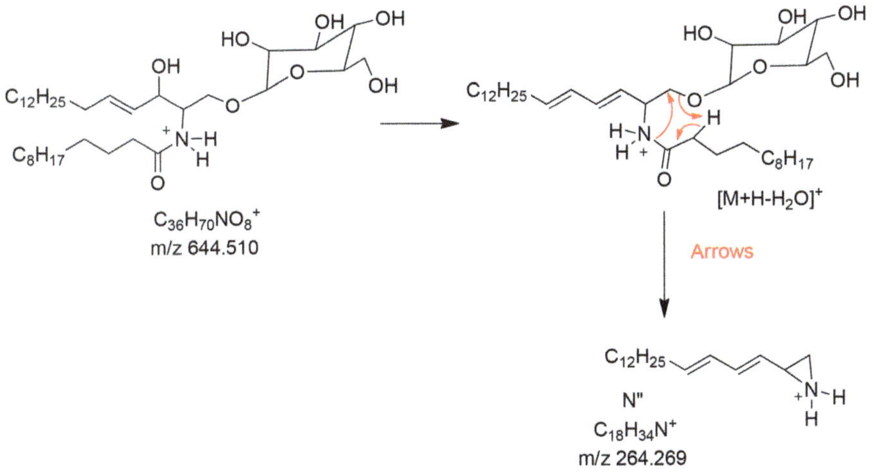

Scheme 6.27

6.4.3 Neutral Glycosphingolipids [M + Li]⁺

The formation of the lithium adducts of neutral glycosphingolipids has been a strategy introduced initially with fast atom bombardment ionization followed by high energy collisional activation and more recently extended to low energy tandem mass spectrometry with electrospray ionization.[7,10] Investigators have suggested improved sensitivity and fragmentation that is quite suitable for detailed structural characterization of these complex lipids extracted from biological matrices.[27,28] Many of the pathways for decomposition of positive ion lithium adducts $[M + Li]^+$ have already been introduced in earlier sections, with structurally less complex sphingolipids. It would appear that the specific ions relevant to the lipid portion of the glycosphingolipid (ceramide) are derived from precursor ions very similar in structure to the much simpler $[M + H]^+$ or $[M + Li]^+$ ions derived from ceramides themselves.

A typical $[M + Li]^+$ product ion spectrum is seen for the neutral glycosphingolipid GalCer(d18 : 1/24 : 0) (Figure 6.11A) and is characterized by a series of high mass product ions that correspond to cleavage of the sugar residue at the sites indicated in Scheme 6.23 as Z_0 and Y_0, identical to those observed for the protonated adduct. A mechanism for formation of Z_0-type ions has been suggested (Scheme 6.24) and the reaction leading to either the Y_0 or the B_0-type ion is suggested by the charge remote mechanism presented in Scheme 6.28. Again, there have not been many detailed studies of the formation of product ions from these metalated species, but when six protons were deuterated by H/D exchange, these carbohydrate cleavage ions shift in mass, consistent with the mechanism suggested by Hsu and Turk, where three exchangeable protons remain on the fragment ion m/z 656 (659) and three on the ion m/z 169 (172).[27] These ions may be formed by the

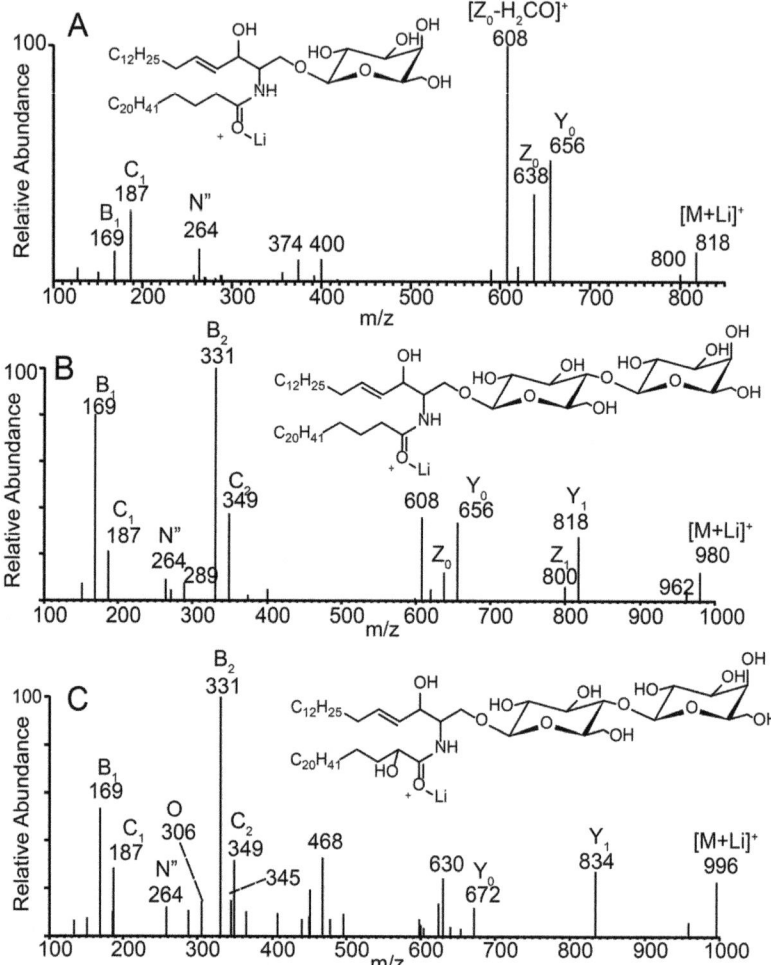

Figure 6.11   Electrospray ionization (positive ions) and tandem mass spectrometry of lithiated cerebrosides and globosides. (A) Product ions obtained following collisional activation of lithiated GalCer(d18 : 1/24 : 0) [M + Li]$^+$ at m/z 818; (B) product ions obtained following collisional activation of lithiated LacCer(d18 : 1/24 : 0) [M + Li]$^+$ at m/z 980; (C) product ions obtained following collisional activation of an α-hydroxy fatty acyl LacCer(d18 : 1/h24 : 0) [M + Li]$^+$ at m/z 996. This figure was redrawn from data presented in ref. 27.

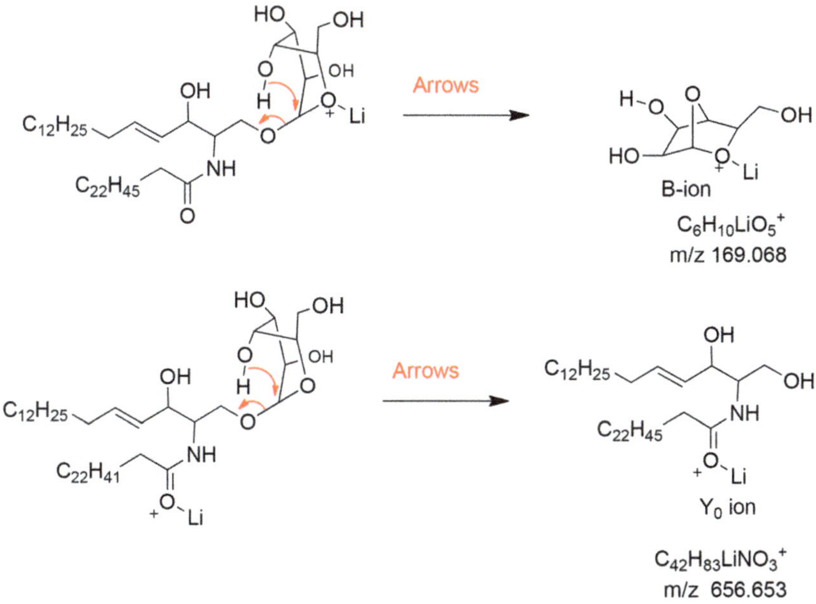

Scheme 6.28

mechanism shown in Scheme 6.28 and the exact ion formed depends on an initial charge site either on the sugar (B ion product) or the ceramide (Y_0 ion product).

The Z_0 ion (observed at m/z 638 in Figure 6.11A) can then undergo the loss of the C-1 carbon as a neutral formaldehyde (CH_2O) to form the abundant product ion observed at m/z 608. These ions all retain the Li^+ atom, revealing the adduct site location on the ceramide portion of this glycosphingolipid for these reaction mechanisms.

Abundant low mass ions retaining the Li^+ charging site on the sugar residue are also observed at m/z 187 and 169 and these are indicated as C_1 and B_1, respectively. These are formed by the same charge remote reactions as indicated for the formation of Z_0 and Y_0 ions, but in these instances the charge is now retained on the sugar fragment, thus charging these as ions, consistent with the lipid portion being lost as neutral species. This is shown in Scheme 6.28 for the Li adduct charge site residing on the sugar portion of the molecule (m/z 169).

The ions observed at m/z 656 (Y_0-type ion) would appear identical to the molecular ion of $[M + Li]^+$ Cer(d18 : 1/24 : 0) and studies similar to MS^3 have revealed identical product ions previously discussed, particularly the N''-type product ion at m/z 264.[19] These observations, combined with several others, have established the value of the $[M + Li]^+$ adduct product ion spectra in structural characterization of the ceramide-lipid portion of the glyco-sphingolipids.[7,10,29,30] With more complex glycosphingolipids, the sugar

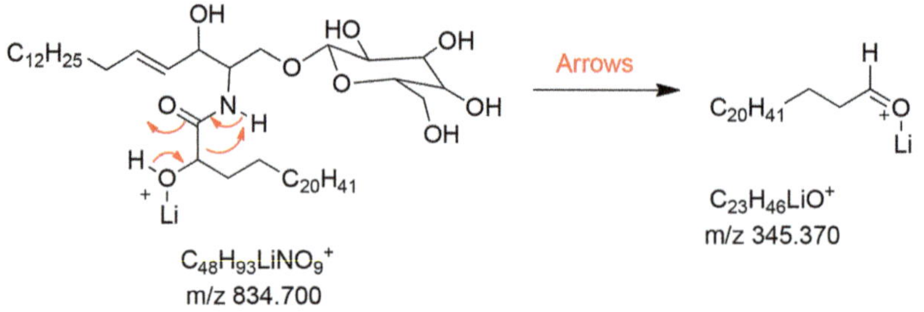

Scheme 6.29

fragment ions containing the charging Li-adduct increase in mass as well as number (C_1, C_2, C_n... and B_1, B_2, B_n...) which are a series of ions that differ by 162 Da, that render a much more complex product ion spectrum. These ion species are observed (Figure 6.11B) at m/z 331 (B_2 ion) and 349 (C_2 ion) as seen for LacCer(d18 : 1/24 : 0).

The presence of an α-hydroxyl group in the N-acyl moiety can also be discerned in the tandem mass spectrum of these glycosphingolipids. The O-type specific fragment ions (m/z 306 for the h24 : 0 fatty acyl group) previously described (Scheme 6.9) are present (Figure 6.11C). Also, the lithiated aldehyde formed by the cleavage of the 1′ and 2′ fatty acyl carbon atoms indicate a population of lithium attachment at the α-hydroxyl group (Scheme 6.29). The aldehydic product ion could be formed directly from the molecular ion along with other product ions, still retaining the N-acyl substituent, by the mechanism suggested for Li^+-adducts that involve formation of CO neutral and amide bond cleavage (Scheme 6.10).

The more complex tetrasaccharide glycosphingolipid generated a tandem mass spectrum presented in Figure 6.12 that reveals many more fragmentation pathways, but most of these are rather specific for the sugar portion of the molecule. This includes internal fragment ions observed at m/z 331 and 349 corresponding to hexoses (galactosyl-galactose or galactosyl-glucose).[27] This tetrasaccharide glycosphingolipid (also called a globoside) has a terminal N-acetyl amino group observed as the abundant products at m/z 210 and 228 corresponding to B_1 and C_1 fragment ions (Scheme 6.23).

6.4.4 Neutral Glycosphingolipids $[M − H]^−$

Negative $[M − H]^−$ ions are readily formed by electrospray ionization of the neutral glycosphingolipids and product ion information about the carbohydrate portion of the sphingolipid predominates as exemplified by Gal-Cer(d18 : 1/12 : 0) and LacCer (d18 : 1/18 : 0) (Figure 6.13). The most abundant product ions formed after collisional activation of the $[M − H]^−$ correspond to cleavage of the sugar residues and formation of the Y-series ions. This was observed at m/z 480 from GalCer(d18 : 1/12 : 0) for the Y_0 ion (Figure 6.13A) and at m/z 536 and 698 from LacCer(d18 : 1/16 : 0) corresponding to

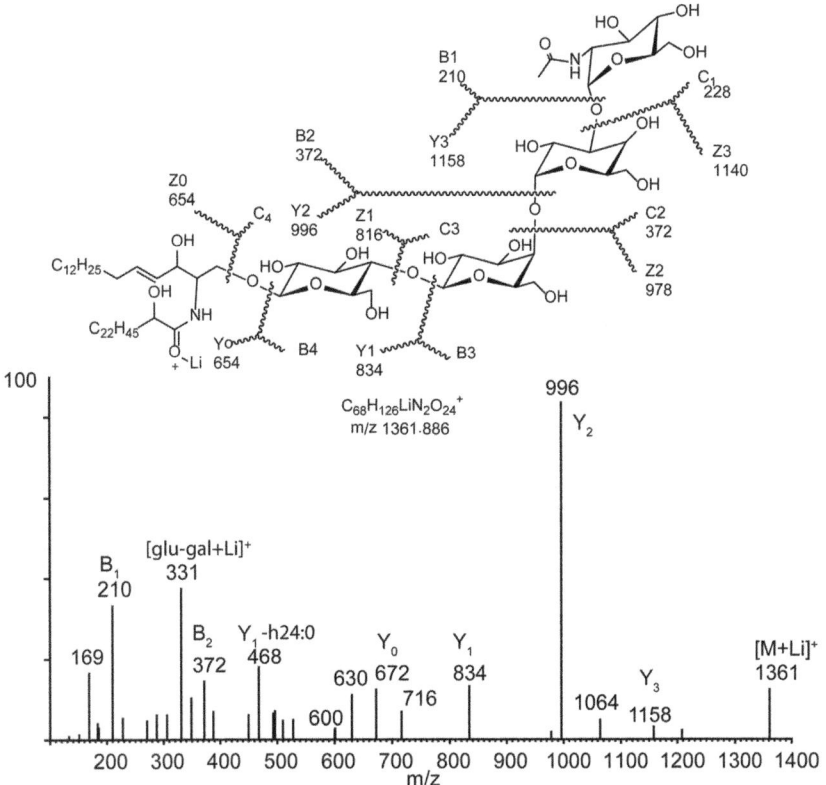

Figure 6.12 Electrospray ionization (positive ions) and tandem mass spectrometry of a tetrahexosegloboside. Product ions obtained following collisional activation of the α-hydroxy fatty acyl Gb₄(d18 : 1/h24 : 0) [M + Li]⁺ at *m/z* 1361. This figure was redrawn from data presented in ref. 27.

Y_0 and Y_1, respectively (Figure 6.13B). This would suggest that the charge site is predominantly located on the sugar portion of the molecule, possibly as an alkoxide anion. The mechanisms of formation of these ions were presented in Schemes 6.25 and 6.28.

6.4.5 Acidic Glycosphingolipids [M − H]⁻

A very abundant acidic glycosphingolipid found in numerous tissues are the 3-sulfogalactosyl ceramide molecular species termed sulfatides. These unique and highly acidic lipids play important and diverse biochemical roles often observed in the white matter of the brain.[31]

Positive ions can be made by electrospray ionization of sulfatides, although not a usual ionization polarity for these highly acidic sphingolipids. The [M + H]⁺ ions lose sulfuric acid [M + H − H₂SO₄]⁺ observed at *m/z* 792 in Figure 6.14A for the ST(18 : 1/24 : 1) molecular species. The other losses

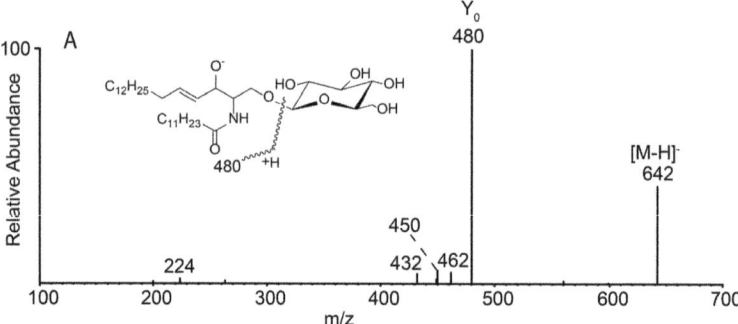

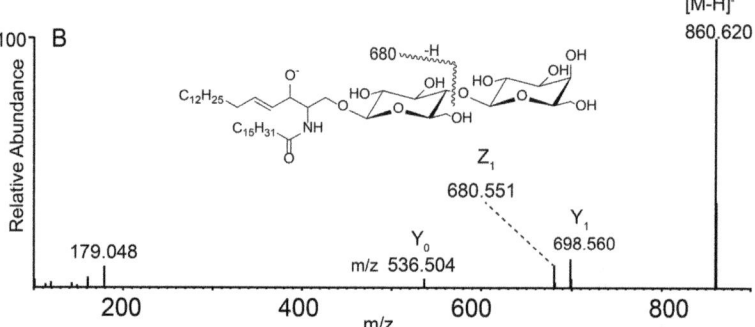

Figure 6.13 Electrospray ionization (negative ions) and tandem mass spectrometry of cerebroside and globoside. (A) Product ions obtained following collisional activation of the anion from GalCer(d18 : 1/12 : 0) [M − H]⁻ at *m/z* 642. This MS/MS spectrum was obtained using a tandem quadrupole mass spectrometer. (B) Product ions obtained following collisional activation of the anion from LacCer(d18 : 1/16 : 0) [M − H]⁻ at *m/z* 860. These high resolution MS/MS spectrum was obtained using a quadrupole time-of-flight mass spectrometer.

correspond to the formation of Z_0 type ions at *m/z* 630 and further loss of H_2O at *m/z* 612 as outlined in Scheme 6.30.

As all sulfuric acid esters, these lipid species readily form negative ions [M − H]⁻ (Figure 6.14B). The example tandem mass spectrum in this figure ST(d18 : 1/24 : 1) corresponds to one of the most abundant sulfatide molecular species found in the mammalian brain. Collisional activation of this sulfatide [M − H]⁻ yields a very dominate product ion at *m/z* 97 corresponding to HSO_4^-, most likely by the process shown in Scheme 6.31. Formation of *m/z* 97 would involve a proton abstraction, perhaps in a charge-driven process involving the sugar C-6 hydroxyl proton, to yield a bicyclic neutral species as indicated in the scheme.[32] By molecular models, this process would appear to be quite favorable as drawn, but other mechanisms engaging the loss of hydroxyl protons from other positions on the sugar ring

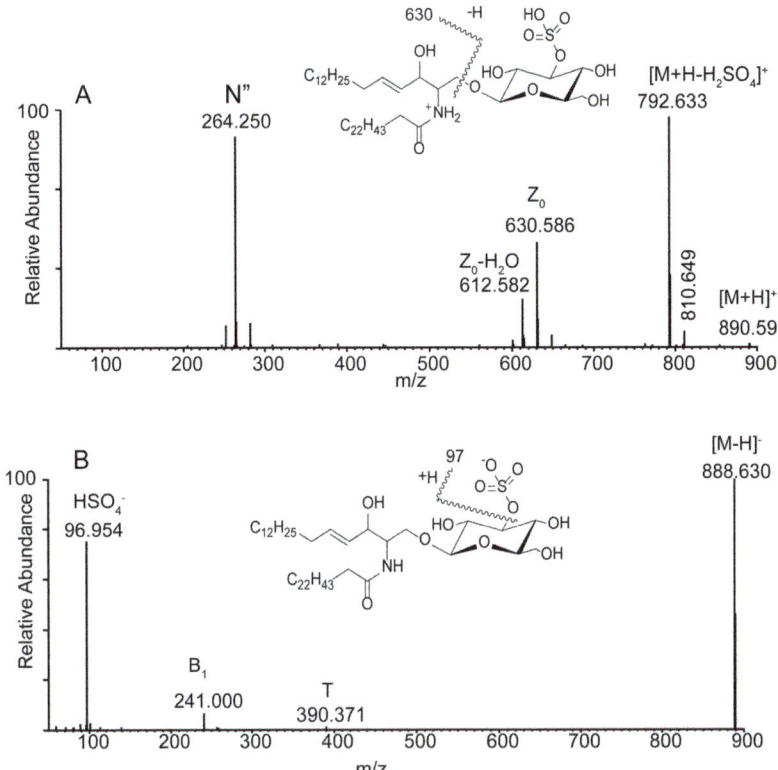

Figure 6.14 Electrospray ionization (positive and negative ions) and tandem mass spectrometry of a sulfatide (ST). (A) Product ions obtained following collisional activation of the protonated ST(d18 : 0/24 : 1) $[M + H]^+$ at m/z 890; (B) product ions obtained following collisional activation of the anion ST(d18 : 1/24 : 1) $[M - H]^-$ at m/z 888. These high resolution MS/MS spectra were obtained using a quadrupole time-of-flight mass spectrometer.

Scheme 6.30

Scheme 6.31

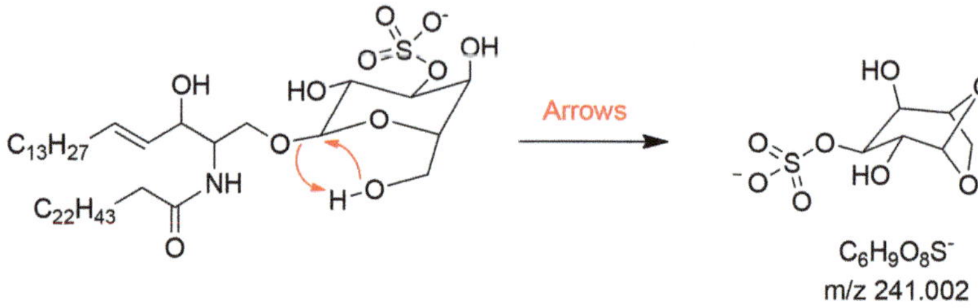

Scheme 6.32

are possible. Likely the formation of m/z 97 is a result of several distinct reaction mechanisms involving all three free sugar hydroxyl moieties.

While other product ions are not particularly abundant, there are unique ions observed at m/z 241 and 390. The formation of the B-series ion by cleavage of the glycosidic bond would lead to the ion at m/z 241 as indicated in Scheme 6.32, as a bicyclic carbohydrate ion containing the sulfate moiety.

A more complex sulfatide product ion generated following collisional activation of $[M - H]^-$ is formed by a charge-driven mechanism that would involve attack of the sulfate anion on the C-3 hydroxyl proton of the sphingoid base. This interaction would initiate formation of an aldehyde fragment from the long-chain base, olefin formation, then abstraction of the amide proton by the glycosidic oxygen to form the anionic fragment retaining the N-acyl group as a T-type ion. The concerted mechanism is outlined in Scheme 6.33.

Hsu and Turk have considered mechanisms for the formation of numerous minor product ions formed following collisional activation of the sulfatide $[M - H]^-$ that reveal many interesting structural features.[32] This includes the potential to be able to glean information about the double location of the various fatty acyl groups.[33]

6.4.6 Acidic Glycosphingolipids Positive Ions

The other major types of acidic sphingolipids are the gangliosides that contain the nine carbon sugar, neuraminic acid (a generic term often used is sialic acid). The structures of these molecules are quite complex and important information required for identification includes defining the sugar isomer present as each hexose, as well as the linkage to each adjacent

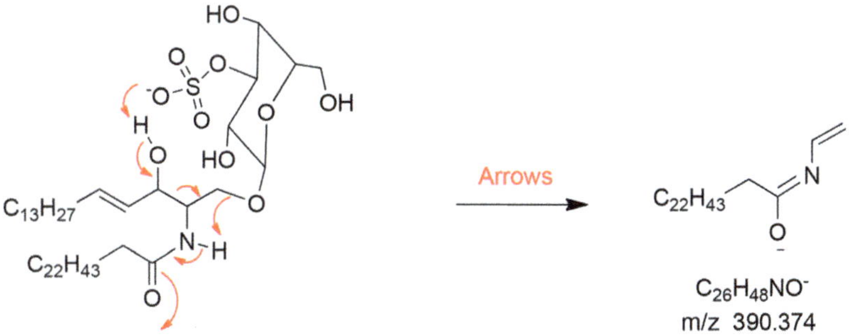

Scheme 6.33

sugar. Thus, detailed information related to carbohydrate structure is one of the important challenges of structure characterization of these molecules. Often mass spectrometry is carried in the positive ion mode with sodium or lithium attachment ions, but negative ion [M − H]⁻ is also used. Furthermore, derivatization of the free hydroxyl groups is often carried out to facilitate linkage analysis. The tandem mass spectrometry of these complex lipids is dominated by the sugar cleavage reactions, rather than fragmentation driven by the lipid portion of the molecule.[34,35] However, many important details of the collisional activation of these complex lipids is available and should be consulted as well as additional experimental approaches to carry out structure elucidation of these molecules.[36]

References

1. A. H. J. Merrill, Sphingolipid and glycosphingolipid metabolic pathways in the era of sphingolipidomics, *Chem. Rev.*, 2011, **111**, 6387–6422.
2. Y. Chen, Y. Liu, M. C. Sullards and A. H. Merrill Jr, An introduction to sphingolipid metabolism and analysis by new technologies, *NeuroMol. Med.*, 2010, **12**, 306–319.
3. B. Lieser, G. Liebisch, W. Drobnik and G. Schmitz, Quantification of sphingosine and sphinganine from crude lipid extracts by HPLC electrospray ionization tandem mass spectrometry, *J. Lipid Res.*, 2003, **44**, 2209–2216.
4. G. T. Kunkel, M. Maceyka, S. Milstien and S. Spiegel, Targeting the sphingosine-1-phosphate axis in cancer, inflammation and beyond, *Nat. Rev. Drug Discovery*, 2013, **12**, 688–702.
5. G. Liebisch and M. Scherer, Quantification of bioactive sphingo- and glycerophospholipid species by electrospray ionization tandem mass spectrometry in blood, *J. Chromatogr. B: Anal. Technol. Biomed. Life Sci.*, 2012, **883–884**, 141–146.
6. C. Bode and M. H. Graler, Quantification of sphingosine-1-phosphate and related sphingolipids by liquid chromatography coupled to tandem mass spectrometry, *Methods Mol. Biol.*, 2012, **874**, 33–44.

7. B. Domon and C. E. Costello, Structure elucidation of glycosphingolipids and gangliosides using high-performance tandem mass spectrometry, *Biochemistry,* 1988, **27**, 1534–1543.

8. V. N. Reinhold and D. M. Sheeley, Detailed characterization of carbohydrate linkage and sequence in an ion trap mass spectrometer: glycosphingolipids, *Anal. Biochem.,* 1998, **259**, 28–33.

9. J. Zaia, Mass spectrometry and glycomics, *OMICS,* 2010, **14**, 401–418.

10. J. Adams, Q. Ann, Structure determination of sphingolipids by mass spectrometry, *Mass Spectrom. Rev.,* 1993, **12**, 51–85.

11. R. t'Kindt, L. Jorge, E. Dumont, P. Couturon, F. David, P. Sandra and K. Sandra, Profiling and characterizing skin ceramides using reversed-phase liquid chromatography-quadrupole time-of-flight mass spectrometry, *Anal. Chem.,* 2012, **84**, 403–411.

12. M. C. Sullards, Analysis of sphingomyelin, glucosylceramide, ceramide, sphingosine, and sphingosine 1-phosphate by tandem mass spectrometry, *Methods Enzymol.,* 2000, **312**, 32–45.

13. F. F. Hsu, J. Turk, M. E. Stewart and D. T. Downing, Structural studies on ceramides as lithiated adducts by low energy collisional-activated dissociation tandem mass spectrometry with electrospray ionization, *J. Am. Soc. Mass Spectrom.,* 2002, **13**, 680–695.

14. M. Fillet, J. C. Van Heugen, A. C. Servais, G. J. De and J. Crommen, Separation, identification and quantitation of ceramides in human cancer cells by liquid chromatography-electrospray ionization tandem mass spectrometry, *J. Chromatogr. A,* 2002, **949**, 225–233.

15. M. Gu, J. L. Kerwin, J. D. Watts and R. Aebersold, Ceramide profiling of complex lipid mixtures by electrospray ionization mass spectrometry, *Anal. Biochem.,* 1997, **244**, 347–356.

16. Q. Ann and J. Adams, Structure-specific collision-induced fragmentations of ceramides cationized with alkali-metal ions, *Anal. Chem.,* 1993, **65**, 7–13.

17. Q. Ann and J. Adams, Structure determination of ceramides and neutral glycosphingolipids by collisional activation of [M + Li](+) ions, *J. Am. Soc. Mass Spectrom.,* 1992, **3**, 260–263.

18. F. F. Hsu and J. Turk, Characterization of ceramides by low energy collisional-activated dissociation tandem mass spectrometry with negative-ion electrospray ionization, *J. Am. Soc. Mass Spectrom.,* 2002, **13**, 558–570.

19. X. Han, Characterization and direct quantitation of ceramide molecular species from lipid extracts of biological samples by electrospray ionization tandem mass spectrometry, *Anal. Biochem.,* 2002, **302**, 199–212.

20. M. H. Lee, G. H. Lee and J. S. Yoo, Analysis of ceramides in cosmetics by reversed-phase liquid chromatography/electrospray ionization mass spectrometry with collision-induced dissociation, *Rapid Commun. Mass Spectrom.,* 2003, **17**, 64–75.

21. L. A. Hoeferlin, D. S. Wijesinghe and C. E. Chalfant, The role of ceramide-1-phosphate in biological functions, *Handb. Exp. Pharmacol.,* 2013, 153–166.

22. F.-F. Hsu and J. Turk, Structural determination of sphingomyelin by tandem mass spectrometry with electrospray ionization, *J. Am. Soc. Mass Spectrom.,* 2000, **11**, 437–449.

23. J. P. Slotte, Biological functions of sphingomyelins, *Prog. Lipid Res.,* 2013, **52**, 424–437.

24. R. C. Murphy and P. H. Axelsen, Mass spectrometric analysis of long-chain lipids, *Mass Spectrom. Rev.,* 2011, **30**, 579–599.

25. M. C. Sullards, Y. Liu, Y. Chen and A. H. Merrill, Jr, Analysis of mammalian sphingolipids by liquid chromatography tandem mass spectrometry (LC-MS/MS) and tissue imaging mass spectrometry (TIMS), *Biochim. Biophys. Acta,* 2011, **1811**, 838–853.

26. X. Jiang, K. Yang and X. Han, Direct quantitation of psychosine from alkaline-treated lipid extracts with a semi-synthetic internal standard, *J. Lipid Res.,* 2009, **50**, 162–172.

27. F. F. Hsu and J. Turk, Structural determination of glycosphingolipids as lithiated adducts by electrospray ionization mass spectrometry using low-energy collisional-activated dissociation on a triple stage quadrupole instrument, *J. Am. Soc. Mass Spectrom.,* 2001, **12**, 61–79.

28. S. B. Levery, M. S. Toledo, A. H. Straus and H. K. Takahashi, Comparative analysis of glycosylinositol phosphorylceramides from fungi by electrospray tandem mass spectrometry with low-energy collision-induced dissociation of Li(+) adduct ions, *Rapid Commun. Mass Spectrom.,* 2001, **15**, 2240–2258.

29. B. Bennion, S. Dasgupta, E. L. Hogan and S. B. Levery, Characterization of novel myelin components 3-O-acetyl-sphingosine galactosylceramides by electrospray ionization Q-TOF MS and MS/CID-MS of Li+ adducts, *J. Mass Spectrom.,* 2007, **42**, 598–620.

30. A. Olling, M. E. Breimer, E. Peltomaa, B. E. Samuelsson and S. Ghardashkhani, Electrospray ionization and collision-induced dissociation time-of-flight mass spectrometry of neutral glycosphingolipids, *Rapid Commun. Mass Spectrom.,* 1998, **12**, 637–645.

31. M. Eckhardt, The role and metabolism of sulfatide in the nervous system, *Mol. Neurobiol.,* 2008, **37**, 93–103.

32. F. F. Hsu and J. Turk, Studies on sulfatides by quadrupole ion-trap mass spectrometry with electrospray ionization: structural characterization and the fragmentation processes that include an unusual internal galactose residue loss and the classical charge-remote fragmentation, *J. Am. Soc. Mass Spectrom.,* 2004, **15**, 536–546.

33. F. F. Hsu and J. Turk, Elucidation of the double-bond position of long-chain unsaturated fatty acids by multiple-stage linear ion-trap mass spectrometry with electrospray ionization, *J. Am. Soc. Mass Spectrom.,* 2008, **19**, 1673–1680.

34. I. Meisen, J. Peter-Katalinic and J. Muthing, Discrimination of neolacto-series gangliosides with alpha2-3- and alpha2-6.linked N-acetylneuraminic acid by nanoelectrospray ionization low-energy collision-induced

dissociation tandem quadrupole TOF MS, *Anal. Chem.,* 2003, **75,** 5719–5725.

35. K. Ikeda, T. Shimizu and R. Taguchi, Targeted analysis of ganglioside and sulfatide molecular species by LC/ESI-MS/MS with theoretically expanded multiple reaction monitoring, *J. Lipid Res.,* 2008, **49,** 2678–2689.
36. S. B. Levery, Glycosphingolipid structural analysis and glycosphingo-lipidomics, *Methods Enzymol.,* 2005, **405,** 300–369.

CHAPTER 7

Steroids

The classification of lipids as steroids restricts family members to products of cholesterol biosynthesis from the cyclization of 2,3-oxidosqualene to various metabolic products with a tetracyclic perhydrophenanthrylenyl basic structural unit. Even so, a large number of very important lipids are included in this designation that serve very diverse roles in biochemistry and physiology. An even larger number of synthetic steroid analogs have been made and marketed as therapeutics. The family member that is found in all cells is cholesterol, which is the single most abundant lipid in all mammalian cells, present in various lipid bilayers as a critical component for proper membrane function. Cholesterol is synthesized from acetic acid by way of the HMGCoA to form squalene. After cyclization of this 30-carbon isoprene hydrocarbon, and approximately 17 steps of hydroxylation and chain-shortening, a 27-carbon product results (Scheme 7.1). About one-half of the cholesterol found in humans is synthesized *de novo* by this pathway, largely taking place in the liver, with the remaining being ingested as a dietary component.[1]

Cholesterol is a major lipid in the brain and it is thought to be largely synthesized within this tissue.[2] Cholesterol is not soluble in aqueous solution and is transported in the blood of higher order animals as free cholesterol or cholesterol esters packaged within lipoprotein particles such as low density lipoproteins (LDL) and very low density lipoproteins (VLDL).[3] The uptake of these cholesterol species into cells takes place primarily by receptors recognizing the LDL particles, such that the process of cholesterol trafficking inside and outside of the cell continues as a highly regulated process.[1]

Cholesterol is metabolized into various steroidal products such as oxysterols, metabolites formed by reaction catalyzed by specific cytochrome P-450s, as well as specific oxidases, leading to several diverse steroid families

New Developments in Mass Spectrometry No. 4
Tandem Mass Spectrometry of Lipids: Molecular Analysis of Complex Lipids
By Robert C Murphy
© Robert C Murphy 2015
Published by the Royal Society of Chemistry, www.rsc.org

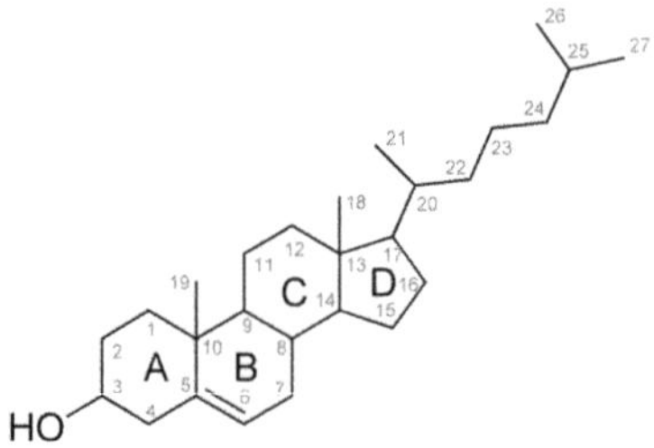

Scheme 7.1

including bile acids, androgens, estrogens, glucocorticoids, and mineralo-corticoids. These metabolites can be further conjugated to very polar entities such as glucuronides and sulfate esters to form more water-soluble, excretable products.

In addition, cholesterol can be oxidized by reactive oxygen species and specific enzymatic reactions.[4] An example is that cholesterol linoleate is the preferred substrate for 15-lipoxygenase to form cholesterol 13-hydro-peroxyoctadecenoate.[5] Such oxidized products and other oxysterols have profound biological activities as well as mediating pathological events. For example, major lipid species that accumulate in arterial plaques of humans in atherosclerosis are oxidized cholesterol esters.[6]

7.1 Cholesterol $[M + NH_4]^+$ and $[M + Na]^+$

Electrospray ionization of cholesterol requires adduct formation with a charging species such as NH_4^+ or alkali metal ion to form $[M + NH_4]^+$ or $[M + Na]^+$. Collision activation of $[M + NH_4]^+$ results in the formation of m/z 369, which is quite characteristic of the cholesterol structure (Figure 7.1A and B). This product ion is the result of an initial hydrogen rearrangement, so that the resulting carbon-centered cationic site emerging at C-3, following the loss of water and NH_3, can be delocalized over an adjacent double bond (Scheme 7.2). None of the deuterium-labeled hydrogen atoms immediately adjacent to the C-3 hydroxyl moiety are lost in this ion struc-ture (Figure 7.1B), supporting Scheme 7.2. The attachment of an alkali metal ion ($[M + Na]^+$) as a charging moiety prevents this decomposition behavior, suggesting the importance of the neutral loss of ammonia driving this process (Figure 7.1C).

7.1.1 Cholesterol and Oxysterol Biochemical Intermediates

Analysis by tandem mass spectrometry of the cholesterol adduct ions and most oxysterol intermediates of cholesterol biosynthesis as $[M + NH_4]^+$ yield very few product ions resulting from carbon–carbon bond cleavages. The stability of the cyclic ring structure renders the vast majority of these

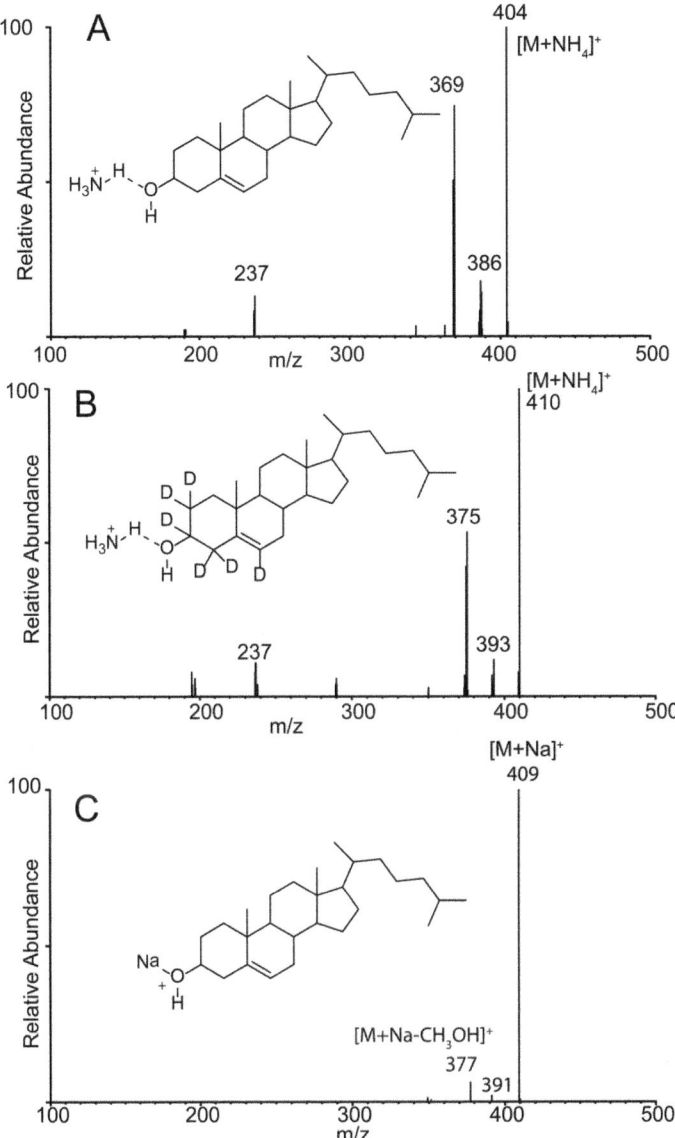

Figure 7.1 Electrospray ionization (positive ions) and tandem mass spectrometry of cholesterol. (A) Product ions following collisional activation of cholesterol $[M + NH_4]^+$ at m/z 404; (B) product ions obtained following collisional activation of D_6-cholesterol $[M + NH_4]^+$ at m/z 410; (C) product ions obtained following collisional activation of cholesterol $[M + Na]^+$ at m/z 409. These MS/MS spectra were obtained using a tandem quadrupole mass spectrometer.

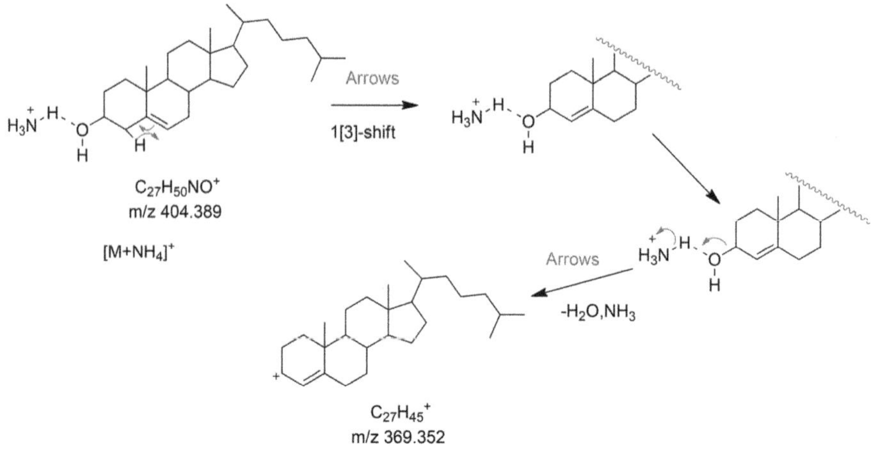

Scheme 7.2

oxysterols to form *m/z* 369 (or analog ion that includes double bonds or hydroxy groups in the basic steroidal structure), by eliminating one or two neutral water molecules, depending upon the number of hydroxy groups on the cholesterol ring. Therefore, little structural information is obtained from the tandem mass spectrometry of simple mono- and dihydroxylated cholesterol species that are involved in the biosynthesis or metabolism of this important lipid. The steroid nucleus product ion is typically the most abundant product ion observed and for an unsaturated cholesterol species, this appears at *m/z* 367 as exemplified for desmosterol (Figure 7.2A) and *m/z* 385 (369 + 16) for 24-hydroxy cholesterol (Figure 7.2B). Very little further fragmentation is evident after collisional activation, even at higher energies. It is possible to determine the mass of the steroid core moiety, but this limited information can be very useful when combined with HPLC retention time as an analytical approach to identify oxysterol intermediates. There are tables published providing molecular weight and HPLC retention time that can be used as guides.[7]

7.1.2 Cholesterol Derivatives

A highly successful strategy to increase structural information that can be obtained from cholesterol and various oxysterols, is derivatization with very polar and in many cases ionized moieties. This was first proposed by Shackleton using Girard's T reagent,[8] followed by Han with dimethylglycine and greatly expanded by Griffith and co-workers using Girard's P reagent.[9–11] This last protocol introduces a quaternary nitrogen group into the derivatized cholesterol and hence facilitates positive ion tandem mass spectrometry. These derivatives also increase electrospray ion yields from these neutral lipids. Many steroids have oxo groups that can covalently react with

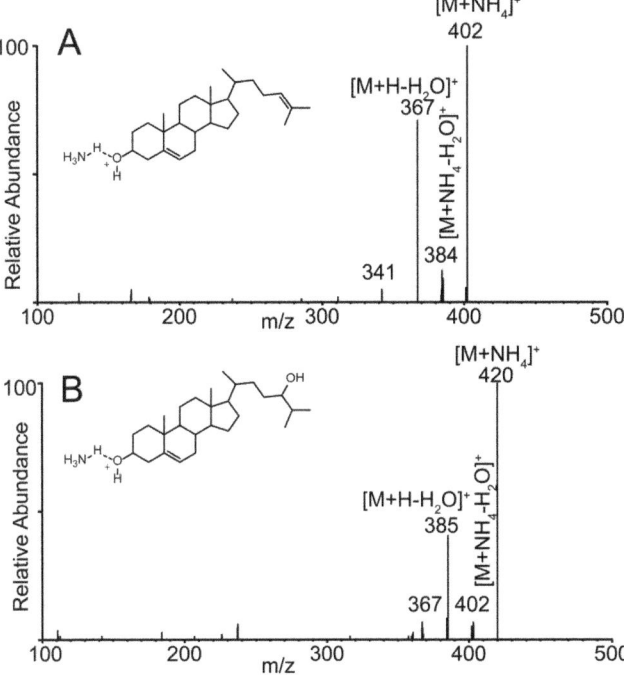

Figure 7.2 Electrospray ionization (positive ions) and tandem mass spectrometry of cholesterol precursors and metabolites. (A) Product ions obtained following collisional activation of desmosterol $[M + NH_4]^+$ at m/z 402; (B) product ions obtained following collisional activation of 24-hydroxycholesterol $[M + NH_4]^+$ at m/z 420. These MS/MS spectra were obtained using a tandem quadrupole mass spectrometer.

Girard's P reagent to form hydrazones such as the very common 3-oxo-steroids testosterone, aldosterone, and cortisol. Some steroids do not have such a structural feature, including cholesterol itself and its many biosynthetic intermediates and metabolites. To overcome this limitation, Griffiths[10] has devised relatively simple biochemical reactions using bacterial-derived cholesterol oxidase to generate the 3-oxo group from the 3-hydroxy moiety of cholesterol. These enzymes can readily convert a 3-β-hydroxy-5-ene structural unit found in cholesterol to the 3-oxo-4-ene structure in high yield and at the microchemical scale. These ketone products can then react readily with Girard's P hydrazine (GP) to form the ionized GP-hydrazone derivative (Scheme 7.3). The electrospray ionization process results in the formation of an abundant $[M]^+$ since the counter anion that would be present in solution to neutralize the charge is removed during the electrospray process.

Collisional activation of the GP-derivatives (Figure 7.3) yields a very abundant product ion that corresponds to the loss of neutral pyridine $[M - 79]^+$. This likely results from a charge-driven cyclization of the hydrazone involving the $\Delta^{4,5}$ double bond of the cholesterol moiety which leaves

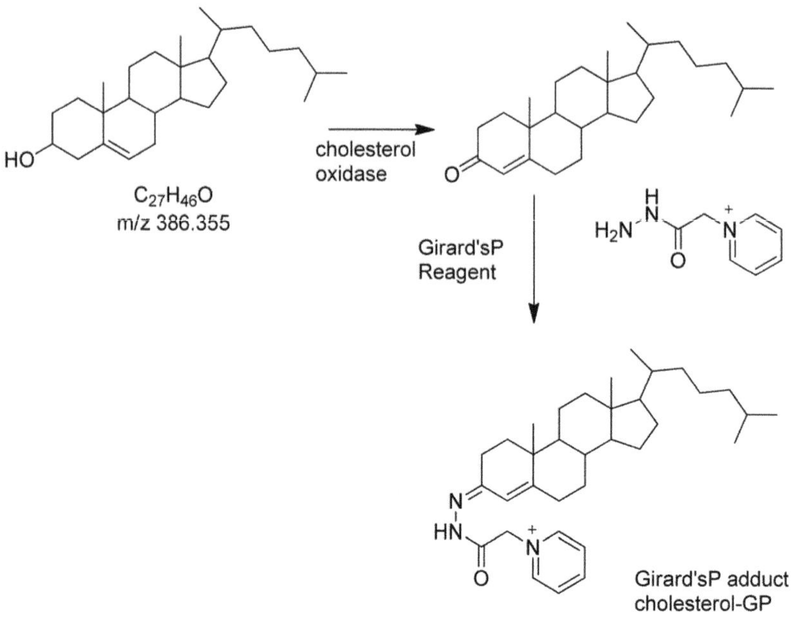

HO

$C_{27}H_{46}O$
m/z 386.355

cholesterol
oxidase

O

Girard'sP
Reagent

H_2N N

O

N

N
HN

O

N

Girard'sP adduct
cholesterol-GP

Scheme 7.3

the charge site on the bridge head carbon-5, as illustrated for 24-hydroxy cholesterol after oxidation with cholesterol oxidase and derivatization (Scheme 7.4). This ion can further undergo a loss of carbon monoxide to form the ion corresponding to $[M - 107]^+$. These ion structures are supported by high-resolution measurements and are typically the two abundant, high-mass ions for such GP-derivatives.[11]

Of particular interest is a series of low-mass ions that can reveal information about substitutions within the steroid A-B rings. These ions are quite common for a large number of steroid GP-derivatives and the origins of the steroid carbon atoms has been suggested and supported, again by high resolution measurements of the product ions.[11] These ions have been designated as b_1-12, b_3-28, b_2-type ions.[10,11] The exact mechanisms responsible for formation of these ions have not been examined in great detail; however, some suggestions are presented here. These product ions appear in large part to be a result of a fragmentation mechanism very similar to the rearrangement termed the Wagner-Meerwein rearrangement (also a pinacol-like rearrangement) driven by a positively charged carbon atom at the bridge-head position between steroid ring A and B.[12] Since the loss of the pyridine neutral from a Girard's P derivative (Scheme 7.5) leads to a favorable cationic site at such a bridgehead, the Wagner-Meerwein rearrangement is facilitated.

The ion at the lowest mass (b_1-12) has eight carbon atoms and could be formed by a migration of the 9-10 bond to the cation at carbon-5. This would form a spiro 5-membered ring leaving the cationic site now at carbon-10 (Scheme 7.5). Since this cyclopentane ring would be perpendicular to the

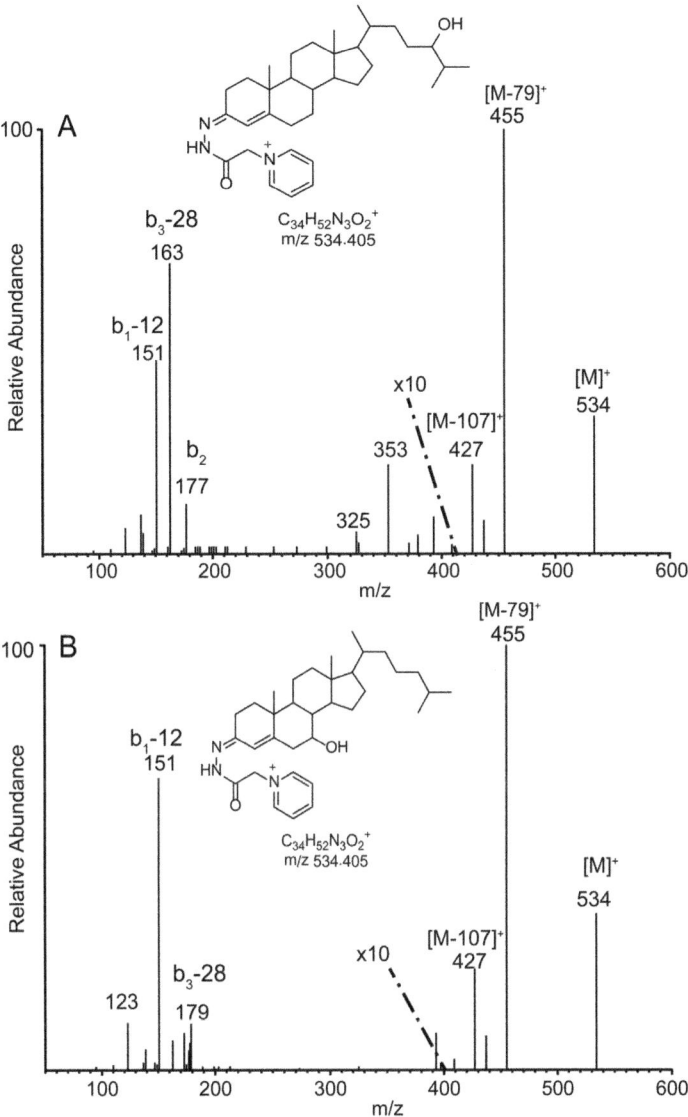

Figure 7.3 Electrospray ionization (positive ions) and tandem mass spectrometry of Girard's P (GP) derivatized hydroxycholesterols. (A) Product ions obtained following collisional activation of the GP-hydrazone of 24-hydroxycholesterol $[M]^+$ at m/z 534; (B) product ions obtained following collisional activation of the GP-hydrazone of 7-hydroxycholesterol $[M]^+$ at m/z 534. This figure was redrawn from data presented in ref. 10.

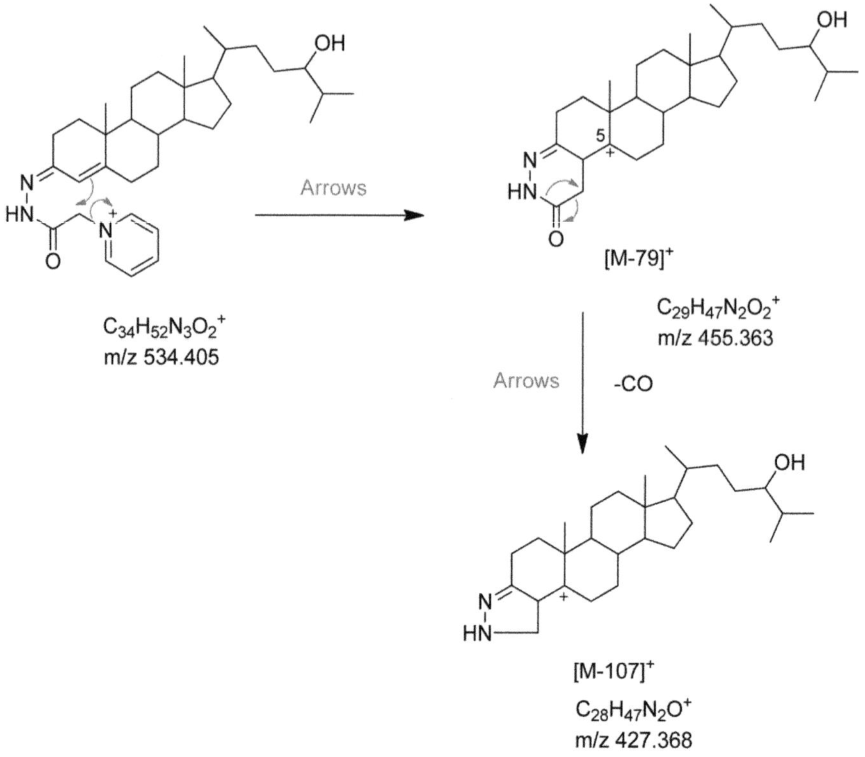

Scheme 7.4

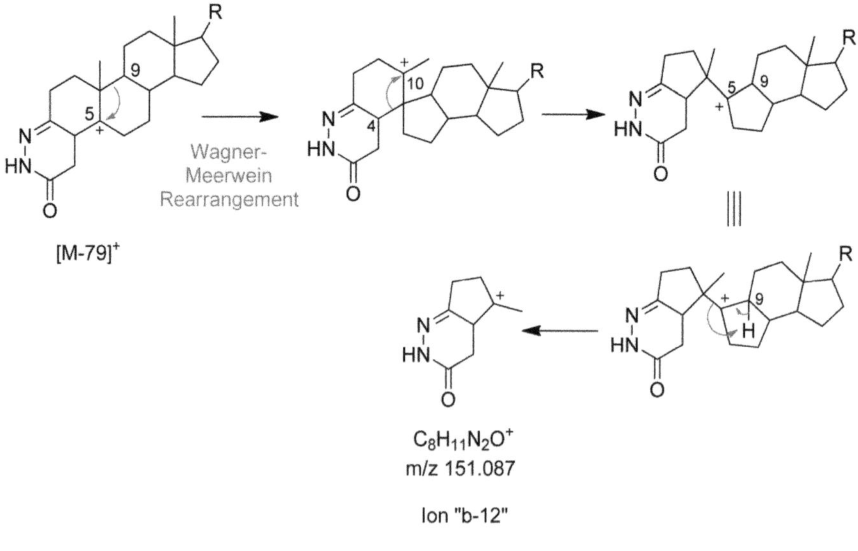

Scheme 7.5

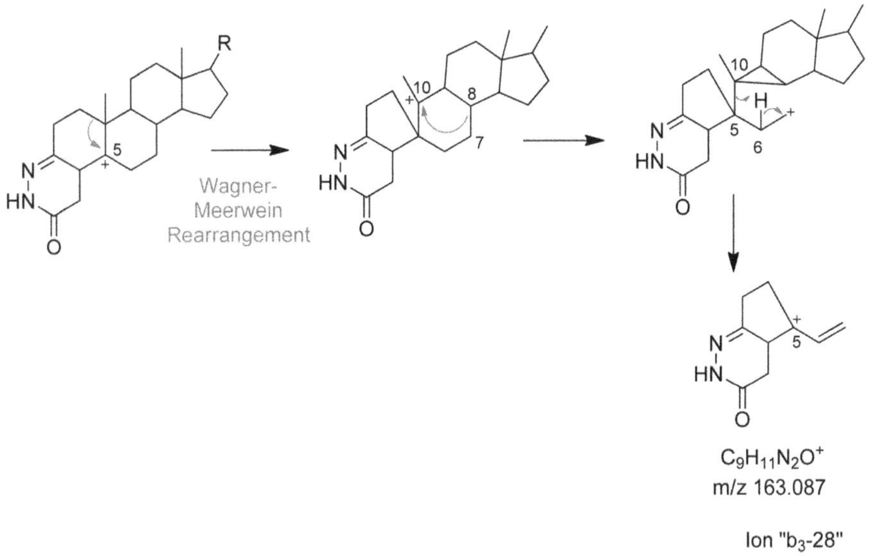

$C_9H_{11}N_2O^+$
m/z 163.087

Ion "b_3-28"

Scheme 7.6

original steroid A-ring cyclohexane ring structure, this would facilitate another cation-driven migration of the C-4/5 bond to C-10, leaving the cationic site now at carbon-5 and loss of the spiro structure. Breaking the new C-5/10 with a concerted proton rearrangement to C-5 from C-9 would leave the ion structure b_1-12 (Scheme 7.5).

Similar serial alkyl carbon migrations by the Wagner-Meerwein rearrangement are outlined in Scheme 7.6 as a mechanism explaining the formation of the *b-28* ion. The reaction pathway is initiated by migration of carbon-1 to the carbon-5 cationic site. This cyclopentane spiro intermediate would have the cationic site also at carbon-10, but it is now in a cyclohexane B-ring. Formation of a C-8/10 bond and breaking the C-7/8 bond during the process would leave the cationic site at C-7 and a new cyclopropane ring from the starting steroid C-ring. Cleavage of the C-5/10 bond with abstraction of a proton from carbon-6 leaves the charge site at C-5 and a 6,7 double bond. This ion, which has been termed *b-28*, would appear at *m/z* 163. A substituent at C-7 such as with 7-hydroxy cholesterol (Figure 7.3B) yields the same *b-28* ion now at *m/z* 179, since this ion includes C-7 from the original structure.

A similar Wagner-Meerwein rearrangement with formation of the cyclopropane fused to the A-ring, leads to the ion called b_2 (Scheme 7.7).

7.1.3 Cholesteryl Esters

Esterification of the 3-hydroxy group of cholesterol with fatty acids is an enzymatic process leading to fatty acyl cholesterol esters (CE). These CE species are packaged into lipoprotein particles such as chylomicrons in the

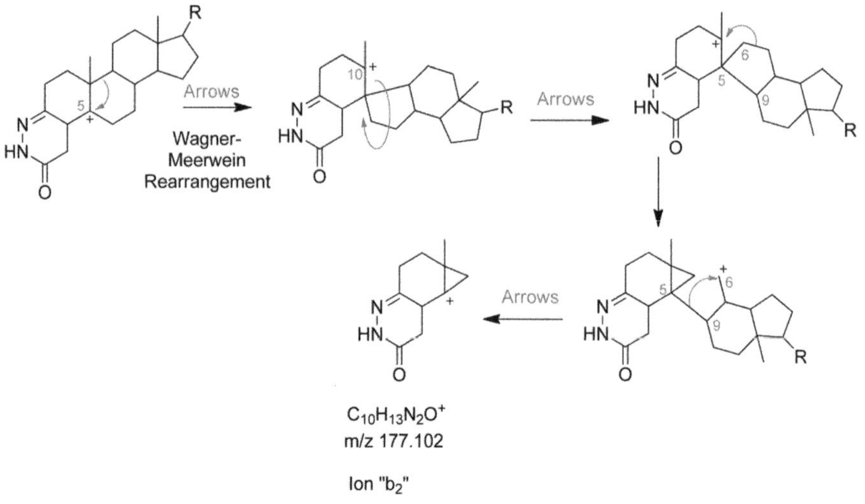

$C_{10}H_{13}N_2O^+$
m/z 177.102

Ion "b$_2$"

Scheme 7.7

intestine as part of food absorption and transport of both fatty acids and cholesterol to the liver as well as all cells in the organism. In the liver the CEs are reformed and packaged into very low density lipoproteins and secreted into the circulation. Blood carries these CEs in concentrations as high as 1–3 mmol L^{-1} within lipoprotein particles, since they are insoluble in aqueous solutions.

7.1.3.1 Cholesterol Esters $[M + NH_4]^+$

The CE lipid esters are neutral lipids and electrospray ionization must form charged adducts such as $[M + NH_4]^+$ or $[M + Na]^+$, depending upon the abundance of cationic buffers in the electrospray solvent system in order to study these molecules by mass spectrometry. Considerable use has been made of the ammonium ion adducts for analysis of cholesterol esters.[13] The activation of $[M + NH_4]^+$ yields a very simple product ion mass spectrum dominated by m/z 369 (Figure 7.4A). This ion corresponds to the cholestene cation seen for cholesterol (Scheme 7.8). It is likely that the 5,6 double bond isomerizes to the 4-5 double bond either before or after loss of a fatty acyl group (as a free fatty acid) and ammonia in order to delocalize the cationic site that is now found at C-3. This cholestene cation does not easily further decompose in the tandem quadrupole mass spectrometer.

Analysis of cholesterol esters typically makes use of this common product ion (m/z 369) to identify molecular species in biological samples. Analysis by either precursors of m/z 369 or MRM analysis of cholesterol esters is widely used to detect these unique lipid esters in biological extracts, using LC-MS/MS assay systems.[13] When the fatty acyl chain is oxidized, such as in CE(9-HODE) (Figure 7.4B) and CE(13-HODE)

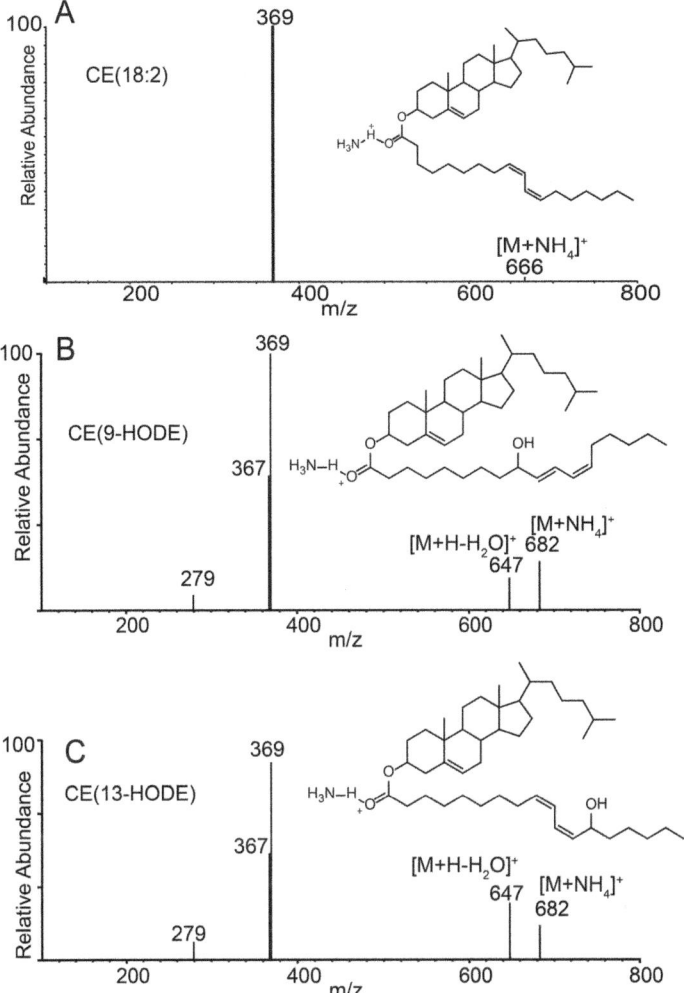

Figure 7.4 Electrospray ionization (positive ions) and tandem mass spectrometry of cholesterol esters and oxidized cholesterol esters. (A) Product ions obtained following collisional activation of cholesterol linoleate $[M + NH_4]^+$ at m/z 666; (B) product ions obtained following collisional activation of cholesterol 9-hydroxy-10,12-octadecadienoate $[M + NH_4]^+$ at m/z 682; (C) product ions obtained following collisional activation of cholesterol 13-hydroxy-9,11-octadecadienoate $[M + NH_4]^+$ at m/z 682. These MS/MS spectra were obtained using a tandem quadrupole mass spectrometer.

(Figure 7.4C), additional product ions are observed such as a significant ion for the loss of water. This ion most likely is due to the formation of a highly conjugated triene in the fatty acyl chain. There are also ions observed at m/z 279 which are acylium ions from the fatty acyl chain. Since non-oxidized CEs do not form such product ions after collisional

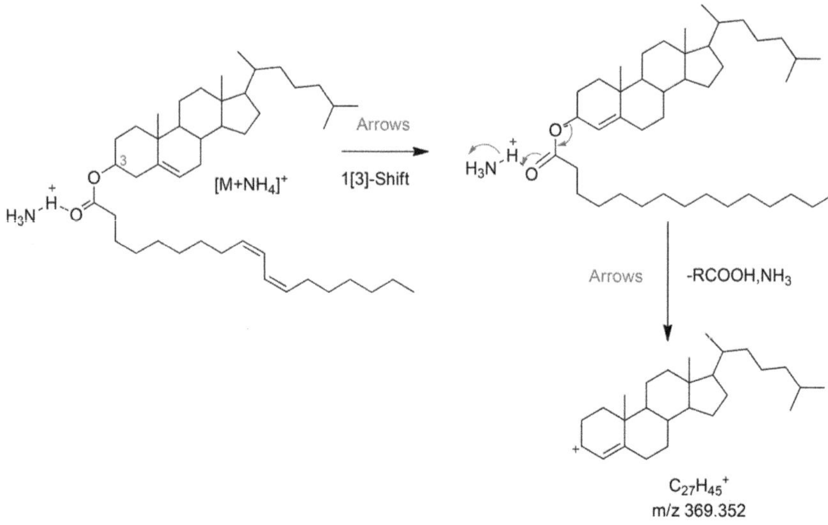

Scheme 7.8

activation, this suggests participation of the conjugated alcohol in this process of the loss of neutral cholesterol, water, and ammonia.

7.1.3.2 Hydroperoxy Cholesterol Esters

An unexpected facile decomposition was observed when CEs were esterified to 9- and 13-hydroperoxyoctadecadienoic acid and collisionally activated.[14] The $[M + NH_4]^+$ adduct ions readily decomposed to product ions that carried specific information about the site of the hydroperoxy group in the fatty acyl chain (Figure 7.5). The cholestene cation was the most abundant product ion (m/z 369) as expected (Scheme 7.9), but many high mass product ions were observed, suggesting that the hydroperoxy group had a fragmentation directing influence that was not observed in the closely related cholesterol esters, cholesterol 9- or 13-hydroxyoctadecadienoate (Figure 7.4B and C).

In most cases, the hydroperoxy group undergoes collision induced scission to form reactive oxygen-centered radicals that can interact with other portions of the molecule and direct product ion formation. Collisional activation of cholesterol 13-hydroperoxyoctadecadienoate will be presented as an example of mechanisms that can operate in formation of these product ions (Figure 7.5). Several unique reactions were identified that involve initial hydroperoxide scission, including reactions leading to the formation of a novel N–O bond, perhaps driven by the loss of neutral water (m/z 680) shown in Scheme 7.10. This product ion could then undergo a proton shift that would set up attack of the ammonium ion nitrogen atom on the acyl carbon-12 with a concerted loss of hexanal to form the unique ion m/z 580 (Scheme 7.10).

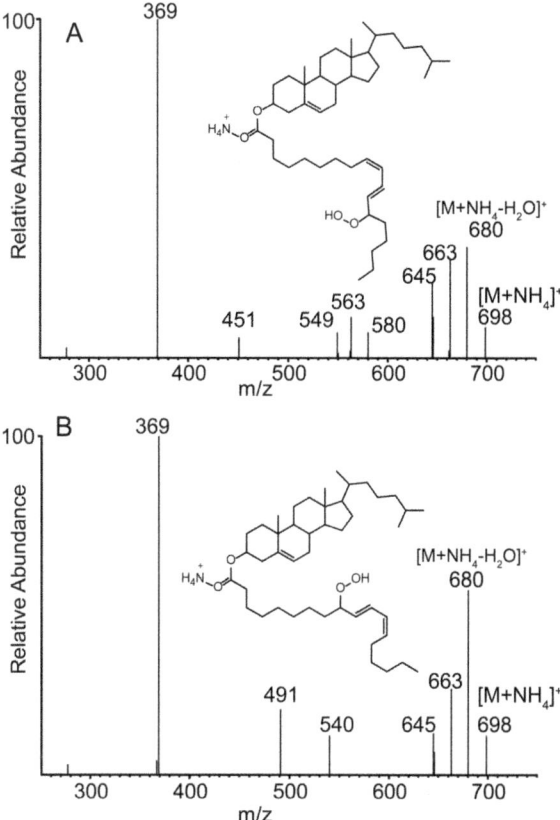

Figure 7.5 Electrospray ionization (positive ions) and tandem mass spectrometry of cholesterol hydroperoxy esters. (A) Product ions obtained following collisional activation of cholesterol 13-hydroperoxy-9,11-octadecadienoate $[M + NH_4]^+$ at *m/z* 698; (B) product ions obtained following collisional activation of cholesterol 9-hydroperoxy-10,12-octadecadienoate $[M + NH_4]^+$ at *m/z* 698. This figure was redrawn from data presented in ref. 14.

A separate hydroperoxide scission pathway involves the oxygen-centered radical at C-13 forming a carbonyl oxide-like bond with the fatty acyl carbonyl to form a cyclic peroxide observed at *m/z* 663. This reaction was likely driven by the favorable loss of neutral ammonia and water (Scheme 7.11).

Two other fragment ions that result in carbon–carbon bond cleavages adjacent to the hydroperoxy moiety could be driven as a result of a sigmatropic proton shift and a charge driven loss of hexanal, water, and ammonia (Scheme 7.12). The ion at *m/z* 549 would come from a second sigmatropic proton shift and loss of the epoxide of heptene. These resulting ions at *m/z* 563 and 549 would have the cationic site on the fatty acyl chain delocalized by a conjugated diene.

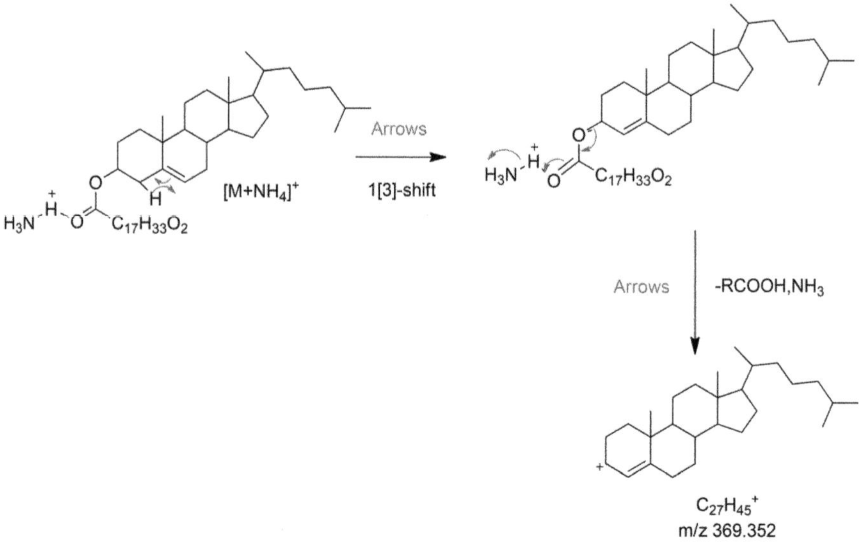

Scheme 7.9

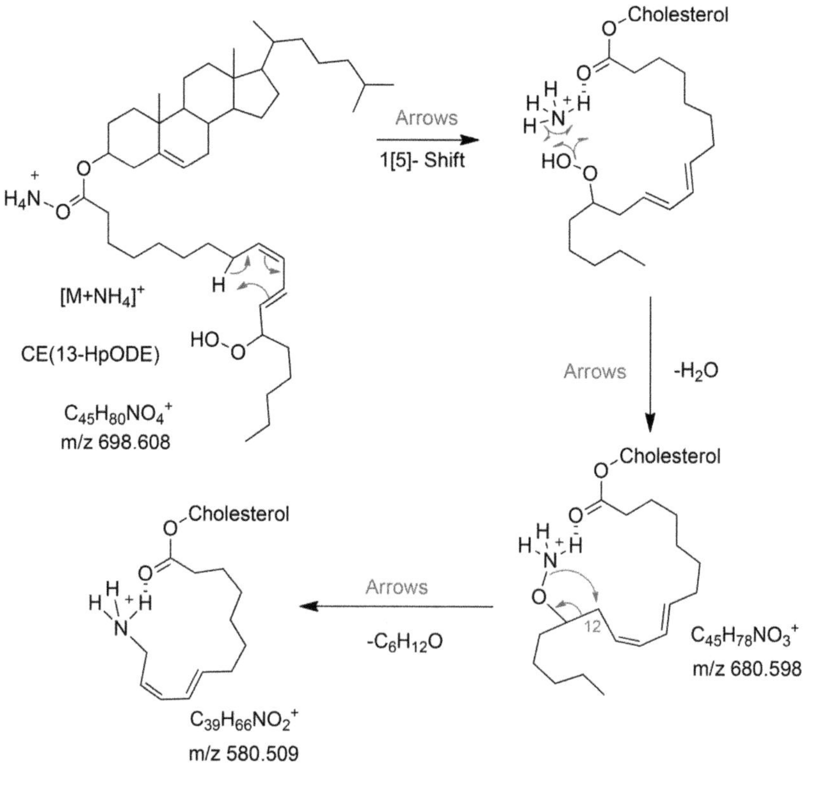

Scheme 7.10

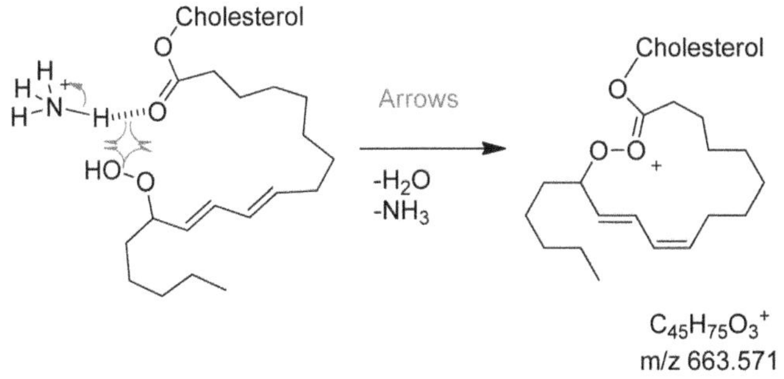

Scheme 7.11

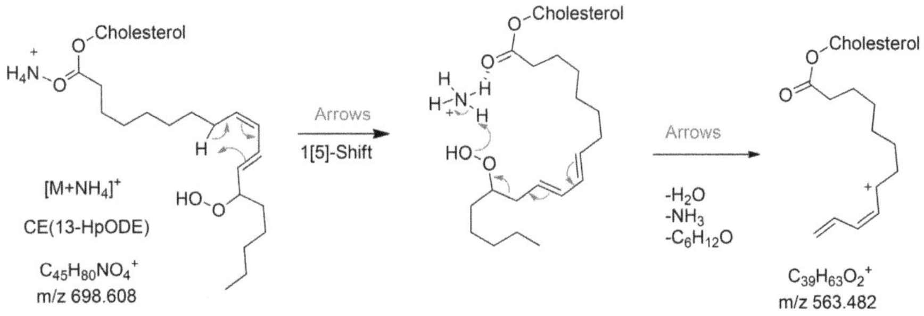

Scheme 7.12

The tandem mass spectra of cholesterol 9-hydroperoxyoctadecenoate also has unique product ions formed from collision induced decomposition and radical cleavage of the hydroperoxy group. Perhaps the most unexpected product ions retained the nitrogen atom from the adducted ammonium ion at m/z 540 (Figure 7.5B), covalently bonded to the alkyl chain (Scheme 7.10). This nitrogen atom-adduct was confirmed using [^{15}N]-labeled ammonium acetate as the electrospray electrolyte.[14]

7.2 Estrogens [M − H]⁻, Estrone, and Estradiol

The naturally occurring steroids that have an 18-carbon backbone include the female sex hormones such as estrone, estradiol, and estriol. These steroids are phenols in the A-ring and have lost the C-19 methyl group that originally was at the bridge head between the A- and B-ring attached to carbon 10. Since these are phenolic lipids, they readily form negative ions and the detailed studies of their collision decomposition have involved analysis of [M − H]⁻.

The product ions obtained following collisional activation of the [M − H]⁻ from these steroids yield, quite remarkably, the same abundant product ions, namely m/z 145, 171, and 183. This suggests very similar origins for these ion species and location of the negative charge, initially at the same place, namely the phenolic anion in ring A. Furthermore, there appears to be little influence of the oxygen atoms in ring D to alter the pathway decompositions. The formation of m/z 145 will be illustrated using estrone (Figure 7.6A). The structure first suggested for m/z 145 from estrone was not based on high resolution measurements or isotope labeling experiments,[15] but has been

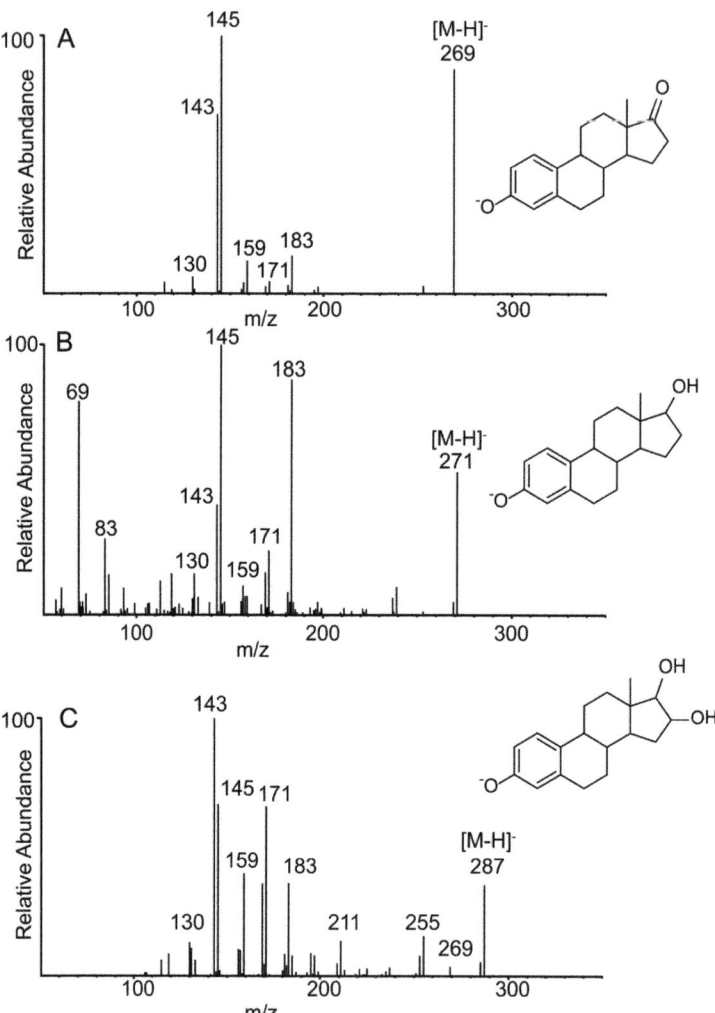

Figure 7.6 Electrospray ionization (negative ions) and tandem mass spectrometry of estrogens. (A) Product ions obtained following collisional activation of estrone [M − H]⁻ at m/z 369; (B) product ions obtained following collisional activation of estradiol [M − H]⁻ at m/z 271; (C) product ions obtained following collisional activation of estriol [M − H]⁻ at m/z 287. These MS/MS spectra were obtained using a tandem quadrupole mass spectrometer.

supported by later studies of both estrone and estradiol using these techniques.[16-18] The charge driven cleavage of the C-ring was suggested to result in formation of neutral ethylene and 2-methyl cyclopentenone in a concerted mechanism from the C and D-rings resulting in the product ion at m/z 145 (Scheme 7.13).[16] The anionic site is illustrated in this scheme at C-8, but resonance would drive formation of a more stable phenolic anion. This ion can then undergo loss of H_2 from C-6/7 to form the aromatic ion at m/z 143.

If this mechanism is not concerted, a first intermediate could involve a charge driven loss of H_2, leading to appearance of the anionic site at C-11. This intermediate has been proposed in the reaction pathway for other abundant ions observed, namely at m/z 171 (Scheme 7.14) from estrone, estradiol, and estriol.[17]

Rannulu and Cole proposed a mechanism for formation of m/z 169 to proceed from m/z 171 with a charged induced loss of a H_2 neutral.[17] This mechanism (Scheme 7.15) would result in the anion charge site being delocalized over much of the ion structure by resonance.

The mechanism responsible for the quite abundant and common product ion for estrone, estradiol, and estriol observed at m/z 183 (Scheme 7.16) was found to be rather complex, which required high resolution measurements as well as three different deuterium and carbon-13 labeled variants of estradiol to fully elucidate.[18] The same intermediate described for estrone (Scheme 7.13) leading to initial cleavage of the C-ring could abstract a proton from carbon-8, which would place the anionic charge at a somewhat more

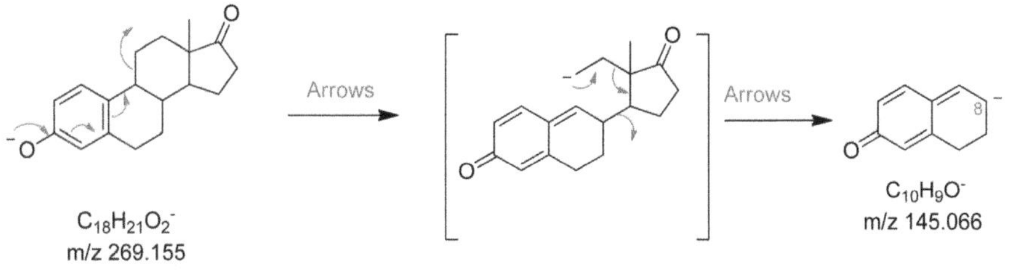

$C_{18}H_{21}O_2^-$
m/z 269.155

$C_{10}H_9O^-$
m/z 145.066

Scheme 7.13

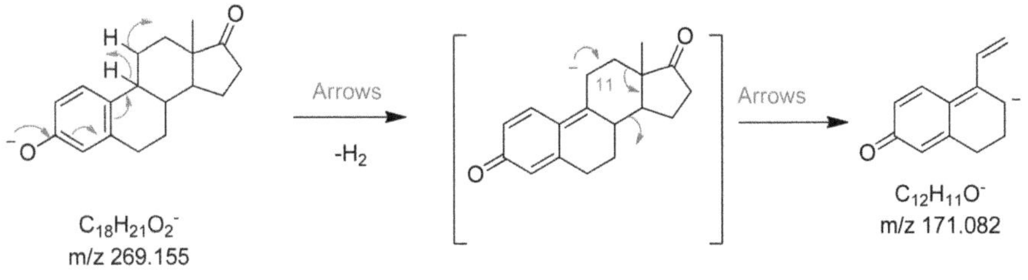

$C_{18}H_{21}O_2^-$
m/z 269.155

$C_{12}H_{11}O^-$
m/z 171.082

Scheme 7.14

stable location by delocalization. When the anionic site is present at C-8, the D-ring could open and by a series of proton abstractions result in a highly conjugated intermediate with a charge site now at C-15 (Scheme 7.16), but still without any loss of mass from the molecular ion (*m/z* 271). This molecular ion isomer could then, in a charge driven reaction, cleave the C-16/17 bond with a proton abstraction from carbon-7, finally yielding the structure indicated in Scheme 7.16 for *m/z* 183.

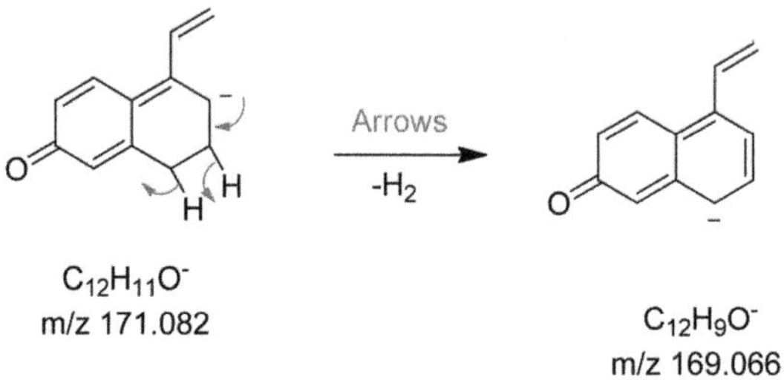
Arrows
-H2
C12H11O-
m/z 171.082
C12H9O-
m/z 169.066

Scheme 7.15

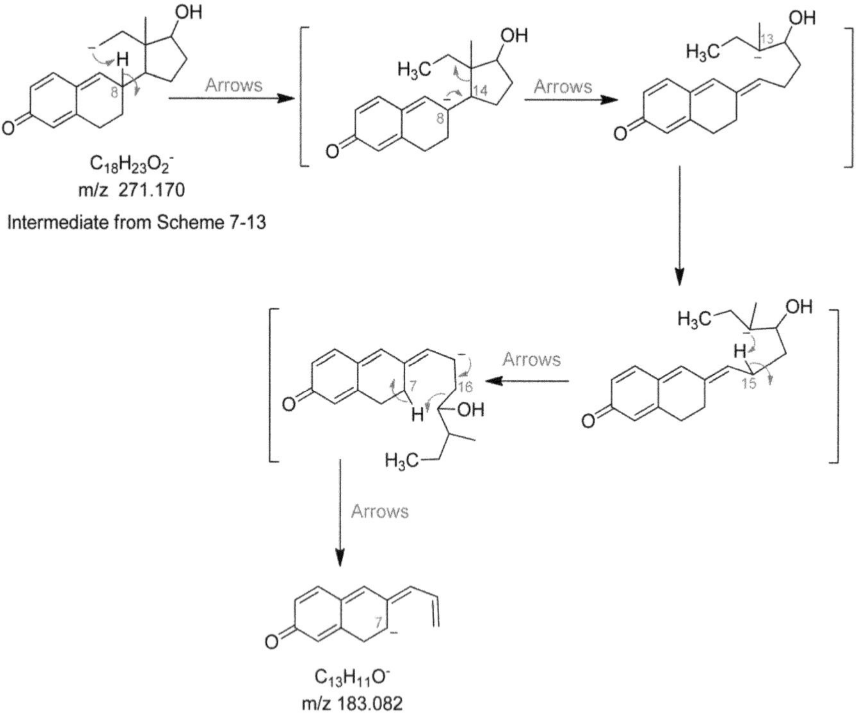
OH
H
8
Arrows
C18H23O2-
m/z 271.170
Intermediate from Scheme 7-13
H3C
14
Arrows
13
15
16
7
OH
C13H11O-
m/z 183.082

Scheme 7.16

7.3 Androgens, Testosterone [M + H]⁺ and Dehydroepiandrosterone Sulfate [M − H]⁻

The androgen steroids, including testosterone and dehydroepiandrosterone sulfate, are examples of quite important biologically active 19-carbon steroids that contain the 3-keto-4-ene structure in the steroidal A-ring. A great number of closely related analogs are known and many of their mass spectra have been studied.[19] These steroids are readily protonated by electrospray ionization to form [M + H]⁺ that undergo rather characteristic collision induced dissociation.[20] The presence of the 3-keto group also was a structural prerequisite to make the Girard's P derivative that had the collision induced decomposition reactions that were previously discussed in Section 7.1.2.

Testosterone will be used as the archetypical example, since a number of detailed studies of the product ions from testosterone [M + H]⁺ have appeared.[21] The tandem mass spectra of testosterone as [M + H]⁺ (*m/z* 289) generated by electrospray ionization is presented in Figure 7.7A. At high

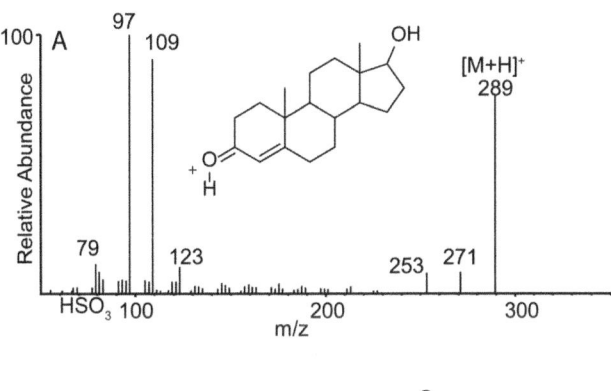

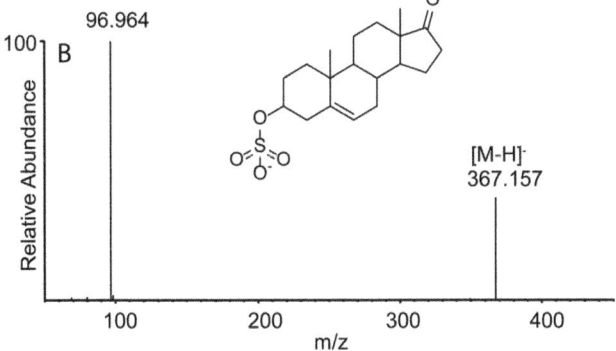

Figure 7.7 Electrospray ionization (positive and negative ions) and tandem mass spectrometry of androgens. (A) Product ions obtained following collisional activation of testosterone [M + H]⁺ at *m/z* 289. This MS/MS spectra was obtained using a tandem quadrupole mass spectrometer. (B) Product ions obtained following collisional activation of the sulfate ester of dehydroepiandrosterone [M − H]⁻ at *m/z* 367. This high resolution MS/MS spectra was obtained using a quadrupole time-of-flight mass spectrometer.

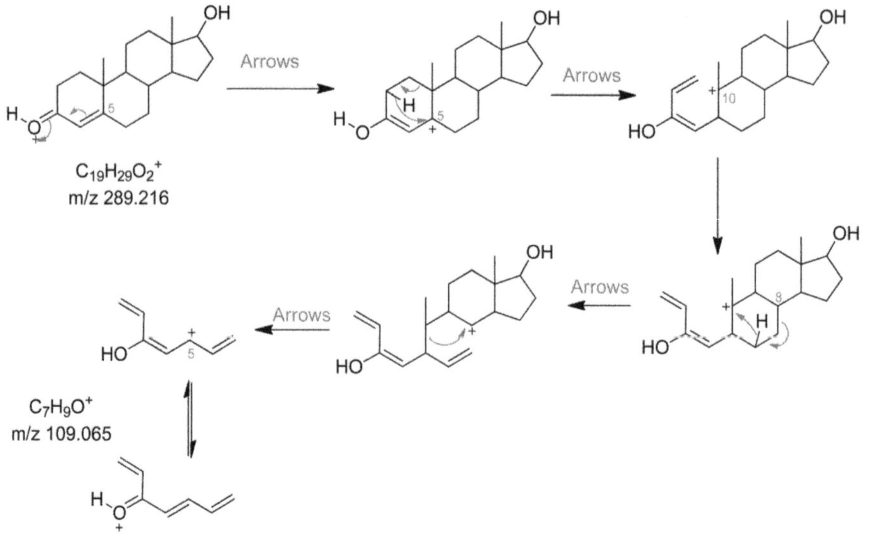

Scheme 7.17

mass the ions are observed at *m/z* 271 and 253, corresponding to the loss of one and two molecules of water, respectively. The dominant product ion abundance is observed as two different species, *m/z* 109 and 97, and both ions are derived from the A-ring.[19]

A mechanism responsible for the formation of *m/z* 109 was first presented in 1999 and it was based on the observed fragment ions of testosterone obtained by electron ionization as well as tandem mass spectrometry of electrospray generated ions.[8,20,22] This pathway proceeds by way of a charge-driven double bond rearrangement that brings the positive charge site to the bridgehead C-5 between the A and B steroid rings (Scheme 7.17). This intermediate structure places a proton close to the C-5 cationic site and the carbon 1-10 bond could break in a 6-membered transition state to cleave the A-ring, leaving the positive charge at C-10. A very similar hydrogen rearrangement is then set up for the B-ring, which transfers the charge site to C-8. With a cationic site now at C-8, there can be one of several hydrogen abstraction reactions that will lead to cleavage of the bond between C-5/10 and formation of the highly conjugated cation shown in Scheme 7.17 at *m/z* 109. This mechanism is supported by high resolution measurements and a stable isotope labeled variant of testosterone.[20]

The second abundant product ion occurs at *m/z* 97. The mechanism responsible for formation of this product ion has been very difficult to rationalize.[20] Using techniques, including density functional theory calculations and infrared multiple photon dissociation spectroscopy, various isotope labeled testosterones, and comparison of the mass spectral data from six potential isomers of *m/z* 97, two pathways were shown to be operating that lead to this ion.[21] The first pathway (Scheme 7.18) involves transfer of the

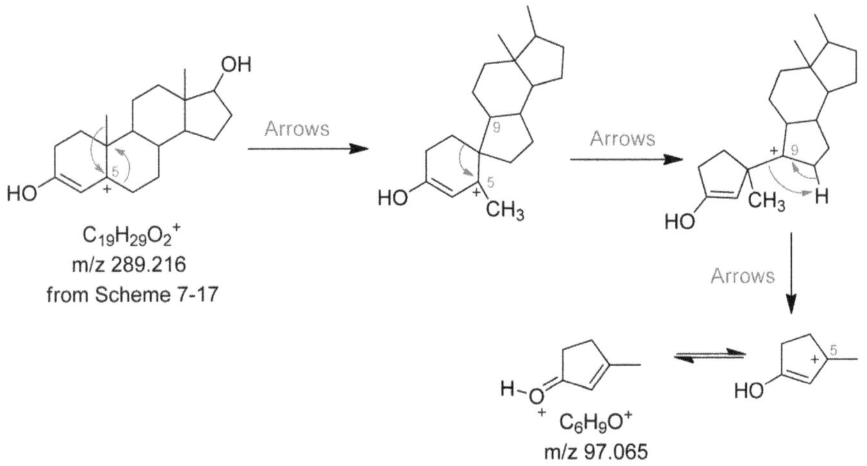

Scheme 7.18

cationic charge site to C-5 just as in the first step of Scheme 7.17. The C-19 methyl group could then migrate in Wagner-Meerwein rearrangement, with formation of a bond between C-6 and C-10. This results in a 6-ring/5-ring spiro structure that can undergo another Wagner-Meerwein rearrangement driven by the cation at C-5 to form a new bond between C-1 and C-5 and leaving the charge site at C-10. This ion structure could then have the bond between C-5 and C-10 break, remove a proton from C-6, which renders a neutral olefin for the steroid BCD ring remnant and formation of a charged cyclopentenal ion at m/z 97.

The second proposed mechanism proceeds from the same initial intermediate with the cationic site at C-5, only with the C-3/4 double bond participating in formation of a more complicated pentacylic structure with a new 5-membered A-ring fused to a cyclic propane/cyclopentane ring (Scheme 7.19).[21] Cleavage of the C-4/5 bond driven by the positive charge center at C-3 opens the cyclopropane ring leaving the charge site now at C-5. In Scheme 7.19 the C10-5 bond is rotated, placing it close to a hydrogen atom on C-6 that would facilitate its abstraction by C-5 while the C-10/5 bond is heterolytically cleaved, leaving carbon-10 the cationic charge site of a methylcyclopentenol cation (m/z 97). This product is identical to that of Scheme 7.18 except for the origin of the carbon atoms in the final ion structure.[21] These carbon atoms are indicated in each of the two schemes.

The sulfate ester of dehydroepiandrosterone is another androgen, but being a sulfuric acid ester, rather abundant $[M - H]^-$ ions are formed by electrospray ionization. The tandem mass spectrometry of this steroid yields essentially only one product ion, namely HSO_4^-, at m/z 97 (Figure 7.7B). Note that the exact mass (and therefore the elemental composition) of this ion is different from the m/z 97 derived from testosterone. The mechanism of sulfate ester decomposition processes is presented in detail in Section 7.7.

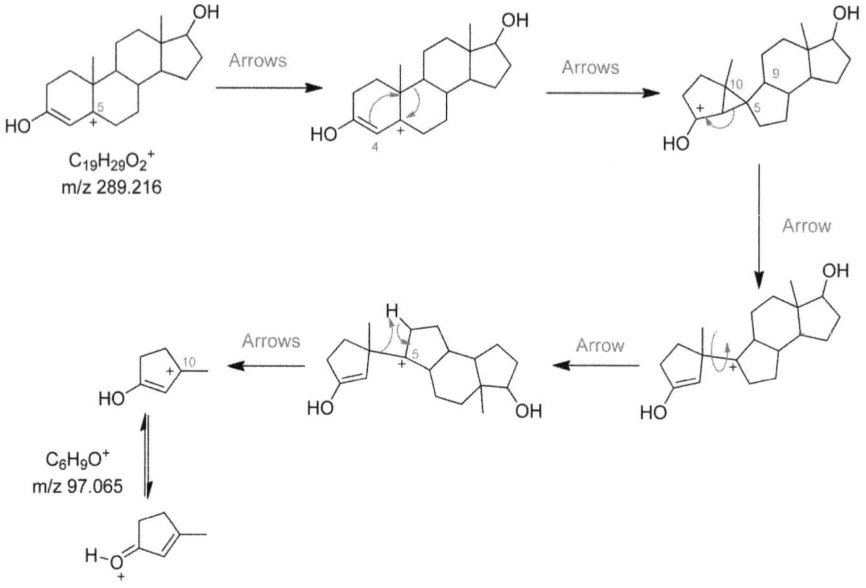

$C_{19}H_{29}O_2^+$
m/z 289.216

$C_6H_9O^+$
m/z 97.065

Scheme 7.19

7.4 Adrenocorticosteroids and Progestins

The steroids with 21 carbon atoms include the glucocorticoids (*e.g.* cortisol) and mineralocorticoids (*e.g.* aldosterone), which are synthesized in the cortex of the adrenal glands. Progesterone has a very similar steroid core structure, but is made in the ovaries as a progestinal steroid ketone. The differences in structure between these steroids center on the number and positions of oxygen substituents as hydroxy and keto moieties. All have the cholest-4-ene-3-one A-ring and form $[M + H]^+$ ions by electrospray ionization.

7.4.1 Progesterone $[M + H]^+$

The electrospray generated $[M + H]^+$ from progesterone occurs at m/z 315 and high mass product ions formed after collisional activation include loss of one (m/z 297) and two molecules of water (m/z 279) (Figure 7.8A). This is consistent with only two oxygen atoms being present at C-3 and C-20 as ketone moieties. The most abundant product ions appear at m/z 97 and 109 and are likely formed by the same mechanisms proposed for testosterone $[M + H]^+$ in Schemes 7.18 and 7.17, respectively.

7.4.2 Cortisol $[M + H]^+$

Cortisol has three additional hydroxy groups at C-11, C-17, and C-21, which are secondary, tertiary, and primary carbinols, respectively, in addition, two keto groups are present at C-3 and C-16. Three water losses are fairly

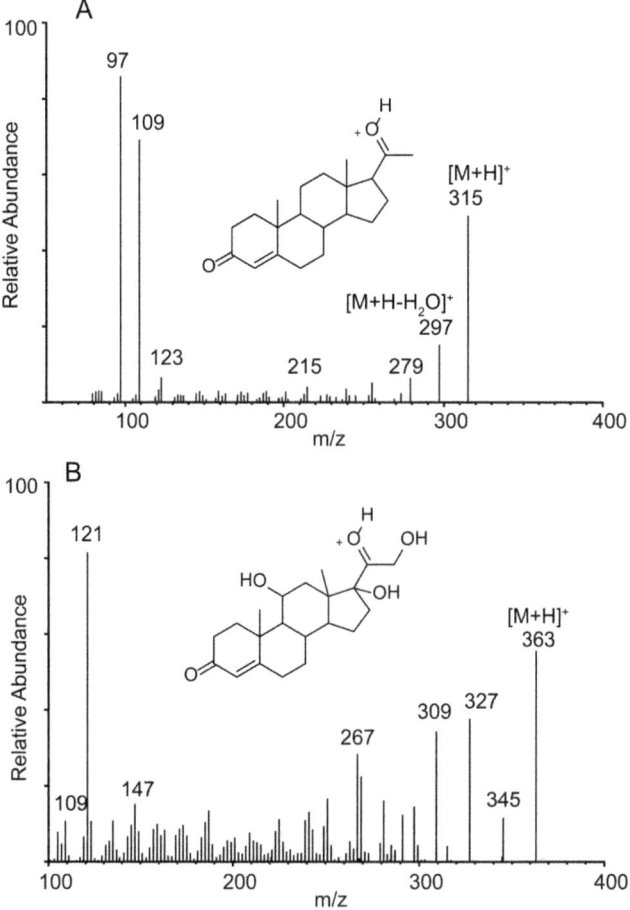

Figure 7.8 Electrospray ionization (positive ions) and tandem mass spectrometry of adrenal corticosteroids. (A) Product ions obtained following collisional activation of progesterone $[M + H]^+$ at m/z 315; (B) product ions obtained following collisional activation of the cortisol $[M + H]^+$ at m/z 363. These MS/MS spectra were obtained using a tandem quadrupole mass spectrometer.

abundant from $[M + H]^+$ at m/z 363 (Figure 7.8B), most likely corresponding to the losses of the three hydroxy groups as neutral water molecules. The ions at m/z 97 and 109 are present at low mass, but are not particularly abundant when compared to that for testosterone or progesterone.

The most abundant product ion occurs at m/z 121. This ion shifts to m/z 125 in the product ions derived from 2,2,4,6,6,12,12-[^{2}H$_7$]-cortisol. Thus, this ion retains four of the seven labeled protons in this deuterium labeled analog. When cortisol is labeled as 9,10,11,11-[^{2}H$_4$]-cortisol, this ion remains at m/z 121.[23] These deuterium atom shifts suggest that m/z 121 is largely composed of elements from the A-ring. The major structural modification

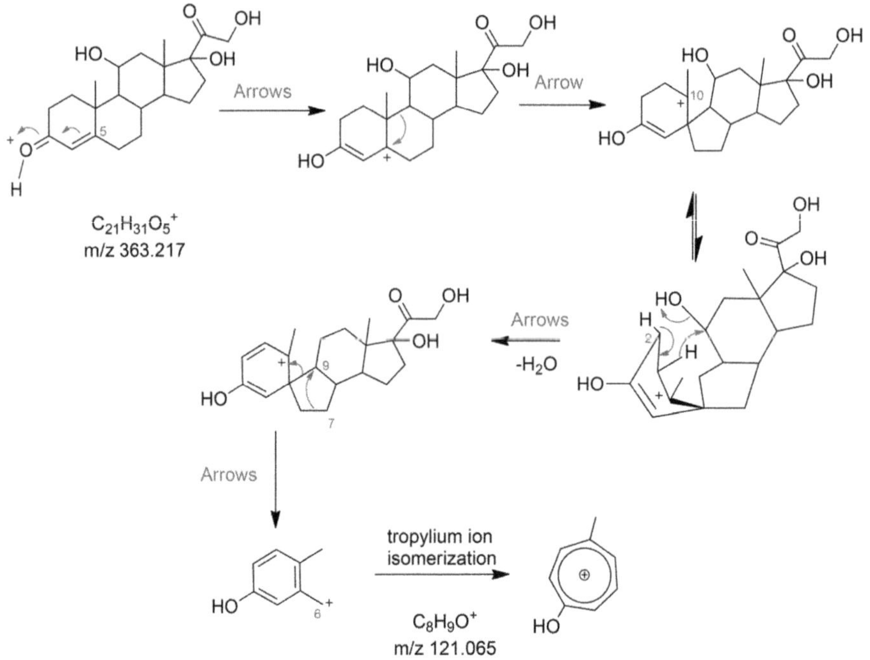

$C_{21}H_{31}O_5^+$
m/z 363.217

tropylium ion
isomerization

$C_8H_9O^+$
m/z 121.065

Scheme 7.20

close to the A-ring (when considering the structural difference between progesterone and cortisol) is the 11-hydroxy group in ring C. Structures for m/z 121 have been proposed to be quinoid-like, but no mechanism for formation had been proposed. This structure could involve formation of $\Delta^{1,2}$ double bond of the A-ring.[23,24]

Considering a Wagner-Meerwein alkyl group migration to C-5, as proposed for testosterone (which also has the 3-oxo-4-ene structural unit in the A-ring), cortisol could form a cation at C-5 and a spiro A-B-ring that would bring into close proximity, the C-11 hydroxy group and a proton at C-2. Loss of water with formation of $\Delta^{1,2}$-double bond with a proton from C-1 migrating to C-11 would likely be facilitated by stabilization of the cationic site at C-10 (Scheme 7.20). Collapse of the spiro-ring structure by movement of C-7 to C-9 would then yield the ion at m/z 121 with a charge site at C-6. Most likely this benzyl cation expands into the highly stable tropylium-like ion shown in Scheme 7.20.[25]

7.4.3 Aldosterone $[M + H]^+$ and $[M - H]^-$

Aldosterone, a mineralocorticoid made in the adrenal gland, can generate abundant positive $[M + H]^+$ as well as negative $[M - H]^-$ species. Measurement of aldosterone by tandem mass spectrometry has been reported using negative ions as a sensitive means to quantitate this important circulating steroid.[26] The most abundant product ion generated by the tandem mass

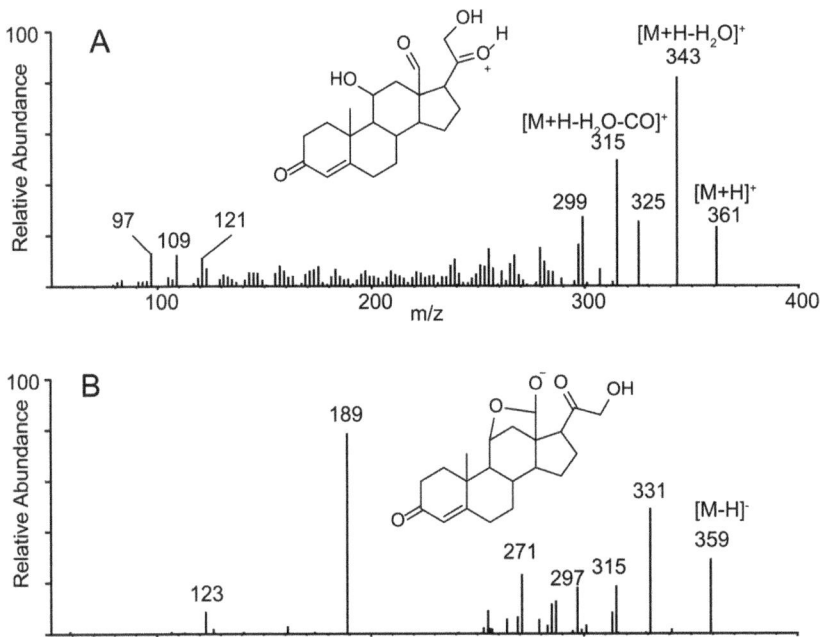

Figure 7.9 Electrospray ionization (positive and negative ions) and tandem mass spectrometry of aldosterone. (A) Product ions obtained following collisional activation of aldosterone [M + H]$^+$ at m/z 361; (B) product ions obtained following collisional activation of aldosterone [M − H]$^-$ at m/z 359. These MS/MS spectra were obtained using a tandem quadrupole mass spectrometer.

Scheme 7.21

spectrometry of [M + H]$^+$ at m/z 361 corresponds to loss of H_2O and H_2O plus carbon monoxide (CO) observed at m/z 343 and 315 (Figure 7.9A). The formation of these ions likely involves protonation of one of the keto groups, C-18 or C-20, and a proposed mechanism is presented in Scheme 7.21 for the sequential loss of these two small neutral molecules. The formation of a highly stable, 5-membered ring fused to the D-ring likely drives the loss of neutral CO from C-20.

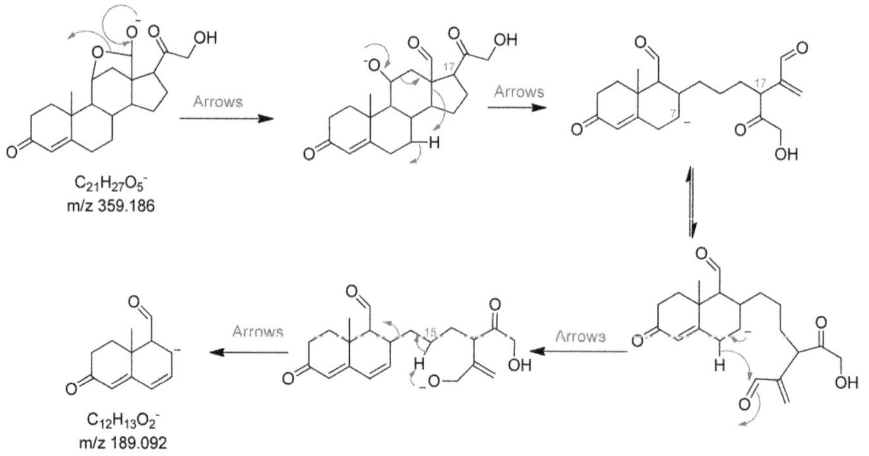

$C_{21}H_{27}O_5^-$
m/z 359.186

$C_{12}H_{13}O_2^-$
m/z 189.092

Scheme 7.22

Product ions are observed at *m/z* 97, 109, and 121, for which mechanisms have been suggested in previous schemes for the related structural steroids cortisol, testosterone, and progesterone with similar A- and B-ring substitutions. The lower abundance of these ions is likely attributable to the more favorable proton attachment (charging site) at the aldehydic moiety and the β-hydroxy-keto group of the D-ring.

The most abundant product ions formed by collisional activation of aldosterone [M − H]⁻ (Figure 7.9B) are observed at *m/z* 189 and is more structurally specific, hence the use of this ion transition as a sensitive MS/MS assay to measure aldosterone in biological fluids. The 2,2,4,6,6,12,12-[²H₇]-aldosterone isotopomer has this ion shifted to *m/z* 193, revealing the retention of only four deuterium atoms of the seven labeled positions during the formation of this product ion.[27] A likely mechanism would involve opening the C-ring by formation of an aldehyde at C-11 that also drives opening of the D-ring and rearrangement of a C-7 proton to C-14 to make a linear keto aldehyde of the C/D-rings. Either carbonyl group could accept the hydride anion from C-6 to form an alkoxide anion. Such an alkoxide anion would readily abstract a proton from C-15, leading to a double bond between C-14 and C-15, and in the process cleaving the C-14/8 bond. This would leave the anionic site at C-8 delocalized by the conjugated keto structure now in the A/B-ring remnant (Scheme 7.22).

7.4.4 Pregnenolone [M + H]⁺ and [M − H]⁻

Pregnenolone is a 21 carbon steroid that is not only a precursor in the biosynthetic pathway for many other glucocorticoids, androgens, and estrogens, but also an active steroid in its own right, especially in the brain. This steroid has the 3-hydroxy-5-ene structure of cholesterol, which is

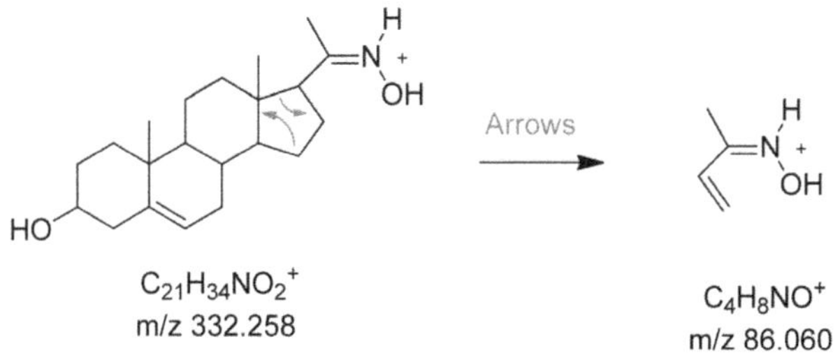

C₂₁H₃₄NO₂⁺
m/z 332.258

C₄H₈NO⁺
m/z 86.060

Scheme 7.23

different from that of many of the biochemical products such as progesterone. The collisional activation of the electrospray positive ions ($[M + H]^+$) have not been reported, but conversion of the 20-keto group to an oxime derivative has been used by several investigators as the basis of a positive ion tandem mass spectrometric assay for pregnenolone. The only abundant product ion formed following collisional activation of pregnenolone oxime $[M + H]^+$ is the ion at *m/z* 86.[28] The structure of this ion has been suggested to involve cleavage of the D-ring and a potential remote site mechanism is presented as Scheme 7.23.[29]

Pregnenolone can also form negative ions as adduct ions with acetate, bicarbonate, and fluoride that display abundant product ions, specifically the fluoride adduct (Figure 7.10A).[17] The collisional activation of $[M + F]^-$ at *m/z* 335 yields a deprotonated anion $[M - H]^-$ that likely involves removal of one C-21 proton or a C-17 proton based on the D_4 analog (Figure 7.10B). All of the abundant product ions appear to be derived from one of these two isomeric ions.

The abundant product ion observed at *m/z* 299 (Figure 7.10A) retains three deuterium atoms when the D_4 analog is collisionally activated (Figure 7.10B). A mechanism from the fluoride adduct involving loss of the C-18 methyl group as methyl fluoride and H_2 is presented in Scheme 7.24, which involves formation of an anion at C-16 that could be stabilized by the adjacent α,β-unsaturated keto group.

This same anion at *m/z* 299 could undergo a retro Diels–Alder reaction within the B-ring, leading to the abundant product ion at *m/z* 161 (Scheme 7.25). The other abundant product ion at *m/z* 123 could originate from an anionic site for $[M - H]^-$ at C-21, which would be derived from the loss of HF from the $[M + F]^-$. A retro Diels–Alder reaction of the C-ring as a charge remote process would lead to *m/z* 123, as indicated in Scheme 7.26.

7.5 Secosteroid (Vitamin D3)

Vitamin D and various related family members are steroid-derived lipids that play important roles in calcium ion homeostasis.[30] These lipids are termed vitamins since we do not synthesize these lipids directly, but rather ingest either vitamin D2 or vitamin D3 precursors from our diet or use sunlight to

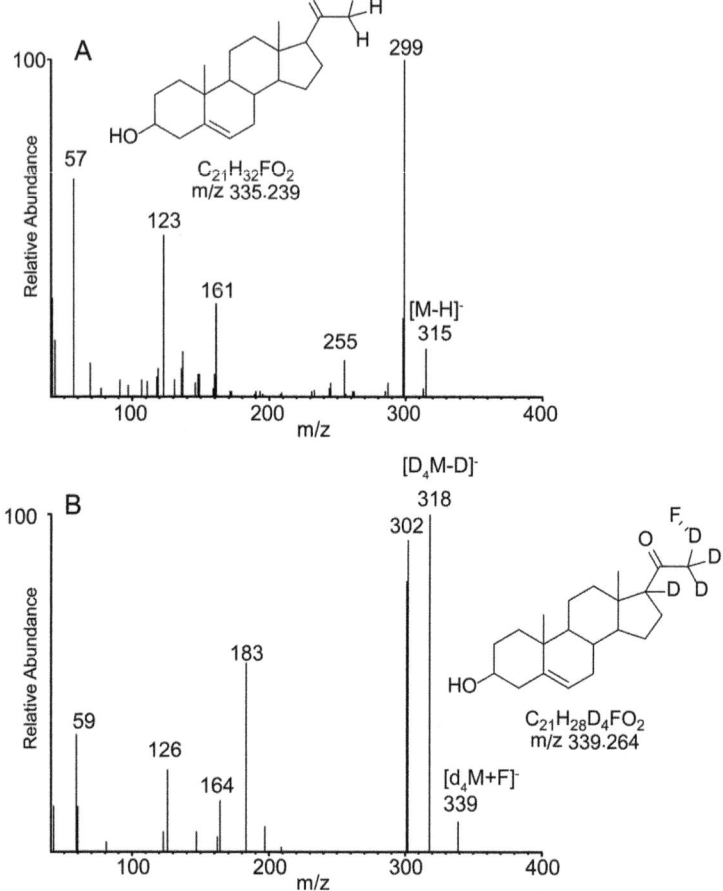

Figure 7.10 Electrospray ionization (negative ions) and tandem mass spectrometry of pregnenolone. (A) Product ions obtained following collisional activation of the fluoride adduct of pregnenolone $[M + F]^{+}$ at m/z 335. The highest mass product ion is observed at m/z 315, corresponding to $[M - H]^{-}$. (B) Product ions obtained following collisional activation of tetradeutero pregnenolone $[D_4M-D]^{-}$ at m/z 339. This figure was redrawn from data presented in ref. 17.

Scheme 7.24

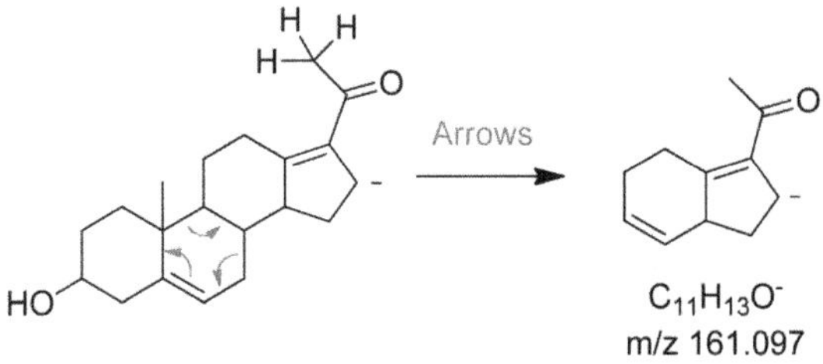

Scheme 7.25

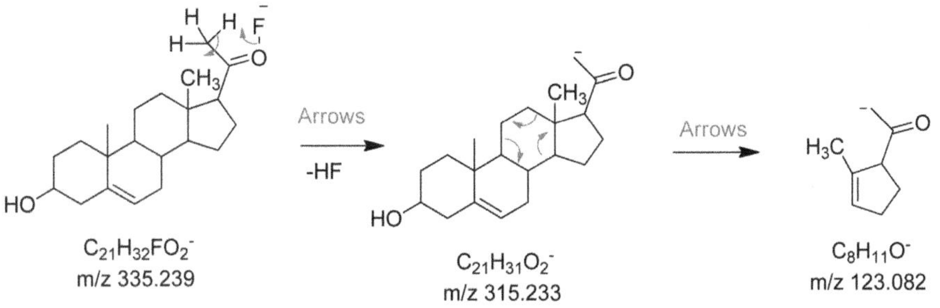

Scheme 7.26

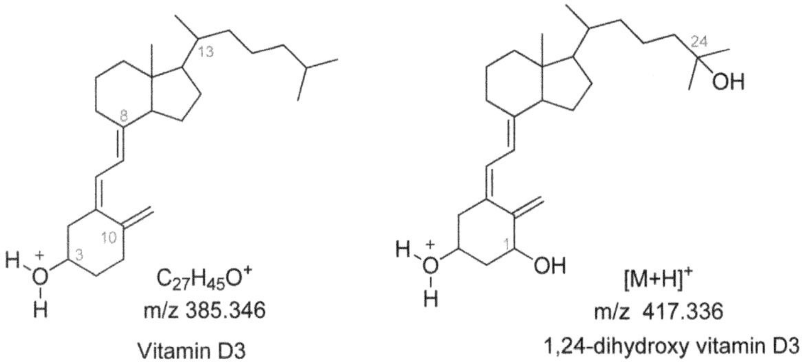

Scheme 7.27

photochemically cleave the B-ring of 7-dehydrocholesterol (a direct precursor of vitamin D3) to form pre-vitamin D3, which isomerizes into vitamin D3, a secosteroid in that it lacks the B-ring. Tandem mass spectrometry and electrospray ionization of vitamin D3 has been used extensively to assay various family members of this series of related lipids such as 1,24-vitamin D3.[30-33]

The collisional activation of the $[M + H]^+$ ion of vitamin D3 (m/z 385) (Scheme 7.27) yields an ion corresponding to the loss of water (m/z 367) and a host of product ions of rather high abundance that appear between m/z 100

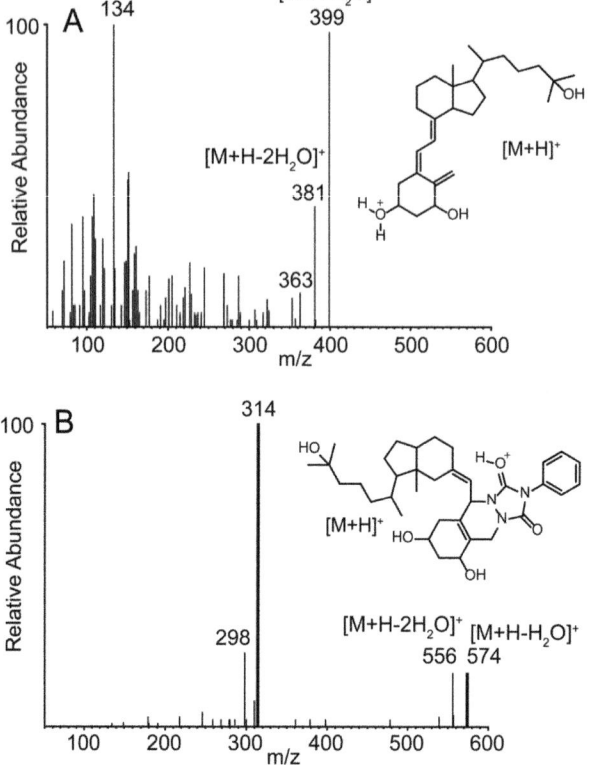

Figure 7.11 Electrospray ionization (positive ions) and tandem mass spectrometry of vitamin D3 and derivatized vitamin D3. (A) Product ions obtained following collisional activation of 1,24-vitamin D3 $[M + H]^+$ at *m/z* 417; (B) product ions obtained following collisional activation of the 4-phenyl-1,2,4-triazoline-3,5-dione (PTAD) derivative of 1,24-dihydroxy vitamin D3 $[M + H]^+$ at *m/z* 592. This figure was redrawn from data presented in ref. 31.

and 300. This large number of product ions greatly reduces the overall sensitivity of any MRM based assay of a precursor ion of the $[M + H]^+$ and any one of the single product ions. This is due to the abundance of any precursor-product ion pair being reduced by the additional reaction pathways. This behavior can be seen for the product ions of 1,24-dihydroxy vitamin D3 (Figure 7.11A),[31] where the most abundant ions correspond to the loss of one and two molecules of water at *m/z* 399 and 381.

For this reason, derivatization of vitamin D3 and its metabolites has been explored as a means to improve the sensitivity for a mass spectrometric-based assay. One interesting approach was found to capitalize on the unique *cisoid* arrangement of the double bonds as a structural remnant of the photochemical cleavage of the B-ring and to use a dienophile reagent to carry out a favorable Diels–Alder chemical reaction. One such reagent reported is

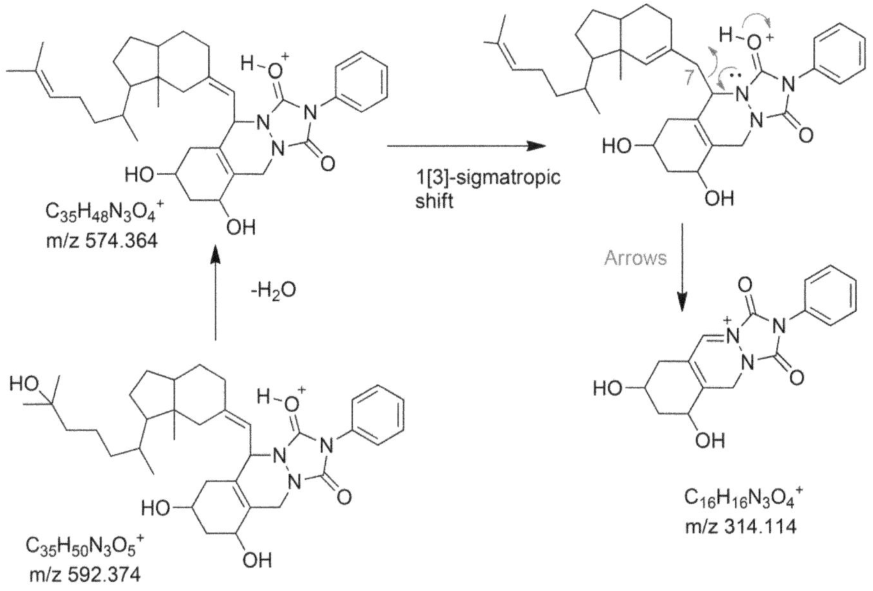

Scheme 7.28

4-phenyl-1,2,4-triazoline-3,5-dione (PTAD) that readily forms an adduct with vitamin D3.[32] The tandem mass spectrum of the PTAD derivative of 1,24-dihydroxy-vitamin D3 is presented in Figure 7.11B, which reveals formation of a very abundant product ion at m/z 314.

The formation of this product ion has been suggested to involve cleavage of the carbon 6–7 bond based on deuterium labeled analogs.[31] This could be rationalized as involving a 1[3]-sigmatropic shift after the loss of water from the C-24 hydroxyl group (Scheme 7.28), that moves a double bond from an exocyclic to an endocyclic position of the C-ring remnant. Following this rearrangement, the nitrogen atom attached to C-6 could participate through a non-bonded electron pair to drive neutralization of the protonated charge site at the adjacent amide, in the process breaking C-7 and forming a product ion corresponding to m/z 314.

7.6 Bile Acids

Bile acids are a family of steroidal molecules derived from the oxidative metabolism of cholesterol in the liver, which is a major metabolic pathway for the removal of cholesterol from the body. Bile acids are readily conjugated into many different products with the major forms excreted into the bile being glycine and taurine conjugates, although a large number of conjugates are known, including sulfate esters, glucuronides, and N-acetyl glucosamines.[34]

Most bile acids and bile acid conjugates are acidic molecules that form negative ions $[M - H]^-$ by electrospray ionization. While tandem mass spectrometry has been used as a basis of many quantitative assays for this diverse steroidal class, there has been little interest in probing the mechanistic details involved in the formation of these product ions.[34–37] In large part, the lack of information about the actual mechanism by which ions are formed is due to the rather straightforward behavior of bile acids and bile acid conjugates (Figure 7.12).

Free bile acids (non-conjugated) do not readily fragment after collisional activation of their $[M - H]^-$ anion[31–34,36,37] as seen for chenodeoxycholic acid $[M - H]^-$ observed at m/z 391 (Figure 7.12A). There are losses for one or more water molecules, depending upon the total number of hydroxyl groups attached to the steroid nucleus of the bile acid.

The tandem mass spectrometric behavior of glycine conjugates of bile acids can be exemplified by the glycine conjugate of chenodeoxycholic acid (Figure 7.12B). The glycine carboxylate anion is typically considered the site of charge localization responsible for $[M - H]^-$ that likely loses water and CO_2 or the concerted loss of carbonic acid to form the product ion observed at m/z 386. The most abundant product ion is found at very low mass, namely m/z 74 (Figure 7.12B), corresponding to the carboxylate anion of intact glycine. The origin of this ion can be a charge remote formation of a ketene-like structure that cleaves the glycine-amide bond and transfer a proton to the amine nitrogen atom as the amide bond cleaves (Scheme 7.29).

Taurine conjugates of bile acids such as taurochenodeoxycholic acid (Figure 7.12C) generate abundant $[M - H]^-$ ions by electrospray ionization, since they are strong acids containing a sulfonate moiety. Collisional activation leads to loss of water in relatively minor abundance, but the majority of the product ions are derived from the taurine moiety and seen at m/z 80, 107, and 124. The same mechanism suggested in Scheme 7.29 likely operates in the formation of m/z 124, which corresponds to the loss of a neutral ketene of the chenodeoxycholic acid and the sulfonate anion of taurine. This ion has been observed for many other taurine conjugates, including taurine conjugates of PGE_2 and leukotriene B_4.[38,39]

Another charge remote mechanism involving double bond formation discussed previously for many lipids that contain amide groups, leads to the ion at m/z 107. A mechanism proposed in Scheme 7.30 is consistent with the appearance of this ion.

The most abundant product ion following collisional activation of taurine conjugates of bile acids is found at m/z 80 derived from the sulfonate moiety following carbon–sulfur bond cleavage. This is a rather interesting ion in that it must be a radical anion, most likely involving a simple homolytic cleavage of the carbon–sulfur bond yielding two radical products (Scheme 7.31). The strong electron withdrawing properties of the oxygen atoms of the sulfonate moiety likely stabilize this radical product ion. The loss of hydroxyl radical from the bisulfate anion has been previously proposed, leading to the sulfonate radical anion at m/z 80.[40]

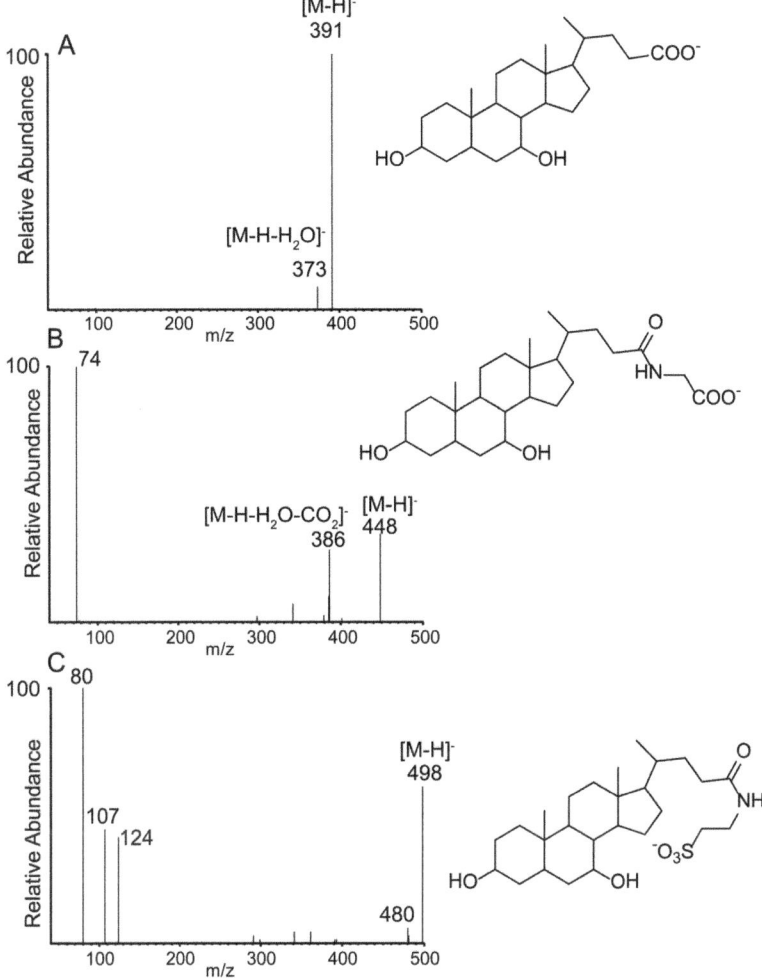

Figure 7.12 Electrospray ionization (negative ions) and tandem mass spectrometry of bile acid and bile acid conjugates. (A) Product ions obtained following collisional activation of chenodeoxycholic acid [M − H]⁻ at *m/z* 391; (B) product ions obtained following collisional activation of the glycine conjugate of chenodeoxycholic acid [M − H]⁻ at *m/z* 448; (C) product ions obtained following collisional activation of the taurine conjugate of chenodeoxycholic acid [M − H]⁻ at *m/z* 498. This figure was redrawn from data presented in ref. 34.

7.7 Steroid Sulfate Conjugates

A common metabolic reaction is the conversion of steroids into sulfuric acid esters to increase their water solubility. While detailed analyses of pathways operating in the tandem mass spectrometer leading to the observed product ions from steroid sulfates are few,[41,42] it is clear that most of the product ions

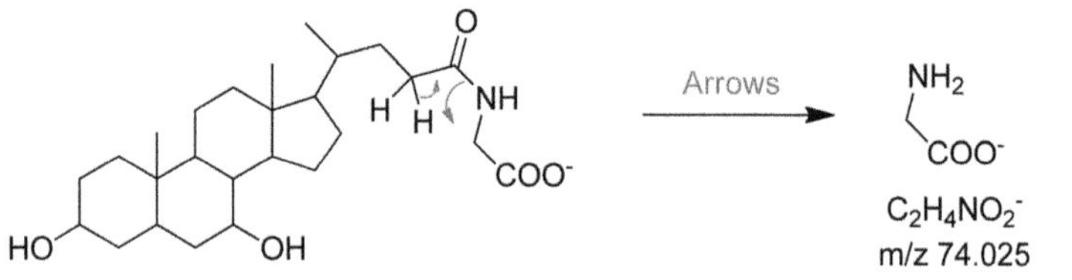

Scheme 7.29

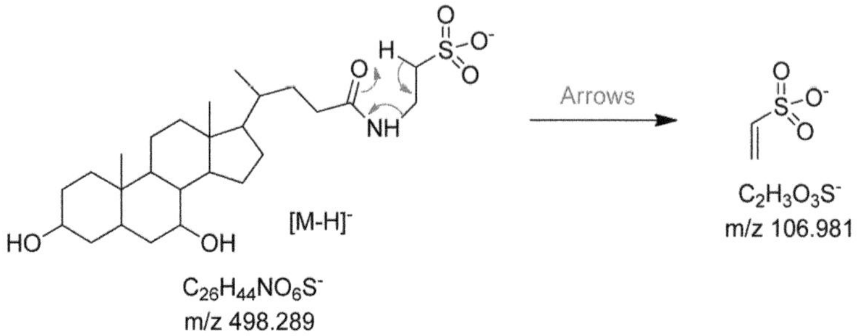

Scheme 7.30

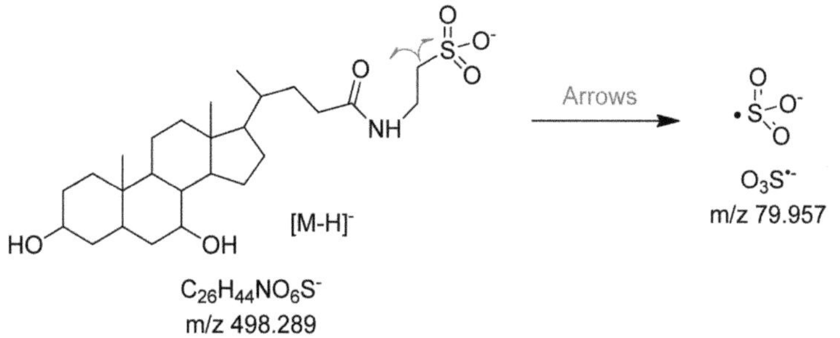

Scheme 7.31

derive from the sulfate ester moiety and are seen at m/z 97, 81, and 80. The other product ions appear to differ in abundance depending upon the instrument (tandem quadrupole *versus* ion trap) and energy employed to activate these ions.[42]

The mechanism of formation of m/z 80 and 97 has been studied in detail with various sulfate esters, including steroid sulfates.[40] For many steroids the reaction leading to m/z 97 involves removing a proton from a beta carbon atom (Scheme 7.32) similar to many β-elimination reactions observed in the gas phase.[40,43]

However, for some steroid sulfates such as cortisone-21-sulfate, the proton attacked by the sulfate anion is an exchangeable proton as revealed when the hydroxyl group (R–OH) at C-17 is labeled by H/D exchange using D_2O. For these labeled molecules the ion at m/z 97 is shifted to m/z 98 (Figure 7.13), suggesting operation of the mechanism outlined in Scheme 7.33.

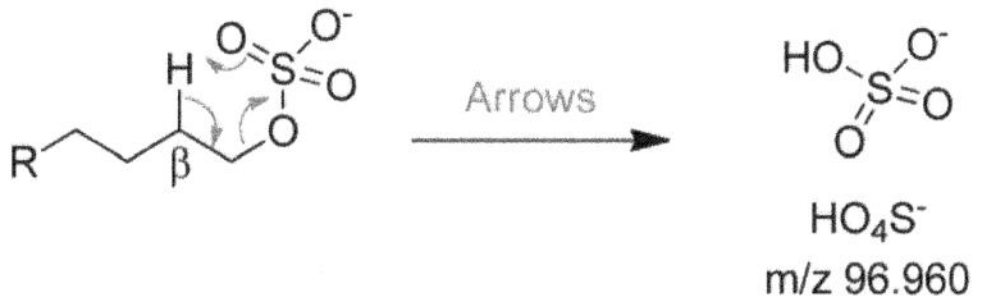

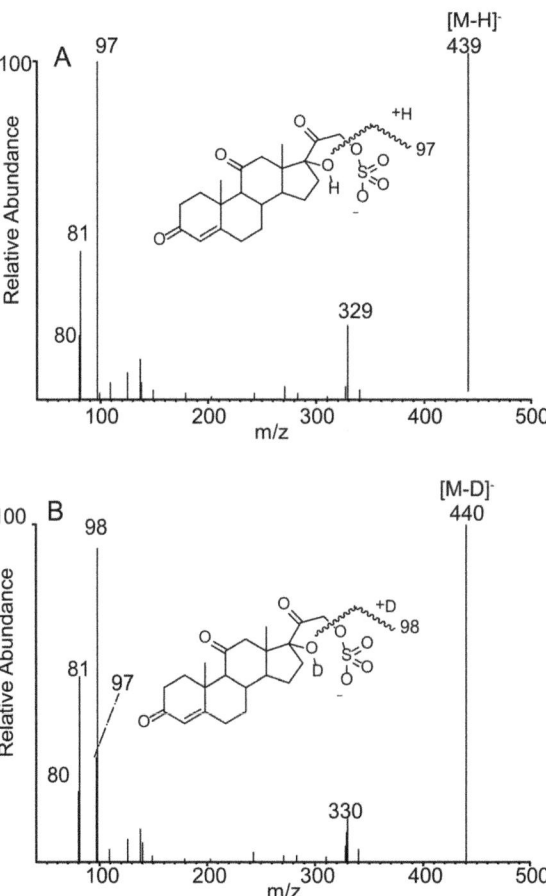

Scheme 7.32

Figure 7.13 Electrospray ionization (negative ions) and tandem mass spectrometry of cortisone sulfate. (A) Product ions obtained following collisional activation of cortisone-21-sulfate [M − H]⁻ at *m/z* 439; (B) product ions obtained following collisional activation of the H/D-exchanged proton of cortisone-21-sulfate [M-D]⁻ at *m/z* 440. This figure was redrawn from data presented in ref. 40.

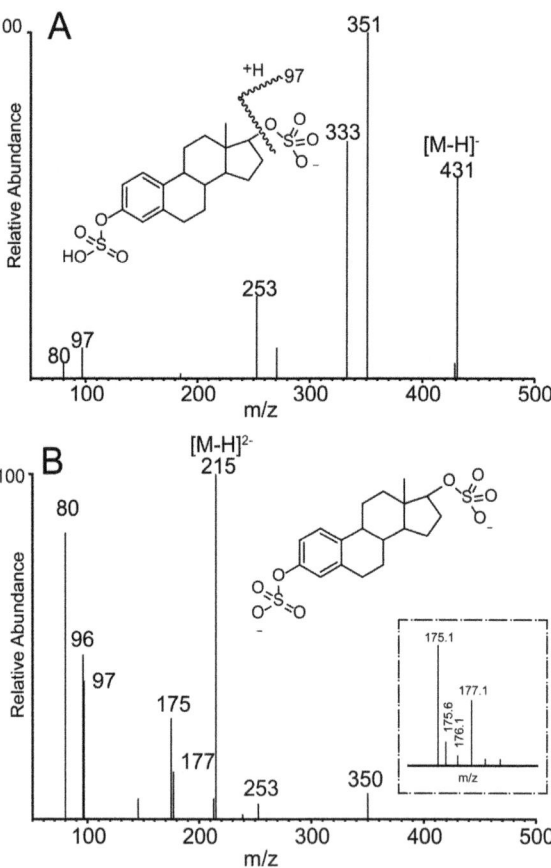

Scheme 7.33

Figure 7.14   Electrospray ionization (negative ions) and tandem mass spectrometry of estradiol-3,17-disulfate. (A) Product ions obtained following collisional activation of the singly charged estradiol-3,17-disulfate [M − H]⁻ at m/z 431; (B) product ions obtained following collisional activation of the doubly charged estradiol-3,17-disulfate [M − H]²⁻ at m/z 215. This figure was redrawn from data presented in ref. 40.

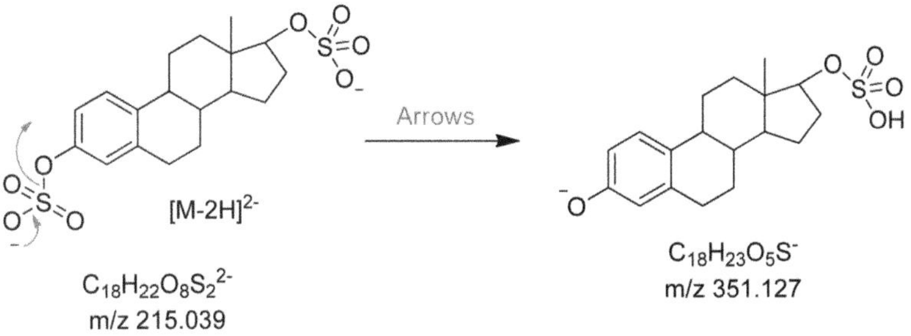

Scheme 7.34

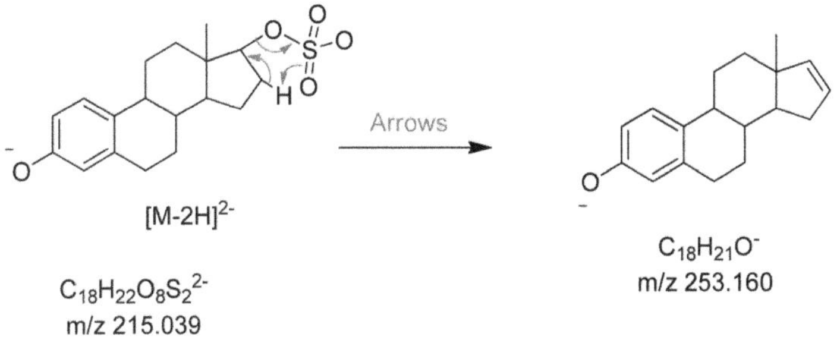

Scheme 7.35

Estradiol can be biochemically sulfated in various tissues on both the phenolic A-ring and the D-ring hydroxyl group to yield estradiol-3,17-disulfate. Electrospray ionization yields both a singly charged species $[M − H]^-$ observed at m/z 431 and a doubly charged $[M − 2H]^=$ at m/z 215. Collisional activation of either the singly or doubly charged ion can be carried out and the observed product ion spectra are quite different (Figure 7.14).[40] Collisional activation of m/z 431 yields abundant ions corresponding to the loss of neutral H_2SO_4 (m/z 333) and SO_3 (m/z 351). The site for the loss of H_2SO_4 is most likely from the aliphatic sulfate ester.

The ion at m/z 351 has been proposed to be a result of the charge driven loss of SO_2, from the singly charged species, leaving the phenolate anion on the A-ring as the charge site (Scheme 7.34).[40] This ion still retains the sulfate ester at C-17, which can also undergo a subsequent charge remote loss of H_2SO_4, forming the ion observed at m/z 253 (Scheme 7.35). For these steroid esters, the observed product ions are driven by the unique chemistry of the sulfate esters.

References

1. M. S. Brown and J. L. Goldstein, Cholesterol feedback: from Schoenheimer's bottle to Scap's MELADL, *J. Lipid Res.*, 2009, (50 Suppl.), S15–S27.
2. J. M. Dietschy and S. D. Turley, Thematic review series: Brain Lipids, Cholesterol metabolism in the central nervous system during early development and in the mature animal, *J. Lipid Res.*, 2004, **45**, 1375–1397.
3. I. Ramasamy, Recent advances in physiological lipoprotein metabolism, *Clin. Chem. Lab. Med.*, 2013, 1–33.
4. R. C. Murphy and K. M. Johnson, Cholesterol, reactive oxygen species, and the formation of biologically active mediators, *J. Biol. Chem.*, 2008, **283**, 15521–15525.
5. J. Belkner, H. Stender and H. Kuhn, The rabbit 15-lipoxygenase preferentially oxygenates LDL cholesterol esters, and this reaction does not require vitamin E, *J. Biol. Chem.*, 1998, **273**, 23225–23232.
6. P. M. Hutchins, E. E. Moore and R. C. Murphy, Electrospray MS/MS reveals extensive and nonspecific oxidation of cholesterol esters in human peripheral vascular lesions, *J. Lipid Res.*, 2011, **52**, 2070–2083.
7. J. G. McDonald, D. D. Smith, A. R. Stiles and D. W. Russell, A comprehensive method for extraction and quantitative analysis of sterols and secosteroids from human plasma, *J. Lipid Res.*, 2012, **53**, 1399–1409.
8. C. H. Shackleton, H. Chuang, J. Kim, X. de la Torra and J. Segura, Electrospray mass spectrometry of testosterone esters: potential for use in doping control, *Steroids*, 1997, **62**, 523–529.
9. X. Jiang, D. S. Ory and X. Han, Characterization of oxysterols by electrospray ionization tandem mass spectrometry after one-step derivatization with dimethylglycine, *Rapid Commun. Mass Spectrom.*, 2007, **21**, 141–152.
10. Y. Wang, K. Karu and W. J. Griffiths, Analysis of neurosterols and neurosteroids by mass spectrometry, *Biochimie*, 2007, **89**, 182–191.
11. W. J. Griffiths and Y. Wang, Analysis of oxysterol metabolomes, *Biochim. Biophys. Acta*, 2011, **1811**, 784–799.
12. H. Meerwein, Uber den Reaktionsmechanismus der Umwandlung von Borneol in Camphen, *Justus Liebigs Ann Chem*, 1914, **405**, 129–175.
13. G. Liebisch, M. Binder, R. Schifferer, T. Langmann, B. Schulz and G. Schmitz, High throughput quantification of cholesterol and cholesteryl ester by electrospray ionization tandem mass spectrometry (ESI-MS/MS), *Biochim. Biophys. Acta*, 2006, **1761**, 121–128.
14. P. M. Hutchins and R. C. Murphy, Peroxide bond driven dissociation of hydroperoxy-cholesterol esters following collision induced dissociation, *J. Am. Soc. Mass Spectrom.*, 2011, **22**, 867–874.
15. T. R. Croley, R. J. Hughes, B. G. Koenig, C. D. Metcalfe and R. E. March, Mass spectrometry applied to the analysis of estrogens in the environment, *Rapid Commun. Mass Spectrom.*, 2000, **14**, 1087–1093.

16. S. Bourcier, C. Poisson, Y. Souissi, S. Kinani, S. Bouchonnet, M. Sablier, Elucidation of the decomposition pathways of protonated and deprotonated estrone ions: application to the identification of photolysis products, *Rapid Commun. Mass Spectrom.*, 2010, **24**, 2999–3010.

17. N. S. Rannulu and R. B. Cole, Novel fragmentation pathways of anionic adducts of steroids formed by electrospray anion attachment involving regioselective attachment, regiospecific decompositions, charge-induced pathways, and ion-dipole complex intermediates, *J. Am. Soc. Mass Spectrom.*, 2012, **23**, 1558–1568.

18. K. M. Wooding, R. M. Barkley, J. A. Hankin, C. A. Johnson, A. P. Bradford, N. Santoro and R. C. Murphy, Mechanism of formation of the major estradiol product ions following collisional activation of the molecular anion in a tandem quadrupole mass spectrometer, *J. Am. Soc. Mass Spectrom.*, 2013, **24**, 1451–1455.

19. O. J. Pozo, E. P. Van, K. Deventer, S. Grimalt, J. V. Sancho, F. Hernandez and F. T. Delbeke, Collision-induced dissociation of 3-keto anabolic steroids and related compounds after electrospray ionization. Considerations for structural elucidation, *Rapid Commun. Mass Spectrom.*, 2008, **22**, 4009–4024.

20. T. M. Williams, A. J. Kind, E. Houghton and D. W. Hill, Electrospray collision-induced dissociation of testosterone and testosterone hydroxy analogs, *J. Mass Spectrom.*, 1999, **34**, 206–216.

21. M. Thevis, S. Beuck, S. Hoppner, A. Thomas, J. Held, M. Schafer, J. Oomens and W. Schanzer, Structure elucidation of the diagnostic product ion at *m/z* 97 derived from androst-4-en-3-one-based steroids by ESI-CID and IRMPD spectroscopy, *J. Am. Soc. Mass Spectrom.*, 2012, **23**, 537–546.

22. Z. V. Zaretskii, *Mass Spectrometry of Steroids*, John Wiley and Sons, Inc., New York, 1976.

23. S. Fustinoni, E. Polledri and R. Mercadante, High-throughput determination of cortisol, cortisone, and melatonin in oral fluid by on-line turbulent flow liquid chromatography interfaced with liquid chromatography/tandem mass spectrometry, *Rapid Commun. Mass Spectrom.*, 2013, **27**, 1450–1460.

24. M. A. Jensen, A. M. Hansen, P. Abrahamsson and A. W. Norgaard, Development and evaluation of a liquid chromatography tandem mass spectrometry method for simultaneous determination of salivary melatonin, cortisol and testosterone, *J. Chromatogr. B: Anal. Technol. Biomed. Life Sci.*, 2011, **879**, 2527–2532.

25. S. Meyerson, Tropylium, chlorine isotopic abundances, monomeric metaphosphate anion, and conestoga wagon theory, *J. Am. Soc. Mass Spectrom.*, 1993, **4**, 761–768.

26. U. Turpeinen, E. Hamalainen and U. H. Stenman, Determination of aldosterone in serum by liquid chromatography-tandem mass spectrometry, *J. Chromatogr. B: Anal. Technol. Biomed. Life. Sci.*, 2008, **862**, 113–118.

27. E. Hinchliffe, S. Carter, L. J. Owen and B. G. Keevil, Quantitation of aldosterone in human plasma by ultra high performance liquid chromatography tandem mass spectrometry, *J. Chromatogr. B: Anal. Technol. Biomed. Life. Sci.*, 2013, **913–914**, 19–23.

28. P. Keski-Rahkonen, K. Huhtinen, M. Poutanen and S. Auriola, Fast and sensitive liquid chromatography-mass spectrometry assay for seven androgenic and progestagenic steroids in human serum, *J. Steroid Biochem. Mol. Biol.*, 2011, **127**, 396–404.

29. S. Liu, J. Sjovall and W. J. Griffiths, Neurosteroids in rat brain: extraction, isolation, and analysis by nanoscale liquid chromatography-electrospray mass spectrometry, *Anal. Chem.*, 2003, **75**, 5835–5846.

30. J. M. van den Ouweland, M. Vogeser and S. Bacher, Vitamin D and metabolites measurement by tandem mass spectrometry, *Rev. Endocr. Metab. Disord.*, 2013, **14**, 159–184.

31. P. A. Aronov, L. M. Hall, K. Dettmer, C. B. Stephensen and B. D. Hammock, Metabolic profiling of major vitamin D metabolites using Diels-Alder derivatization and ultra-performance liquid chromatography-tandem mass spectrometry, *Anal. Bioanal. Chem.*, 2008, **391**, 1917–1930.

32. T. E. Lipkie, A. Janasch, B. R. Cooper, E. E. Hohman, C. M. Weaver and M. G. Ferruzzi, Quantification of vitamin D and 25-hydroxyvitamin D in soft tissues by liquid chromatography-tandem mass spectrometry, *J. Chromatogr. B: Anal. Technol. Biomed. Life Sci.*, 2013, **932**, 6–11.

33. T. Higashi, Y. Shibayama, M. Fuji and K. Shimada, Liquid chromatography-tandem mass spectrometric method for the determination of salivary 25-hydroxyvitamin D3: a noninvasive tool for the assessment of vitamin D status, *Anal. Bioanal. Chem.*, 2008, **391**, 229–238.

34. M. Maekawa, M. Shimada, T. Iida, J. Goto and N. Mano, Tandem mass spectrometric characterization of bile acids and steroid conjugates based on low-energy collision-induced dissociation, *Steroids*, 2014, **80**, 80–91.

35. Y. Alnouti, I. L. Csanaky and C. D. Klaassen, Quantitative-profiling of bile acids and their conjugates in mouse liver, bile, plasma, and urine using LC-MS/MS, *J. Chromatogr. B: Anal. Technol. Biomed. Life Sci.*, 2008, **873**, 209–217.

36. M. Scherer, C. Gnewuch, G. Schmitz and G. Liebisch, Rapid quantification of bile acids and their conjugates in serum by liquid chromatography-tandem mass spectrometry, *J. Chromatogr. B: Anal. Technol. Biomed. Life Sci.*, 2009, **877**, 3920–3925.

37. T. Murai, K. Oda, T. Toyo, H. Nittono, H. Takei, A. Muto, A. Kimura and T. Kurosawa, Determination of 3-β-hydroxy-Δ^5-bile acids and related compounds in biological fluids of patients with cholestasis by liquid chromatography-tandem mass spectrometry, *J. Chromatogr. B: Anal. Technol. Biomed. Life Sci.*, 2013, **923–924**, 120–127.

38. J. A. Hankin, P. Wheelan and R. C. Murphy, Identification of novel metabolites of prostaglandin E_2 formed by isolated rat hepatocytes, *Arch. Biochem. Biophys.*, 1997, **340**, 317–330.
39. M. A. Shirley and R. C. Murphy, Metabolism of leukotriene B_4 in isolated rat hepatocytes. Involvement of 2,4-dienoyl-coenzyme A reductase in leukotriene B_4 metabolism, *J. Biol. Chem.*, 1990, **265**, 16288–16295.
40. A. B. Attygalle, S. Garcia-Rubio, J. Ta and J. Meinwald, Collisionally-induced dissociation mass spectra of organic sulfate anions, *J. Chem. Soc., Perkin Trans.*, 2001, 2498–2506.
41. Y. Yan, D. L. Rempel, T. E. Holy and M. L. Gross, Mass spectrometry combinations for structural characterization of sulfated-steroid metabolites, *J. Am. Soc. Mass Spectrom.*, 2014, **25**, 869–879.
42. F. F. Hsu, F. Nodari, L. F. Kao, X. Fu, T. F. Holekamp, J. Turk and T. E. Holy, Structural characterization of sulfated steroids that activate mouse pheromone-sensing neurons, *Biochemistry*, 2008, **47**, 14009–14019.
43. C. A. Grob von and P. W. Schiess, Die heterolytische fragmentierung als reaktionstypus in der organischen chemie, *Angew. Chem.*, 1967, **79**, 1–14.

Subject Index